The Correspondence of John Tyndall

VOLUME 2

THE CORRESPONDENCE OF JOHN TYNDALL

VOLUME 2

The Correspondence, September 1843–December 1849

Edited by
Melinda Baldwin and Janet Browne

General Editors
James Elwick, Bernard Lightman, and Michael S. Reidy

UNIVERSITY OF PITTSBURGH PRESS

Published by the University of Pittsburgh Press, Pittsburgh, Pa., 15260

Manufactured in the United States of America
Printed on acid-free paper
10 9 8 7 6 5 4 3 2 1

Cataloguing-in-Publication data is available from the British Library

ISBN 13: 978-0-8229-4471-3
ISBN 10: 0-8229-4471-5

Frontispiece: A sketch of Queenwood College, where John Tyndall taught from 1847–8, from a prospectus of the school. From F. Podmore, *Robert Owen: A Biography*, 2 vols (London: Hutchinson & Company, 1906), vol. 2, p. 542. Image courtesy of Baker Old Class Collection, Baker Library, Harvard Business School

CONTENTS

ACKNOWLEDGEMENTS

The editors of this volume wish to thank the many people who offered us their advice and expertise as we worked on this volume, especially Ruth Barton, Geoffrey Cantor, Gowan Dawson, Evan Hepler-Smith, Peter Wirzbicki, and the general editors of *The Correspondence of John Tyndall*. The editors would particularly like to thank Julie Hipperson, who assisted with research in London, and Mark Tynan, for his research in the archives of the Ordnance Survey of Ireland (deposited in the National Archives of Ireland). The wonderful library resources at Harvard University were also indispensible as we prepared our Introduction and foonotes.

Finally, the editors would like to express our very great appreciation for the work done by our research assistants. Geoff Belknap was invaluable in the early stages of this volume's existence, and much of our Biographical Register information is owed to his work. Katherine Baca's research skills and keen editorial eye are reflected on every page of this volume, from the Introduction to the Appendix and all of the letters and footnotes in between.

MELINDA BALDWIN AND JANET BROWNE

LIST OF ILLUSTRATIONS

LIST OF ABBREVIATIONS

Journal	John Tyndall's Journal (RI MS JT/2/1–22)
LT	Louisa Tyndall, John Tyndall's widow
LT, 'Biography'	Louisa Tyndall's typescript biography of her husband (RI MS JT/5/9–11)
NA OS	National Archives, Kew, Ordnance Survey Records
NA WO	National Archives, Kew, War Office Records
NAI OS	National Archives of Ireland, Dublin, Ordnance Survey Records
ODNB	*Oxford Dictionary of National Biography*
OED	*Oxford English Dictionary*
RI MS JT	Royal Institution, London, John Tyndall Papers

INTRODUCTION TO VOLUME 2

The 161 letters in this volume encompass a period of dramatic change for the young John Tyndall. They begin in September 1843, in the midst of a fiery public conflict with the Ordnance Survey of England, and end in 1849 with him as a student of mathematics and experimental science at the University of Marburg, pursuing a doctoral degree in a language he had begun to learn only a year and a half earlier. In between, Tyndall was fired from his position in the Ordnance Survey, worked as a railway surveyor at the height of British railway mania, read the works of Ralph Waldo Emerson, Thomas Carlyle and Lord Byron, taught mathematics, and seriously contemplated emigration to the thriving new city of Cincinnati, USA. He began lasting friendships with Thomas Archer Hirst and Edward Frankland, who, along with Tyndall, would eventually become influential figures in nineteenth-century science. The letters between Tyndall and his correspondents also alluded to some of the most important events of the 1840s, including the Irish famine, political agitation for the repeal of the Acts of Union, the arrest of the Irish leader Daniel O'Connell and the railway boom. This introduction provides an overview of Tyndall's complicated path between 1843 and 1849: a period of professional uncertainty and personal transformation that led to Tyndall's decision to devote himself to natural philosophy.

Controversy on the Survey

As a general rule, *The Correspondence of John Tyndall* tries to begin and end volumes with full years. However, the enormous amount of correspondence for the year 1843 prompts a split between Volumes 1 and 2. The reason for the large number of letters in 1843 was Tyndall's conflict with his then-employer, the Ordnance Survey of England. The Ordnance Surveys of England and of Ireland were bodies of the state and mostly staffed and operated by the Royal Engineers. The two Surveys had been tasked with creating maps of townships for administrative and taxation purposes. Tyndall began work as a civil assistant on the Ordnance Survey of Ireland and transferred to the Survey of England in Preston, Lancashire in 1842, just before this volume begins.[1] However, Tyndall quickly became aware that his prospects of earning a comfortable living on the English

Survey were limited. His pay, according to a later recollection, was less than twenty shillings per week.[2] Furthermore, there was increasing feeling among the civil assistants that their prospects were being undercut.

A central issue at hand was that Colonel Thomas Colby, the Superintendent of both the English and Irish Ordnance Surveys, had proposed that in addition to the Survey's regular mapping work, the Ordnance Survey should also take charge of preparing maps for the Tithe Survey, a separate undertaking meant to enable the Church of England to collect appropriate tithes on local landowners. This meant that the Ordnance Survey would have to drastically change its mapping scale; Tithe maps were drawn at 26 2/3 inches to the mile, versus just 6 inches to the mile for an Ordnance Survey map. Furthermore, Tithe mapping work had previously been offered to civil assistants and to private surveyors by individual parishes. Colby's proposal meant that the publicly-funded Royal Engineers – the branch of the British military tasked with, among other things, supervising the creation of the Ordnance Survey maps – would be able to (allegedly) pocket two payments for a single map, while undercutting both the private surveyors and the civil assistants like Tyndall who hoped to supplement their income with Tithe work.[3] It was also clear that the civil assistants could not expect a salary increase from the Survey. On 16 September 1843, Tyndall and the other civil assistants were asked to sign a document saying they were satisfied with their current rate of pay and that they would remain on the Survey even without a raise; it was strongly implied that those who did not sign would be dismissed.[4]

Tyndall, who had chafed under the Irish Survey's poor treatment of its civil assistants, found the situation on the English Survey increasingly intolerable. He felt the Survey's leaders, especially Colby and Captain Henry Tucker, the local Director of the Preston Ordnance Survey, were mistreating the civil assistants and mismanaging public funds in order to line the pockets of the Royal Engineers. He was far from alone in these feelings. Tyndall's good friend William Ginty (whom Tyndall had met on the Irish Survey) and his fellow civil assistant Archibald McLachlan exchanged dozens of letters with Tyndall decrying the mismanagement of the Survey and the poor treatment of the surveyors. These letters are notable for their frequent use of code names; Tyndall was 'Alpha', Ginty 'Omega', McLachlan simply 'B'. McLachlan also adopted the nickname 'Carbon' for Thomas Colby and 'Acid' for another superior, likely Tucker. The three seem to have used these pseudonyms to avoid detection should fellow surveyors or their superiors intercept their mail. Nicknames and pseudonyms – often satirical or humorous – were to become a prominent feature of Tyndall's correspondence. The 1843 letters also stand out due to their rushed, unpolished style and expressions of frustration and anger – a marked contrast to the later letters in this volume, which tend to be elegant missives of friendship and encouragement. The intense tone of the 1843 letters speaks to the sense of anxiety and drama that surrounded Tyndall and his colleagues' protests against the Ordnance Survey leadership.

It was Tyndall who first made the group's accusations public with a series of five open letters to Prime Minister Robert Peel published in the *Liverpool Mercury,* which appeared between September 1843 and February 1844.[5] In these letters, Tyndall called attention to current conditions on the Survey and to previous mismanagement of Survey funds in the 1830s, evidently hoping to have a lucrative proposed survey of London taken out of his opponents' hands. Tyndall signed these letters with the pseudonym 'Spectator'. The sharp, defiant tone of the 'Spectator' letters would become characteristic of Tyndall's prose when he engaged in future conflicts in print. While Tyndall never addressed the Prime Minister with anything other than careful formality and respect, his remarks about the leaders whom he felt were responsible for the Survey's mismanagement were scathing. 'A sum of £200,000 was wasted in the production of this "Comedy of Errors"', he said of the surveying undertaken in the 1830s.[6] His final open letter, printed in February 1844, accused Henry Tucker by name of pompously taking credit for the hard work of the mistreated civil assistants.[7]

Tyndall did not confine himself to anonymous attacks on the Ordnance Survey management. He was also one of the authors of a September 1843 appeal addressed to the Master General of the Ordnance, George Murray, which protested a recent order by Tucker indicating that the Surveyors would not receive a pay increase.[8] Unfortunately for the signatories, Tucker himself intercepted 'the Memorial' (as the surveyors came to call this appeal) and returned it to the group unseen by Murray.[9] Tyndall and his allies, undaunted, wrote again to Murray to call his attention to Tucker's order.[10] The letter informed Murray that many on the Survey worked long hours and yet still lived in near poverty; Tyndall and his colleagues urged Murray to recognize the 'necessity of amending a system which subjects meritorious and talented men to such poverty and humiliation'.[11]

The result of these missives, however, was not what Tyndall had hoped. On 7 November 1843, Tucker signed a notice dismissing nineteen civil assistants and fifteen labourers, all under the command of Lieutenant Charles Stanley, from their positions on the Ordnance Survey. Among the civil assistants fired from Stanley's division were Tyndall himself, plus his close friends Philip Evans, George Latimer and Edmond Sayers.[12] Other divisions were similarly culled of troublemakers. On 8 November, Ginty, who had been stationed at Liverpool, wrote to Tyndall to say that he too had been dismissed from the Ordnance Survey.[13]

A Transition for Tyndall

Tyndall had anticipated that there might be consequences for his criticism of the Ordnance Survey, but as he told his father in an 8 November 1843 letter, 'I did not anticipate so sudden a blow, coming as it did without previous signal'.[14] He recounted the event with a mixture of pride and outrage in his journal, writing,

> Nov. 8 An eventful day – I have been dismissed from the Ordnance Survey … During my connexion with it a complaint was never made of me, either in point of morals or ability – my name was never brought before my superintendent but in terms of commendation – all useless however, Capt. Tucker, by an arbitrary exercise of the power committed to him, has dismissed the flower of his division, because they had the hardy honesty to state their grievances.[15]

Tyndall was not entirely sanguine about this prospect. That month he wrote another letter to Murray, this one personally signed, recounting his experiences on the Survey and asking the Master General to meet with him about his dismissal. He made this request, he said, in 'the hope that when my case is laid before you I shall be rescued from many disagreeable circumstances which are naturally consequent on being suddenly dismissed'.[16] Tyndall followed this with a second and third letter to Murray, suggesting that his initial letter did not receive a personal reply.[17] The appeals to Murray failed to have their desired effect. While some of the discharged civil assistants did return to work on the Survey, Tyndall's dismissal proved permanent.

Tyndall had formed many close friendships while on the Survey, however, and these would endure well beyond the end of his employment. His correspondence in 1844, the year after his dismissal, includes many letters to and from former surveyors. McLachlan and Ginty are particularly notable correspondents in this period. McLachlan followed his mathematical and scientific interests after his dismissal from the Survey, becoming absorbed in tide surveying and statistical analysis.[18] The exchanges with McLachlan hint at the intellectual environment Tyndall had found among his fellow civil assistants, many of whom were accomplished mathematicians with strong scientific interests. Although Tyndall himself seems to have had little thought of a long term career in mathematics at this time, correspondence with friends like McLachlan seems to have made an impression on him and helped him gain familiarity with the world of mathematics beyond the Survey.

Ginty and Tyndall remained close well after their dismissal from the Survey. They regularly exchanged original poetry, a custom that had been important to the social life of the civil assistants on the Ordnance Survey.[19] Ginty also continued to take great joy in ribbing Tyndall about his romantic life. 'I hear she is the wrong side of 30 – Oh damn your taste', he wrote about one love interest.[20] Unlike the harried and secretive style of correspondence that characterizes many of the letters from 1843, by 1844 Ginty and Tyndall had adopted a generally effusive, literary form of communication. For example, expressing his distaste that a former colleague had rejoined the Survey, Ginty noted: 'Oh degenerate sons of "Erin's Isle" how I blush for your spiritless conduct! a dive into the depths of the Shannon were preferable'.[21] Ginty was a source of encouragement and contacts during the months when Tyndall struggled to find employment after his work on the Survey. Indeed, just days after their dismissal from the Survey, Ginty told Tyndall of a position at the Liverpool Athenaeum, a gentleman's club.[22]

Later accounts of Tyndall's life often gloss over the period after Tyndall was obliged to leave the Survey; even his wife Louisa's unpublished biography has little to say about November 1843–September 1844. It appears, however, that this was an important period of struggle and uncertainty for the newly-unemployed young man. He returned to his hometown of Leighlin Bridge, Ireland in December 1843, where he renewed contact with old friends and mentors. He earned a little money by taking on small local surveying jobs, but his desire for consistent employment proved difficult to fulfil.

Tyndall found himself increasingly frustrated by the poor prospects open to a young Irishman in the United Kingdom. Like many Irish men and women in the 1840s, Tyndall began to consider emigrating to the United States of America to find better employment and a wider scope for advancement. Even before his dismissal from the Ordnance Survey, Tyndall told his father that 'The thought of going to America has more or less occupied my mind for years, but within the last nine months it has become a settled determination on my part'.[23] William Wright, a friend from Leighlin Bridge who had emigrated to Cincinnati, enthusiastically encouraged Tyndall's American scheme. Wright assured Tyndall that

> A young active enterprising fellow like you need have no doubts about the propriety of emigrating for there are a number of ways by which a man can live comfortably in this country and which do not exist in Ireland.[24]

After their dismissal from the Ordnance Survey, Tyndall and his friend George Latimer began to make plans to emigrate together. Tyndall's plans were near completion by March 1844; Ginty even wrote his friend a farewell letter, admitting, 'I never sat before a letter with bitterer feelings' and admonishing Tyndall, 'oh do not fail to write when you cast your anchor in the strangers clime!'[25] However, in late March Latimer decided not to make the trip to America after much pleading from his family.[26] Latimer's change of heart appears to have discouraged Tyndall. It seems too that Tyndall became increasingly aware how uncertain his prospects might be in America. Either way, he changed his plans.

Tyndall's hard luck continued. He applied for a position surveying the lands of a local landowner in County Carlow, but the local government committee that was supposed to choose the surveyor eventually determined that they had no legal power to appoint anyone. 'There was an end to all my airbuilt speculations!', Tyndall lamented in his journal.[27] A few weeks later Tyndall learned of an opportunity to do some privately commissioned work in England for a railroad agent named Mr. Wren. The position took Tyndall back to Preston, Lancashire, where he was happy to be reunited with some of his old friends from the Survey. By September 1844, Tyndall and Wren realized that Tyndall was not quite the right person for the position and (according to Tyndall's journal) 'came to the determination of parting friends'.[28] Tyndall was, naturally, alarmed at being unemployed again, but quickly determined his next course of action: he would go to Manchester to see if he could find work as a railroad surveyor.

Railroad Surveying and Halifax

The 1840s were a period of tremendous innovation and expansion for railways in Great Britain and the country seemed to be almost consumed with railway mania. Tyndall would later write that

> It was a time of mad unrest – of downright monomania. In private residences and public halls, in London reception-rooms, in hotels and in the stables of hotels, among gipsies and costermongers, nothing was spoke of but the state of the share market, the prospects of projected lines, the good fortune of the ostler or pot-boy who, by a lucky stroke of business, had cleared ten thousand pounds.[29]

The mid-1840s were a particularly fortuitous time to look for railway work, as Parliament had passed forty-eight railway bills in 1844 authorizing the construction of over 800 miles of new railways, and was considering another 220 railway bills for the 1845 session.[30] It is interesting to recall that Alfred Russel Wallace and Herbert Spencer, also impecunious figures who became noted men of science, worked as railway surveyors in the mid-nineteenth century.

Tyndall's employment was chaotic during this period, as most surveying and engineering firms experienced feast-and-famine cycles in the fast-paced railway market, especially after the Bank of England raised the bank rate in 1845 and brought a halt to the wildest period of railway speculation.[31] His most important employer was a railroad surveyor named Richard Carter, who first hired Tyndall in December 1844 to help prepare plans for a proposed West Yorkshire Railway. Tyndall spent most of 1845 and 1846 travelling between Carter's London and Halifax offices; in fact, he spent much of 1845 in London helping to prepare testimony before the House of Lords in a dispute over whether Carter's West Yorkshire Railway or a rival plan would be given Parliamentary approval.[32] However, in September 1846 Tyndall was called back to Halifax permanently when both railway plans were rejected by Parliament. Tyndall's frequent change of location during the mid-1840s may explain one of the most striking oddities in the Tyndall correspondence: although Tyndall's journal records him sending and receiving many letters, we have found no surviving letters from the year 1846.

It was while working for Carter that Tyndall met Thomas Archer Hirst, a future X Club member who would become one of Tyndall's closest friends. Hirst was then a fifteen-year-old apprentice engineer working for Carter's firm. Tyndall did not appear to take much note of their first meeting; his journal entry for 25 August 1845 says simply 'Mr Hirst came'.[33] However, Tyndall soon took an interest in the young man's development, impressed by his intelligence and intellectual curiosity. Tyndall began mentoring Hirst, along with another young friend from

Halifax, James 'Jemmy' Craven. Using his own self-education as a guide, Tyndall encouraged the pair to pursue independent studies of mathematics and to read the works of nineteenth-century literary luminaries such as Carlyle and Emerson.[34]

Tyndall's surviving letters from 1845 through 1847 also touch on troubles in his home country of Ireland. Ireland's impoverished rural population had rapidly grown in the nineteenth century and potatoes were the staple diet of the poor (although corn was also grown and exported). In 1845, more than half of the Irish potato crop was destroyed by potato blight, followed by near total failure of the crop in 1846 and a poor harvest in 1847. The Great Famine, as the crop failure came to be known, began as a natural catastrophe of extraordinary magnitude but its effects were intensified by the inaction of the government in London, headed by Lord John Russell. The deficiency in food was very real and the British government made some attempts to assist. Yet the economic orthodoxy of the day held that there should be as little government interference with the economy as possible.Some public work schemes and road building projects were established to provide employment, but the Famine's effects were still devastating.

Well-attested figures indicate that about one million people died of starvation or diseases associated with poverty and famine, and more than a million paupers emigrated under dreadful conditions to North America or other parts of Britain, such as Liverpool and Glasgow. Another consequence was mass eviction of families from their cottages, as landlords sought to rid their estates of pauperized farmers and labourers. Tyndall, dismayed by the suffering and malnutrition he observed when he visited Ireland in 1847, wrote to his former teacher Henry McClintock to ask if there was anything he might do for McClintock's impoverished parish in Macroom, County Cork.[35] Other letters from Ireland hinted at the social unrest bubbling up as a result of the Famine.[36]

The controversial Irish nationalist and reformer, Daniel O'Connell (1775–1847), is also referred to in Tyndall's letters during this period. O'Connell had won international esteem by obtaining the right of Catholics to sit in the British Parliament in 1829, and as a crusader against slavery. He was repeatedly elected to Parliament as a candidate supporting the repeal of the Act of Union. To push for this, he held a series of well-attended political rallies, dubbed 'Monster Meetings' by the press. The Prime Minister, Sir Robert Peel, banned one proposed monster meeting at Clontarf, just outside Dublin, in October 1843. O'Connell was arrested, charged with conspiracy and sentenced to a year's imprisonment in London with a fine of £2,000, but was released after three months by the House of Lords, which quashed the conviction and criticized the unfairness of the trial. Tyndall's Irish correspondents followed the news of O'Connell's arrest and trial; while Tyndall and his father had previously expressed a strong dislike of O'Connell, many of Tyndall's friends hoped for his release.[37]

Queenwood College

1847 was an eventful year for Tyndall. Family tragedy struck when Tyndall's father died in March. Tyndall recorded the event in his journal entry for 1 April 1847:

> Received a letter from my uncle Ned acquainting me with the death of my father; one of the strongest links between me and life is thus broken, one kind tongue for ever silent, one fountain of boundless sympathy for ever dried up, my father, my father – I have long shrunk from making this entry, I knew it would be thus, tears, bitter, hopeless tears.[38]

Although Tyndall Sr had not always approved of his son's agitations on the Ordnance Survey, Tyndall respected his father deeply and relied on him for advice and support, and his death was a blow.

The next change for Tyndall was a far more fortunate one. In July 1847, another former civil assistant named Robert Martin wrote to Tyndall to urge him to contact George Edmondson (1798–1863), a Quaker schoolmaster, who was looking for a man to teach engineering and surveying at his new college in Stockbridge, Hampshire.[39] The school, Queenwood College, would be located in a building that had once been home to a school run by the Owenite socialist movement (see frontispiece to this volume). Robert Owen, the socialist leader, had called the building Harmony Hall and intended for it to be a communitarian school where pupils would to be educated according to his utopian principles. However, Owen's lavish spending on this and other buildings in Hampshire all but bankrupted the Owenite movement.[40] In 1846, the Society of Friends (Quakers) took a lease on Harmony Hall for a relative pittance of £600, renamed it Queenwood College and installed Edmondson as Headmaster.

Edmondson envisioned Queenwood as a school that would provide superior scientific and mathematical education. Harmony Hall's eight hundred acres of grounds also provided the opportunity for Queenwood to offer practical education in agriculture and surveying. Edmondson and Martin apparently felt that Queenwood would be a good match with Tyndall's background and abilities. On 4 August 1847, Tyndall wrote in his journal that Edmondson had written to him to offer '£150 a year, rooms, board &c in the College'.[41] Tyndall accepted the position and moved to Queenwood later that month. This would prove to be one of the most significant steps in his early life.

At Queenwood Tyndall met Edward Frankland, a future natural philosopher and X Club member, who would have a profound influence on Tyndall's career path.[42] Frankland had previously been an assistant to the chemist Lyon Playfair in his laboratory at Putney College of Engineering, and had just returned to England after a long visit to the University of Marburg, Germany, where he had pursued research in organic chemistry. Edmondson had recruited Frankland to teach chemistry at Queenwood and Tyndall was immediately impressed with his new colleague. After their first meeting, Tyndall wrote in his journal that Frankland was 'evidently all Mr Edmondson taught me to expect him. A deeply thoughtful and instructive young man'.[43] Although at the time Tyndall had little knowledge of experimental chemistry or physics, he became a regular attendee of Frankland's school classes on chemistry and took careful notes on them in his journals.

Sometime that fall, Frankland and Tyndall decided that they would collaborate in a demanding course of self-instruction. In September 1847, Tyndall wrote to his friend Jack Tidmarsh explaining that he and Frankland

> make an exchange of knowledge – he teaches me the mysteries of his art while I teach him mathematics. We meet together every night and read for a certain time: afterwards each questions the other, and thus the matter is planted in our minds.[44]

The two also joined forces to teach an innovative laboratory-based science course, possibly one of the first in England.[45] At some point, however, Tyndall and Frankland began to wonder if they could truly fulfil their educational aspirations while at Queenwood. They began to aspire to university-level instruction. In May 1848, Frankland wrote in his journal that he had written to his friend Heinrich Debus (later a great friend of Tyndall's as well) to ask his opinion of the University of Marburg, where Debus was studying for a doctorate in chemistry.[46] Meanwhile, Tyndall began learning German in September 1847, suggesting that he was thinking of pursuing scientific study abroad.[47]

Other factors may also have contributed to Tyndall and Frankland's growing feeling that they should move on from Queenwood and undertake doctoral work. Despite his close friendship with Frankland and his fondness for his students, Tyndall was not finding Queenwood an entirely congenial atmosphere. While none of the existing letters contain any criticism of Queenwood's headmaster, Tyndall's journal records several complaints about Edmondson's management of the school and his social treatment of Tyndall. Most significantly, Tyndall disapproved of Edmondson's conduct during a conflict between two Queenwood teachers, Josiah Singleton (Edmondson's brother-in-law) and John Yeats. Tyndall held Yeats in low regard; in May 1848, Tyndall's journal records that Yeats beat two students 'black and blue' for swimming, and a later entry describes Yeats as 'a shallow hypocrite and unprincipled serf'.[48] When Edmondson, following Yeats's advice, seemed to be on the verge of dismissing Singleton, Frankland and Tyndall attempted to intervene on Singleton's behalf. Their efforts were met with indifference from Edmondson and with a rebuke from his wife, Anne Singleton Edmondson.[49] Singleton ultimately left Queenwood. This uncomfortable relationship with the Edmondsons appears to have provided an additional motivation for the decision Tyndall and Frankland reached in the spring of 1848: they would leave Queenwood at the end of the school year and go to the University of Marburg.

Marburg and Mentorship

It might seem strange that Tyndall would choose to pursue a doctoral degree far from home, in a language that he had only begun studying in earnest a few months earlier. However, Tyndall was far from the only young Briton to pursue scientific study in the German states in the mid-nineteenth century. As one historian has observed, 'By 1900 there was hardly a chemist of any standing in Britain who had not a German PhD'.[50] In the 1840s Germany was not yet a unified country; it was a collection of small principalities, duchies, electorates and other types of states. Many of these states regarded an impressive university as an essential part of building their province's reputation, and competition for intellectual prestige led to tremendous investment in both teaching and research. This sense of rivalry also prompted important innovations in university education. German universities had been the first to require doctoral students to produce a dissertation that made an original contribution to their field of study. In 1847 there was no analogous requirement in Great Britain. Tyndall hoped to master a great deal of science and mathematics as quickly as possible. Germany was, for him, an obvious choice despite the language barrier.

Tyndall and Frankland quickly settled on the University of Marburg in the electorate of Hesse[51] as their doctoral home. Marburg was founded in 1527, making it one of the oldest universities in the German states. It was also famous as the first German university to establish a professorship in chemistry, which it did in 1609. However, Tyndall and Frankland's choice of Marburg had less to do with its past than with its present. Marburg was home to Robert Bunsen, a brilliant experimentalist and one of the most respected teachers of chemistry in the world.[52] The pair wrote to professors at Marburg and were each invited to come and pursue their work beginning in the fall of 1848.[53]

After resigning from their positions at Queenwood, Tyndall and Frankland made a summer trip to France and Germany. The pair were thus on hand for some of the most dramatic events of the Revolutions of 1848. Much of Europe experienced republican uprisings against monarchies in early 1848. The unrest was arguably most successful in France, where the rebellions ended the Orleans monarchy and established the Second Republic in February 1848. Shortly after their trip began in June 1848, Tyndall was called back to London to resolve some outstanding business. He left Frankland alone in Paris on the morning of 23 June 1848, and learned the next day that another rebellion had broken out, this time against a conservative turn in the Second Republic's government. He quickly returned to Paris to find Frankland and described the scene there in a letter to Thomas Hirst:

> All was tranquil when I arrived, groups of military were encamped in the streets. … Of course the scenes of the hottest encounter engaged my attention first. Paris will long bear the mark of the affray. The Rue St. Jacques; Rue St. Antoine, Faubourgh, St Antoine, Clos St. Lazare, &c will long be memorable.[54]

Some three thousand people died during the June insurrection in Paris; fortunately, Frankland had come through the events unharmed. The June rebellions were quickly crushed, and in December of 1848 France elected the conservative Louis-Napoleon Bonaparte as President.[55]

Tyndall and Frankland began their studies at Marburg in October 1848. Tyndall described his typical day in a November 1848 letter to Hirst:

> I work tolerably hard – up at 5 oC every morning – work at German until 7½ breakfast, from 8 to 9 a lecture on Physics – 9 to 10 Professor Bunsen on Chemistry – 4 to 5 a private lesson from the professor of mathematics – I could not possibly be better suited than I am – the day I spend in the laboratory – have tea at 6 oC and study mathematics until 10 at which hour I punctually go to bed. I'm not likely to get fat on this work for so far I have stood it very well.[56]

In order to complete a doctoral degree at Marburg, Tyndall would have to pass three *vivâ voce* examinations. He chose mathematics as his principal subject, and physics and chemistry as his secondary subjects. Tyndall pursued his study of chemistry under Bunsen, and found that the famed chemist lived up to his reputation as a teacher. In later life, Tyndall would reflect that Bunsen 'merged himself with his subject; his exposition was lucid, and his language pure ... after some experience of my own, I still look back on Bunsen as the nearest approach to my ideal of a university teacher'.[57] Tyndall's supervisor for mathematics was Christian Ludwig Gerling (1788–1864). Finally, Karl Hermann Knoblauch (1820–1895) guided Tyndall's study of physics after Knoblauch came to Marburg in 1849.

A major component of the doctoral degree at Marburg was to present an original dissertation, written in German, on which Tyndall began to work in late 1849. He decided to focus his dissertation on the mathematical properties of screw surfaces. Tyndall was eager to finish his degree relatively quickly, as he was supporting himself with a limited savings of £300. The desire to complete his doctorate before his savings ran out likely influenced his decision to pursue mathematics and not physics or chemistry as his primary subject, since at the time Tyndall's knowledge of mathematics was far more advanced than his knowledge of the experimental sciences.

However, Tyndall's experience in Marburg ultimately seems to have convinced him to pursue the experimental sciences as his main field of inquiry. Frankland attributed this choice largely to Knoblauch, whom he said was 'probably the main cause of [Tyndall] devoting himself, for the future, chiefly to physical science'.[58] Under Knoblauch's supervision Tyndall pursued his first investigations in physics. The two collaborated on a study of the diamagnetic properties of crystals, work that would one day form the basis of Tyndall's first published physics paper, lead to his extended study of glaciers and would gain him admission to the Royal Society of London.

Tyndall continued mentoring both Hirst and Craven by letter during his time at Marburg. His two protégés were, like their mentor, pursuing the study of chemistry. They relied on two textbooks by the German chemist Carl Remigius Fresenius to guide their work. Tyndall wrote to them with advice on how best to read Fresenius (including a table of elements that he felt would aid the pair as they worked through the books) and also outlined elementary chemical experiments they could conduct in their home laboratory.[59]

Tyndall and Hirst became particularly close while Tyndall was at Marburg. In June 1849, Tyndall wrote what seems to be his first letter to his great philosophical idol Thomas Carlyle, not to introduce himself, but to ask that Carlyle consider sending Hirst some advice: 'a word of counsel when you have time to give it – I have strong faith that he will one day prove himself worthy of your pains'.[60] Later that year, Hirst made the trip to Marburg to visit Tyndall, likely because he was considering pursuing a doctorate himself.

At the end of 1849, Tyndall and Hirst's relationship began shifting from a protégé-mentor exchange to something more like a friendship between equals. The catalyzing event was the September 1849 death of Hirst's mother. The death was not only a personal blow, but left Hirst's finances in some question. Tyndall, no stranger to uncertain financial prospects, assured Hirst that 'long as I have a hand and brain and nine years extra experience, these goods and chattels and whatever marketable value they can bring is yours as well as mine'.[61]

However, Hirst soon learned that his mother had in fact left her children a small inheritance. His mother's bequest would give Hirst an income of around £150 per year. He offered to share his good fortune with Tyndall so that his friend might continue studying in Europe: 'Now I know well your desire to remain longer in Germany Berlin or France & that this affair of money is your sole hindrance. Will you accept of part of this for the purpose?'[62]

The offer rendered Tyndall near-speechless (an unusual condition for him). 'I dont know what to say ... I will decide nothing now', he replied. He seemed torn between joking refusal and sincere consideration of the prospect of a loan: 'The offer you have made me is not a thing common in the world. It argues a degree of courage which I hardly contemplated as belonging to your pulpy texture! What you dog!'[63] Although Tyndall made no decision about the proposed loan before December 1849, by the close of this volume of correspondence Hirst had assumed the role he would play for the rest of Tyndall's life: that of Tyndall's closest friend and confidante.

Conclusion

Tyndall's correspondence from 1843 through 1849 covers a fascinating and dramatic period in European history and a critical turning point in Tyndall's life. These letters trace Tyndall's path through unemployment, railroad surveying and schoolteaching, and show him developing new interests in physics and chemistry. The early letters reveal a young man preoccupied with simply making a living. Indeed, at beginning of the volume, there is no indication that Tyndall was considering a career in science, much less that he would become one of Victorian Britain's most famous physicists. However, by the late 1840s, Tyndall was determined to find a path that suited his temperament and engaged his intellect. Inspired in part by his new friend Frankland, Tyndall began to believe that natural philosophy might be that path and he left a stable position teaching at Queenwood College to spend his life's savings in pursuing a professional science education and PhD degree in Germany. Tyndall remained at Marburg until 1851, and would return to England with a firm goal in mind: to obtain a position where he could pursue research in physics and make a name for himself in natural philosophy.

Notes

1. See 'Introduction', *Correspondence of John Tyndall, Volume 1* (London: Pickering & Chatto, 2015), pp. xxxix–liii, on p. xlvi. See also M. Tynan, 'Appendix: John Tyndall's Role in the Ordnance Survey of Ireland', *Correspondence of John Tyndall, Volume 1*, pp. 451–7.
2. Tyndall, 'Address Delivered at the Birkbeck Institution on October 22, 1884', *New Fragments* (London: Longmans, Green, and Co., 1892), pp. 224–47, on p. 228.
3. See 'Introduction', *Correspondence of John Tyndall, Volume 1*, p. xlvi. See also R. Oliver, 'The Ordnance Survey in Great Britain 1835–1870' (PhD thesis, University of Sussex, 1985).
4. See letter 0236 of this volume, in which a copy of this document was enclosed.
5. See letter 0228, *Correspondence of John Tyndall, Volume 1*; and letters 0248, 0252, 0261 and 0287 of this volume.
6. See letter 0252 of this volume.
7. See letter 0287 of this volume.
8. See letter 0236 of this volume.
9. See letter 0245 of this volume.
10. See letter 0246 of this volume.
11. See letter 0246 of this volume.
12. See letter 0262 of this volume. See also the Appendix in this volume for Tyndall's own chronology of the events leading up to his dismissal.
13. See letter 0263 of this volume.
14. See letter 0264 of this volume.
15. RI MS JT/2/13a/2.
16. See letter 0276 of this volume.
17. See letters 0277 and 0278 of this volume.
18. See letters 0283 and 0301 of this volume.
19. See 'Introduction', *Correspondence of John Tyndall, Volume 1,* p. xlix. For examples of verse in this volume, see letters 0274, 0303.
20. See letter 0259 of this volume.
21. See letter 0288 of this volume.
22. See letter 0266 of this volume.
23. See letter 0234 of this volume.
24. See letter 0237 of this volume.
25. See letter 0298 of this volume.
26. See letter 0299 of this volume.
27. RI MS JT/2/13a/46.
28. RI MS JT/2/13a/62.
29. Tyndall, 'Address Delivered at the Birkbeck Institution on October 22, 1884', p. 230.

30. P. J. G. Ransom, *The Victorian Railway and How It Evolved* (London: Heinemann, 1990), pp. 82–4.
31. Ransom, *The Victorian Railway*, p. 88.
32. See letter 0323 of this volume.
33. RI MS JT/2/13a/98.
34. See, for example, letters 0346, 0348, 0365 of this volume.
35. See McClintock's reply, letter 0341 of this volume.
36. See letter 0360 of this volume.
37. For exchanges between Tyndall and his father about O'Connell, see, for example, *Correspondence of John Tyndall, Volume 1*, letters 0045, 0047, 0063 and 0106. For mentions of O'Connell in this volume, see letters 0268, 0269, 0270 and 0289.
38. RI MS JT/2/13a/191.
39. RI MS JT/2/13a/230.
40. G. Claeys, 'Owen, Robert (1771–1858)', *ODNB*.
41. RI MS JT/2/13a/240.
42. On Frankland, see C. Russell, *Edward Frankland: Chemistry, Controversy and Conspiracy in Victorian England* (Cambridge: Cambridge University Press, 2003); R. Barton, 'The X Club: Science, Religion, and Social Change in Victorian England' (PhD thesis, University of Pennsylvania, 1976); and R. Barton, '"Huxley, Lubbock, and Half a Dozen Others": Professionals and Gentlemen in the Formation of the X Club, 1851–1864', *Isis*, 89 (1998), pp. 410–44.
43. RI MS JT/2/13a/252.
44. See letter 0336 of this volume.
45. C. A. Russell, 'Frankland, Sir Edward (1825–1899)', *ODNB*.
46. LT, 'Biography', vol. 1, p. 212.
47. See letter 0336 of this volume.
48. RI MS JT/2/13a/320; and RI MS JT/2/13a/328.
49. See letter 0349 of this volume. Josiah Singleton was Anne Edmondson's brother; the Edmondsons' reasons for siding with Yeats rather than Singleton are unclear. Tyndall himself wrote in his journal that 'I am unable to account for his [Edmondson's] anxiety to prove John Yeats innocent and his own brother a rascal' (RI MS JT/2/13a/329).
50. W. V. Farrar, 'Science and the German University System, 1790–1850', in M. Crosland (ed.), *The Emergence of Science in Western Europe* (London: The Macmillan Press, 1997), pp. 179–92, on p. 189.
51. Many of the correspondents in this volume refer to the electorate by its older name, Hesse-Kassel.
52. See, e.g., the testimonials from his students in this obituary: 'Professor Robert W. Bunsen', *The Journal of the American Chemical Society*, 23 (1900), pp. 89–107.
53. LT, 'Biography', vol. 1, p. 212.
54. See letter 0356 of this volume.
55. On the Revolution of 1848 in France, see M. Agulhon, *The Republican Experiment, 1848–1852,* trans. J. Lloyd (Cambridge: Cambridge University Press, 1983).
56. See letter 0365 of this volume.
57. Tyndall, 'Address Delivered at the Birkbeck Institution on October 22, 1884', p. 238.
58. E. Frankland, 'Obituary Notices of Fellows Deceased: John Tyndall', *Proceedings of the Royal Society of London* 55 (1894), pp. xviii–xxxiv, on p. xix.
59. See letters 0370 and 0372 of this volume.
60. See letter 0376 of this volume.
61. See letter 0381 of this volume.
62. See letter 0382 of this volume.
63. See letter 0384 of this volume.

EDITORIAL PRINCIPLES

Our aim is to include all letters to and from John Tyndall that are currently extant. Some of the letters are from published sources, such as letters written by Tyndall to newspapers or journals, but the majority of them are either the originals or typescripts produced by Louisa Tyndall from originals. One central editorial principle has been to transcribe, wherever possible, from original letters rather than Louisa's typescripts. This was because of Louisa's conscious and unconscious editorial interventions, a point discussed in more detail below. We have also aimed to reproduce, as accurately as possible, the text of the letters as they were written. Spelling mistakes that have appeared in the handwritten letters have therefore been preserved; mistakes appearing where only the typescript letter has survived are silently corrected as in our judgement they are almost always typographical errors. But, in general, non-standard spellings (e.g. the use of dialect) has been retained even when in a typescript only letter (though there is the possibility that the spelling is Louisa's, not her husband's). We have included illustrations that are part of the letter.

The letters are presented in chronological order of writing. To indicate the beginning of a new year, the date of that year has been inserted in a large font before the first letter from that year.

Both the transcription of non-English language letters and the English translations are presented. The transcription of the non-English letter will appear first, followed by the English translation in a smaller font. Some foreign letters are in an original hand while others exist only as transcriptions by Louisa or others who often did not have a full command of the language. In the latter case these letters contain frequent spelling, grammatical and interpretation errors that do not seem to have been committed by the original author. In these typescript-only cases, and only in these cases, spelling and grammatical errors have been corrected, as have poor interpretations where a better word choice was obvious from the context. If the transcription and translation are based on a typescript transcription, this will be indicated by 'LT Transcript Only'.

There is a standard format for each letter. The first line lists the author, if it is to Tyndall, or the recipient, if it is from Tyndall; the date; and the letter number. The second line lists the place the letter was written (if known), and any information supplied by the writer in their salutation, such as the date or time of day when the letter was written. Information from stationery – for instance from institutional letterhead – is also supplied in the second line if extant. The opening salutation, the text of the letter, and closing then follow. Then come the source and editorial notes, the latter indicated in the text by Arabic numerals.

Editorial notes are intended to provide the contextualization necessary to understand the contents of each letter. There are basically six types of editorial notes. Many of them are informative notes on persons, real or mythical, mentioned in the letters. If a person referred to in a letter is not described in a note, they have been mentioned more than twice in the volume and they will have an entry in the Biographical Register at the back of the volume. For the rest – those mentioned once or twice – biographical information will appear in the editorial note appended to the letter in which he or she is first mentioned. When there is a second reference to that person a note will refer back to the letter in which he or she was first mentioned. Biographical notes on well-known figures such as Faraday or Huxley stress what that person was doing at the time when the letter was written rather than providing a full overview of the person's life.

A second type of editorial note identifies allusions and quotations. In the case of quotations, where possible, the first edition of the work is cited unless a different edition is mentioned in the letter. If a quote is not in English a translation is given. Glosses on obscure places, abstruse words, and words used in an unfamiliar sense comprise the third type of editorial note.

The fourth type provides information about the context of the letter, drawing on such sources as contemporary publications, scholarly sources, Tyndall's personal journals and field notes, and other letters in the Tyndall Correspondence. The fifth type of editorial note is cross-references to other letters in the Correspondence, referring to the letter number. The last type is information on significant textual changes that Tyndall made to the original that is thought to be relevant to the meaning of the letter.

There are a series of conventions for indicating when editors have given additional information about the text of a letter:

When Louisa Tyndall has made an annotation or marginal insertion that the editors have deemed important enough to retain, we have included her annotation in an editorial footnote.

When editors other than LT have made insertions they are indicated by unitalicized square brackets, e.g. [word].

When certain words are ambiguous and we have had to conjecture their meaning they are indicated by italicized square brackets, e.g. *[*words*]*; when we have considered such words illegible the number of illegible words are indicated in full italicization and square brackets, e.g. *[3 words illeg]*.

If a word or series of words are not present or have been destroyed– by inkblots, holes, or mould, for example– they are indicated as missing in italicized angular brackets, e.g. *<3 words missing>*.

In extremely rare cases where it is clear that Louisa herself has intentionally destroyed a few words, they are indicated as excised in italicized angular brackets, e.g. *<3 words excised>*.

Finally, a brief word about Louisa. A Tyndall correspondence project would be very difficult, perhaps even impossible, without the work done by Louisa, much of which seems to have been done in order to write a biography of her late husband. This biography was not completed. We have therefore treated Louisa as a *de facto* member of the editorial team. Some of her handwritten comments on the typewritten letters are extremely helpful and thus have been retained as footnotes to the relevant passage. Conversely, many of these handwritten annotations seem to have been inserted to help write her biography – for instance, summaries of passages or interesting quotes. When the editors have deemed these notes of Louisa's add no new information, they have not been included for publication.

It is also important to note that Louisa was not always reliable as a transcriber. She wasn't always accurate and may have gone back and edited portions of Tyndall's writing to make his letters seem more consistent or grammatically correct. She also left out details that she may have considered too private or too salacious for public consumption. In cases where there is an original, handwritten letter, therefore, we have considered that as taking precedence over a typed transcription.

NOTE ON MONEY

Money features prominently in these letters, which include discussions of the pay received by civil assistants and the overall cost of the Ordnance Surveys. In the 1840s the currency used in both England and Ireland was the pound sterling, usually abbreviated to pound (£). A pound consisted of 20 shillings and each shilling contained 12 pennies. In the notes to this volume the editors have used the following format: £3 4*s*. 5*d*. to express three pounds, four shillings and five pence. The letter L (for livres) has sometimes been used in place of the £ sign and the *d*. (derived from denarius, a small coin (Latin)) indicates pennies.

Several forms of representation were used in these letters. Sometimes the numbers for pounds, shillings and pence were separated by dots, forward slashes or spaces; thus £3 4*s*. 5*d*. could be written as £3.4.5, £3/4/5 or £3 4 5. A zero could be represented by an en-dash; thus 7 shillings was often written as 7/– or just 7/.

A farthing was a quarter of a penny (¼*d*.).

A crown was 5 shillings and half a crown equalled two shillings and sixpence, often written as 2/6.

A guinea was 21 shillings, that is £1 1*s*.

TIMELINE OF JOHN TYNDALL'S LIFE

Year	Event
c. 1822	John Tyndall born at Leighlin Bridge, Carlow County, Ireland
c. 1836	Attends John Conwill's National School in Ballinabranna
1839	Begins employment as civil assistant with Ordnance Survey of Ireland in Carlow
1840	Becomes civil assistant in the Ordnance Survey Office at Youghal, Cork County
1841	First appearance in print (poem in Carlow *Sentinel* under pseudonym 'Walter Snooks')
1842	Joins English Ordnance Survey in Preston
1843	Leads written criticism of Ordnance Survey in Liverpool Mercury; dismissed November
1844	Home in Ireland, mostly unemployed, until obtaining a position with the firm of Nevins and Lawton, Surveyors, of Manchester
1845–7	Works for Richard Carter, Surveyor, of Halifax
1845	Meets Thomas Archer Hirst in Halifax
1847	Begins teaching mathematics and surveying at Queenwood College, Hampshire
1848–50	At University of Marburg, working for PhD with Robert Bunsen and Hermann Knoblauch
1850	Returns to England: meets Michael Faraday and attends his first British Association meeting, before returning to Marburg
1851	Moves from Marburg to Berlin (April–June); then returns to Queenwood
1851	Meets Thomas Henry Huxley
1852	Elected Fellow of the Royal Society
1853	First lecture at Royal Institution (RI)
1853	Meets Herbert Spencer
1853	Appointed Professor of Natural Philosophy at RI
1853	Awarded Royal Medal for magnetic work; declines when controversy arises over award
1857	Climbs Mont Blanc for first time with Hirst
1859	Demonstrates existence of greenhouse gases and climatic implications
1859	Joins Royal School of Mines as Professor of Natural Philosophy
1860	*The Glaciers of the Alps* published
1860	Meets John Lubbock and his wife, Ellen Lubbock

Year	Event
1861	First ascent of the Weisshorn
1861	Delivers his first Christmas Lectures at RI
1862	*Mountaineering in 1861* published
1862	Begins to champion Mayer's priority in discovery of conservation of energy
1863	*Heat Considered as a Mode of Motion* published
1864	Awarded Royal Society's Rumford Medal
1864	First meeting of X Club
1865	*On Radiation* published
1865	Engages in public controversy over the efficacy of prayer
1866	Succeeds Michael Faraday as scientific adviser to Trinity House
1867	Michael Faraday dies; Tyndall becomes Superintendent of the RI
1867	*Sound: A Course of Eight Lectures* published
1868	Successfully climbs Matterhorn
1868	Discovers the cause of light scattering, to be known as 'Tyndall Effect'
1868	*Faraday as a Discoverer* published
1868	Lecture on 'Scientific Materialism' at the British Association
1868	Resigns from Royal School of Mines
1869	Joins Metaphysical Society
1870	'On the Scientific Use of the Imagination' lecture at British Association
1870	*Researches on Diamagnetism* published
1870	*Three Scientific Addresses* published
1870	*Researches on Diamagnetism and Magne-Crystallic Action* published
1871	Meets Louis Pasteur for first time while in Paris
1871	*Hours of Exercise in the Alps* published
1871	*Light and Electricity* published
1871	*Fragments of Science* published
1872	Debates over how to measure the efficacy of prayer, aka the 'Prayer-Gauge Debate'
1872	*Contributions to Molecular Physics in the Domain of Radiant Heat* published
1872	*The Forms of Water in Clouds and Rivers, Ice and Glaciers* published
1872–3	USA Lecture Tour
1873	*Six Lectures on Light* published
1874	As President of the BAAS, delivers the 'Belfast Address'
1876	Marries Louisa Charlotte Hamilton
1877	Develops 'Tyndalliszation' (lengthy sterilization to destroy heat-resistant spores)
1877	Tyndall and Louisa build summer cottage 'Alp Lusgen' at Belalp, northern side of Valais, above Brieg
1877	*Fermentation and its Bearings on Phenomena of Disease* published
1879	*Fragments of Science*, which had gradually expanded, first published in two volumes
1881	*Essays on the Floating Matter of the Air* published
1885	Tyndall and Louisa build Hind Head retreat, Surrey Downs

Year	Event
1887	Resigns from RI
1890	Fight with Gladstone in *The Times* over Irish Home Rule
1892	Hirst dies
1892	*New Fragments* published
1893	Dies Haslemere, Surrey, accidentally poisoned

TIMELINE OF EVENTS IN JOHN TYNDALL'S LIFE SPECIFIC TO VOLUME 2

Year	Month	Tyndall Biographical Details	Notable social, political and cultural events
1843–4	September–February	Five letters by Tyndall about the Ordnance Survey appear in the *Liverpool Mercury*; Tyndall signs these letters with the pseudonym 'Spectator'	In September, Britain claims the Boer republic of Natal as a colony.
1843	16 September	Captain Henry Tucker issues a District Order requiring the Civil Assistants on the English Ordnance Survey to sign a document indicating that they are either satisfied with the rate of pay and will remain on the Survey without a raise, or that they are dissatisfied and wish to return to Ireland	
	23 September	Tyndall and other Preston surveyors submit 'the Memorial' to George Murray, complaining about the employment conditions on the Survey and about Tucker's District Order	
	October		Arrest of Daniel O'Connell, Irish nationalist and reformer
	7 November	Henry Tucker issues an order dismissing fifteen labourers and nineteen civil assistants from Lt. Charles Stanley's division. Tyndall is among those fired, along with his friends Edmond Sayers, George Latimer, Phillip Evans, and Robert Martin	
	21 November	Tyndall's dismissal from the Survey goes into effect	

Year	Month	Tyndall Biographical Details	Notable social, political and cultural events
	December	Tyndall returns to Leighlin Bridge	*A Christmas Carol* by Charles Dickens is published
1844	February		
	March	Tyndall cancels his plans to emigrate to Cincinnati, USA	The Factory Act is introduced to Parliament and proposes limiting children's work to 6 1/2 hours a day to allow time for schooling. Women would be brought under the same protective restrictions. After much debate and criticism, the Act passes in March.
	April		Texas became a US territory
	May		Samuel Morse sends the first telegraph message
	July	Tyndall fails to obtain a hoped-for applotment job	US signs the Treaty of Wanghia with the Chinese government, the first ever diplomatic agreement between China and the US
	August	Tyndall returns to Preston to work for Mr. Wren, who needs an assistant skilled in mathematics and accounting	
	September	Tyndall leaves Wren's employment and is hired as a railroad surveyor by the Manchester firm of Nevins and Lawton	
	October		*Vestiges of the Natural History of Creation* is published; the Great Disappointment of the American Christian Millerite movement occurs when Jesus Christ does not return on the prophesied date
	December	Tyndall moves to Halifax to work for Richard Carter, a surveyor preparing plans for a West Yorkshire Railway	
1845			Michael Faraday discovers that magnetism affects the polarization of light travelling through glass, also called the 'Faraday effect'; Faraday also discovers diamagnetism

Year	Month	Tyndall Biographical Details	Notable social, political and cultural events
1845	January or February	Tyndall returns to Manchester to work for Nevins and Lawton	
	March	Tyndall returns to Halifax and resumes working for Richard Carter	
	June–August	Tyndall travels to London to help the West Yorkshire Railway scheme prepare to make its case before Parliament	In July and August, English engineer Isambard Kingdom Brunel's SS *Great Britain* crosses the Atlantic
	August	Tyndall meets Thomas Hirst in Halifax	The Bank of England raises interest rates, bringing the wildest period of 'railway mania' to a close; *Scientific American* is founded
	September		Onset of the Great Famine in Ireland; the potato blight destroys more than half of the Irish potato crop until 1850. It also kills approximately one million people and causes up to two million people to emigrate from Ireland
	December		US President James Polk speaks in favour of enforcing the Monroe Doctrine, endorses Manifest Destiny
1846	February		US President James Polk annexes Texas
	April		Outbreak of the Mexican-American War
	June		Prime Minister Robert Peel resigns after the repeal of the Corn Laws. He is replaced by the liberal cabinet of Lord John Russell who encourages Ireland to rely on its own resources thereby worsening the effects of the potato famine. Britain and the US sign the Oregon Treaty, dividing Oregon Country along the 49th parallel.

Year	Month	Tyndall Biographical Details	Notable social, political and cultural events
	September	Parliament rejects both the West Yorkshire Railway scheme and a rival railway plan; Carter is obligated to fire many employees, but keeps Tyndall	The planet Neptune is discovered by German astronomer Johann Gottfried Galle with assistance from Heinrich Louis d'Arrest
1847			William Makepeace Thackeray publishes *Vanity Fair*; Charlotte Brontë publishes *Jane Eyre*; Emily Brontë publishes *Wuthering Heights*; Anne Brontë publishes *Agnes Grey*
	March	Death of John Tyndall, Sr	
	June		Irish Poor Law Extension Act is passed by Parliament
	July		Hermann von Helmholtz reads "On the Conservation of Forces" at the Physical Society in Berlin
	August	Tyndall accepts a position at Queenwood College in Hampshire	
	September	Tyndall moves to Hampshire to begin teaching at Queenwood; meets Edward Frankland	
	November		Sonderbund War, a civil war in Switzerland, takes place
1848			Foundation of the Pre-Raphaelite Brotherhood; publication of the *Communist Manifesto* by Karl Marx and Friedrich Engels; Gold Rush begins in California
	February		Revolutionary uprisings in most European countries; in France, the revolution results in the formation of the Second Republic; the Treaty of Guadalupe Hidalgo is signed, ending the Mexican-American War.
	March		Beginning of the First Schleswig War, a military conflict in southern Denmark and northern Germany
	April		Second Anglo-Sikh War begins

Year	Month	Tyndall Biographical Details	Notable social, political and cultural events
	June	Tyndall and Frankland decide they will pursue doctorates at the University of Marburg; both give their notice at Queenwood	
	June–July	Tyndall and Frankland travel in Europe	Uprisings take place in Paris protesting a conservative turn in the Second Republic; Matale Rebellion againstBritish rule in Sri Lanka
	July		The Young Irelander Rebellion takes place in County Tipperary; Seneca Falls, a women's rights convention, takes place in New York
	August		English Public Health Act is passed
	October	Tyndall arrives at the University of Marburg and begins his doctoral work	
	December		Louis-Napoléon Bonaparte is elected the first President of the French Second Republic
1849	March		Second Anglo-Sikh War ends, resulting in the annexation of the Punjab by the East India Company
	May		May Uprising in Dresden, one of the last events in the Revolutions of 1848
	June		Constitution of Denmark is signed, making Denmark a constitutional monarchy
	August	Hirst visits Tyndall at Marburg	
	September	Death of Hirst's mother, Hannah Oates Hirst; Hirst offers to lend Tyndall part of his inheritance	

1843

From Archibald McLachlan 6 September 1843 0231

Wednesday – Sep. 6th 1843

Yours of Sunday[1] came in due time – 3 times have I commenced a reply, & interupted by Company, however it is good company, valuable, and will be turned to good account – I will not reply to your last[2] – that will be done tomorrow or next day –

Several 111[3] came to different parties, and the whole hive may be likened to one in harvest, conjecture is alive – it was proposed to convey through the P–[4] a vote of thanks to Spec:[5] this wont do – however I will turn all into a proper channel – every man – old, young, high, low, red, black – all approved, & the general opinion is, 'it will do much good, or harm' Cause a change of action, or, break up the affair altogether, the parties concerned in an answer cannot remain silent, that would not do – an answer cannot be made; – The whole is in a ferment, I mean the Goths & Vandals –

Tell D.[6] his enquiries have not been forgotten – And to Omega,[7] that if he does not exercise more prudence, the whole is injured – I dreaded some indiscretion, 'Discretion is the better part of valour, by which better part I have saved my life'[8] &c &c – And now for a bit of generalship if it can be managed. Carbon[9] is a Tartar[10] on all points touching the honor of the O.S.[11] – It is desirable that a damper be immediately put on Acid[12] – or that he should be made to vanish altogether – & that soon – Weighing all things – Spec: Lumper[13] &c &c – Could not something be done with Carbon in a friendly way, and thereby cause a change – without hazard of Destruction to the whole O.S. – for if Spec: continues, & the deficit in the Revenue continues (it is said) '10 to one but the whole affair of the O.S. is stopped'

Consult – & say what you think of the following –

To – – – Director of the Ord: Svy: of England[14] –

Sir, It has fallen to my lot to write this letter and I begin with a request that you will not misconstrue its meaning – It cannot longer be permitted that public money should be employed to the injury of private enterprise in this Country – nor public confidence violated with impunity. It is clearly ascertained, that money which has been voted by Par:[15] for one purpose, has, to some extent been misapplied, or misappropriated by your officers to an undertaking never contemplated, nor authorised by the Par: who voted that money – and that your officers have gone farther in this matter than the nature of their Commission will warrant – Your off: have for some time pursued a course of action, inconsistent with the act of Par: for the Survey of the Northern Counties of this Country[16] – and it remains to be seen whether they will persevere in this course –

I would direct your attention to the Evidence given before the Com: of the Ho of Com:[17] in relating[18] to the Suvy & maps for the Commutation of the Tithes[19] – To the opinion of that Committee; to the Act. of Par: for the S. of the 6 N. counties[20] – and to a note written by the Master Gen. of the Ordnance[21] in reply to certain memorial from Land Surveyors[22] – And then ask you, are your officers justified by all these or any of them, in making Tithe maps, either by Public Competition or private contract or otherwise – or do you, as director, approve of and sanction such a proceeding –

<u>All</u> your officers <u>are not</u> implicated, there are some exceedingly imprudent men among them, who have <u>committed themselves publicly</u> by entering into competition <u>without authority</u>!!! and afterwards acted, and continue to act, in such a manner as to submit the character of the Mast Gen to suspicion, and as if his note in reply to a Memorial, was equivocal <u>and designed to be so</u> – But lest you should mistake the nature of the charges to be proferred – I shall condense the Third Letter[23] on the Ordnance Survey of England, which is to appear in the Liverpool Mercury, and arrange the substance under the following Items – and –

1<u>st</u> –Your officers employed under your direction on the O.S.E.[24] did enter into competition with certain Land Surveyors – and were successful – on this subject a memorial was presented – and the Master General of the Board of Or:[25] replied nearly in the following words 'The circumstances out of which the Memorial arose were unauthorised by the Mas: Gen: or B of Or: – and the Mast Gen: will give such instructions to the officers in charge of that work, as will prevent the recurrence of a like nature in future'. –

2 – It follows that your officers did enter into competition without authority –

3 – It is to be enquired into, by what, or by whose authority they acted, <u>and act</u>.

4 – That your officers did construct, and are now constructing maps for Tithe purposes –

5 – That Tithe maps are not paid for until they until approved by the Commissioners –
6 – There is no act of Par: authorising Government officers to construct maps for Tithe
7 – No money has been voted by Par. for the construction of Do[26] — Do —
8 – That men employed on the Ordnance S. of 6 N. counties were withdrawn from that work, & employed by your officers in drawing Tithe maps – & that during the time they were so employed, they were paid as formerly out of money voted for the S of the 6 N. Counties
9 – By this it appears that money voted and designed for the S of the 6 N. Counties of this country has been appropriated to another purpose – without the authority of the Par: who voted the money –
10 – That men employed in constructing Tithe maps were entered in your public accounts as if Employed on the O.S.E –
11 – The O.S.E. is authorised by one Act of Par – Tithe by another act: and that in your public accounts these are confounded together –
12 – That proper accounts have not been kept – nor can it be ascertained from your public accounts whether the public gain or lose –
13 – That should it be proved that proper accounts – and returns were duly made, they will show the amount of loss sustained by the Public –
14 – That after the amount of loss was known from your Accounts – your Officers <u>persevered</u> in constructing Tithe maps –
15 – That should it be made to appear from your accounts, that the Public has gained – it is to be examined into, whether the amount of such gain has been placed to the credit of the Public Accounts –
16 – That your officers, if acting under the orders of the Government, or by authority of Parliament, are not responsible for the amount of any loss – and cannot appropriate to their private use, any of the money gained –
17 – That the design of the Legislation is defeated – the Survey of the 6 N Counts delayed, & the Public has sustained a loss of time and money –
18 – That some of the maps constructed by the R.E.[27] were not undertaken by Public Competition or Contract – but by private treaty – or in a clandestine manner; which is unprecedented – & seems to be direct violation of Public confidence

I have now gone over the principal points of complaint relating to Tithe maps – and will repeat my former assertion, that this system cannot be permitted to continue, and however unpleasant an investigation may be to all parties; it is necessary, & unavoidable – commercial embarrassment involves on every side, & if Government officers are allowed to aid & increase that embarrassment without an attempt on our part at a remedy, we deserve to suffer – It is said that the

Director of the O.S.E. is not aware of whole facts of the case – and it is unwise to proceed without giving him a friendly hint; should his authority extend so far, he may correct the evil complained of – give him a chance – if he will not – nor cannot remedy. Then proceed – I end this letter as I began, with a request, that you will not misconstrue its design –

Now what do you think – what is the first thought that strikes you on reading it – My design would be, that you should correct the letter and enlarge it – or make it more pungent – and that each and every one of the Items should be retained similar to the manner I have done – you will see the items depend on each other – and are a consequence of one another to a great extent – when you have done this, send it to me – I will have it copied – and posted in London – it is received by —[28] and sets him & Acid at variance – he will censure Acid, and perhaps order that *[jobbing]* to cease – it might do more – it might ship Acid, and that is the principle design of it – Consult, and determine, whether such a maneuver would be prudent in the present state of affairs – If you think that such a thing would be effective, do it <u>immediately</u> – and send it to me – or – if you think Omega could manage better, send it to him – and post it in Liverpool –

I know he would dread the very Idea of such a letter & such Items getting into the Public press – you observe the Mast Gen: seems to be a party to the *[jobbing]* – and should he see such a letter in the press, he would come down on Carbon with a vengeance – Carbon knows this – and perhaps to prevent a Row, he might immediately look into the affair – you will have an answer to your last tomorrow – Cap B.[29] is now in London – so is Acid

The reason of Cap B. – And Acid being in London just now is matter of general speculation among the bandits – write soon

<u>Post time</u> | <u>B</u>[30] <u>Thursday</u> –

Form of Recommendation for the Pay of Civil Assistants[31]

Sir
I recommend
who is now receiving   per diem
in class   as
deserving of an increase of survey Pay
in that class, or of Promotion to class
District officer

Class A. (Miscellaneous Duties.) Rate of Pay not to exceed 2s. per Diem

---

Superintending Piling Parties and other duties of a responsible nature, requiring more intelligence than is necessary for the duties of a chainman.

Class B. (Clerks.) Rate of Pay not to exceed 2s per diem

Copying and other duties which require persons who can write tolerably well

---

Class C (Clerks) Rate of Pay above 2s and not exceeding 3s per diem.

Duties which require superior qualifications in Writing, considerable knowledge of Arithmetic, and skill in the particular Books and Accounts of the Survey, and in checking computations.

Class E. (Draughtsmen) Rate of Pay above 2s. and not to exceed 3s per Diem

---

1. Road Surveying and Plotting neatly and accurately.
2. Penning in lines of all Kinds either for Roads or Boundaries neatly and accurately

---

Class F. (Draughtsmen) Rate of Pay above 3s and not to exceed 4. per Diem.
Qualified for Road Surveying and Plotting and also capable of penning in lines of all Kinds, either for Roads or Boundaries.

Class G. (Draughtsmen) Rate of Pay above 4s. and not to exceed 6s per diem.
Qualified for Road Surveying, sketching Gardens, and completing the drawing of fair Plans.

Class H. (Draughtsmen) 6s per Diem, and upwards
Qualified to perform the whole of their duties, both in Surveying and drawing, in a rapid and efficient manner

---

Classification of the Civil Assistants

Class								
				E				
A	B	C	D	1	2	F	G	H

This Table is to be inserted in the Monthly Return, by the side of that showing the classification of the Royal sappers and miners,[32] and to be filled up in a similar manner.–

In each recommendation the Division Officer will fill up the Table with the members in the respective classes attached to his Division –

In recommending the Civilians for the respective classes, the District Officer will take care to state definitely the grounds on which he recommends the individual for the class, in order that I may not only put him in the class, but also assign to him the Rate of Pay in that Class to which, on comparison with orders, he seems to be entitled.

Also when a Civilian has acquired higher qualifications, and contributes more to the progress of work, than he did when I assigned his Rate of Pay, the District Officer is to send a fresh recommendation in his favour if he has not already attained the highest Pay of his Class.

(signed) | Thomas Colby | Lieut Col. | Royal Engineers
Ordnance box office, Tower. April 28, 1828

RI MS JT/1/TYP/11/3788–3791
RI MS JT/8/1/4a

1. *Yours of Sunday*: letter missing.
2. *your last*: letter missing.
3. *Several 111*: not identified, possibly a reference to the draft of Spectator's third *Liverpool Mercury* letter.
4. *through the P–*: not identified, possibly 'through the Post'.
5. *a vote of thanks to Spec*: to 'Spectator', Tyndall's *Liverpool Mercury* pseudonym.
6. *D.*: not identified. Louisa Tyndall's unpublished biography mentions 'D', a 'cool head' who was also stationed in Preston; therefore presumably a civil assistant in the 1st Division (LT, 'Biography', vol. 1, p. 76).
7. *Omega*: an occasional alias for William Ginty.
8. *'Discretion is the better part of valour, by which better part I have saved my life'*: W. Shake-

speare, *Henry IV Part I*, V.iv.115.

9. *Carbon*: a nickname for Colonel Thomas Frederick Colby, possibly relating to his experiments with using pure carbon in printing Ordnance Survey maps in the early 1830s (see *The Correspondence of Michael Faraday*, ed. F. James, 6 vols (London: Institution of Electrical Engineers, 1991–2012), vol. 2, p. 90.
10. *Carbon is a Tartar*: i.e., a solid deposit causing decay.
11. *O.S.*: Ordnance Survey.
12. *Acid*: a nickname for an unidentified opponent of Tyndall's on the Ordnance Survey, possibly Henry Tucker.
13. *Lumper*: not identified.
14. *Director of the Ord: Svy: of England*: Thomas Frederick Colby.
15. *Par*: Parliament.
16. *Survey of the Northern Counties of this Country*: the Survey of the six northern counties (of England): Cumberland, Durham, Lancashire, Northumberland, Westmoreland, and Yorkshire.
17. *Com: of the Ho of Com*: Committee of the House of Commons.
18. *in* *relating*: this blank was written into the letter deliberately.
19. *Commutation of the Tithes*: the Tithe Commutation Act of 1836 allowed the Church of England to be paid tithes (grants) in cash instead of in goods. Tithe maps showed the names of all owners and residents of an individual parish. Tyndall and his allies on the Ordnance Survey were upset that officer engineers were being paid to construct tithe maps, as they felt this work should go to civil engineers and surveyors.
20. *the S. of the 6 N. counties*: the Survey of the 6 Northern counties, see n. 16.
21. *the Master Gen. of the Ordnance*: George Murray.
22. *certain memorial from Land Surveyors*: possibly the letter to George Murray that Tyndall was drafting; see letter 0236 of this volume.
23. *the Third letter*: this is a reference to the third letter Tyndall was to write under the pseudonym of 'Spectator' for the *Liverpool Mercury*.
24. *the O.S.E.*: the Ordnance Survey of England.
25. *Board of Or*: The Board of Ordnance was a British governmental body in charge of supplying the Royal Navy and British Army.
26. *D^o^*: ditto.
27. *R.E.*: Royal Engineers.
28. —: McLachlan has omitted this name.
29. *Cap B.*: probably a nickname for Captain John Bailey of the Royal Engineers, who was stationed at Yorkshire; see letter 0195, n. 14, *Correspondence of John Tyndall, Volume 1.*
30. B: the pseudonym McLachlan used to avoid detection.
31. *Form of Recommendation for the Pay of Civil Assistants*: McLachlan included a copy of this form as an enclosure to his letter.
32. *the Royal sappers and miners*: the division of the British military responsible for military engineering. Each Ordnance Survey division consisted of a Royal Engineer, members of the Corps of Royal Sappers and Miners, and a number of civil assistants.

From William Ginty 7 September 1843 0232

7th Sep. 1843.

I received yours of yesterday[1] – after sending my last letter to you.[2] I found the verses alluded to which I now enclose. We tried a little petty agitation here today. W. Hunter wrote a note to Hamley for more pay, saying there were recommendations forwarded this month by other officers and calling his 'especial attention' to his 'miserable' pay. Hamley answered through Scott,[3] saying 'he did not like the tone of the note at all' that 'it was more like a demand than anything else'. That he could not recommend him this month and that he was sorry he wrote him such a note for 'he had his eye upon him'. Willy[4] in reply 'considered every word of the note perfectly correct and proper'. And in an hour after this I made another fellow write to him (I wrote it and he copied it) to which the reply was 'that Lt. H.,[5] wished it to be understood that he considered himself perfectly competent to judge who deserved increase and who did not'.

I thought to get another to write after this but I have not yet succeeded. He has however forwarded 6 or 7 journals this month! He gave us another hour this evening for the good of our health. See what agitation is doing – *[*mind*]* the truckling system the fellow is forced to adopt. This comes of going in squads to the doctor. The Hospital is more than 2 miles from the office – and the usual time lost by this is from 9 until ½ past 11 – he does not check us for this time.

There was a regular hubbub raised about the O'Neill[6] subscription.[7] Many when they heard such a thing was to be done said they would not do it unless privately. This put a settler on my intentions and after calm reasoning I thought there would not be a rascal wanting to say it was out of remuneration and not charity that it was done! And do it privately among all was quite impossible and would only go to prove such!!! The safest way being always the best, I wrote to Corpl. Scott desiring him to ask Lt. H., if he had any objection, saying 'it was not usual to ask leave to do such things'! But 'the circumstances of the present case seem to warrant it' and 'if it seemed odd, why I undertook it' – it was because I knew O'Neill many years longer than any man in the party and Smith[8] as long and that I knew them to be in distress!' Read the answer I got!!!* Read the last few lines again – It seems he is at them again. I went last night to look for him – but I can't find him. He can't hinder us from doing it privately. Mind the impertinence of the ruffian 'the men do what they please with their money'. By the word 'individuals' he means himself as for the 'injury of the whole survey Department' 'tis all stuff! 'leather prunello'![9] And this is from the man that said O'Neill was to come to him if he was discharged for 'sympathy'! Poor spite and he will catch it for it. I will send the fellow one of these days an anonymous scribble in the shape of an advice that will spoil his appetite for a week. I can counterfeit my hand if it was to take 5 minutes to every word! It shall merely bear upon domestic affairs. Some of his petty tyrannies and the like!!! And some of his lies and false, premeditatedly false, promises! to many of us!!!

I this day computed the spoiled work of the Borough of Liverpool. The borough alone, tis upwards of seventeen hundred acres, completely resurveyed and the original books thrown aside – to say nothing of the expense – the alarming expense of corrections of bad work, notwithstanding the severe checking that was practised at the time. (I think it was one third of their pay that was stopped while correcting the *[*results*]*). As for the portion resurveyed – some of them could not be checked, for they were discharged, and one of them, the worst, was a sapper[10] (Trimble)[11] and he is off the survey! I can't say whether those of these botches who are still on the survey were all checked. Some of them I believe were checked something! This is the facts of the case. I might add more acres if I was to include that portion of each suburban triangle that was outside the Borough B.V.[12] Mark you – The Borough of Liverpool thickly studded with houses and so much more difficult than work in the country!!! See what an expense!!! Again I can prove that the remeasurement – that is the correction of the bad work that 8[13] was at in Yorkshire – cost a vast deal more than the original survey. It was not much, of course, it is tho' nothing less the fact. What was 5 of us and our 5 subs., doing up here for 3 months – Just remeasuring! And it is not all done yet. Yes I could swear that it cost, if I would include the wet days, 4 times, aye 6 times the expense of the original miscalled survey! The worst of it was done by a sapper (Corpl. Bishop[14] since broke) on the cloud-cap'd fells of Yorkshire and Westmoreland! It would be damned folly to let these fellows (the sappers) off clean. It would be inferred that they are were[15] the most proficient and then might it not be said why not get it done by sappers alone!!! Again there is more expense lost by the examiners labours to correct the bad survey of the Town of Prescott than if good surveyors were employed at double the pay in the commencement! And this argument may be carried to an alarming extent thro' every square inch done by this party!

Again what means that system of trying surveys on common cartridge paper and having the corrections done before plotting it on the diagrams or content plots. It means this! They have no confidence in the surveys. They know them to be bad before they try them (exceptions of course) and they are therefore obliged to do so – knowing that if they subjected the plots to the scoring of the compasses and rubbing of India rubber, they would tell tales and speak volumes to their disgrace! This was done all thro' L. Pool[16] thro' all large scale work and many small scale plots!!! in this division.

Facts are stubborn things.
'Anti-Humbug'[17]

The circumstances under which Messrs O'Neill and Smyth[18] were discharged were such that I can do nothing to promote their interest in any way and therefore I shall forbid that any subscription be made during working hours.

The men can do as they please with their own money but I do not think it creditable to any of them to make open expression of sympathy with persons who have endeavoured to make themselves extensively mischievous not only to individuals but to the whole department and one of whom is still understood to be continuing conduct of this kind.

W.G.H.[19]

RI MS JT/1/TYP/11/3609
LT Transcript Only

---

1. *yours of yesterday*: letter missing.
2. *my last letter to you*: letter missing.
3. *Scott*: Corporal William Scott.
4. *Willy*: William Hunter.
5. *Lt. H.*: Lieutenant William George Hamley.
6. *O'Neill*: John O'Neill, a civil assistant who had worked in the 5th Division in the C District on the Irish Survey. He joined the Survey on 7 October 1839 and worked in the office party along with Tyndall. His primary duty was registering. He was transferred to the British Survey on 5 August 1842. He was dismissed from the Survey at some point in 1843.
7. *the O'Neill subscription*: a collection being taken up to aid two recently dismissed surveyors, including John O'Neill.
8. *Smith*: a recently dismissed surveyor, first name unknown.
9. *'leather prunello'*: i.e., vanity. A reference to A. Pope, *Essay on Man, Epistle IV*, 203–204: 'Worth makes the man, and want of it the fellow; | The rest is all but leather or prunello'. Prunello was a heavy woollen fabric used in clerical robes and fine shoes.
10. *a sapper*: a member of the Royal Sappers and Miners, the division of the British military responsible for military engineering. Each Ordnance Survey division consisted of a Royal Engineer, members of the Corps of Royal Sappers and Miners, and a number of civil assistants.
11. *Trimble*: Henry Trimble (*c.* 1807–?) was a Corporal in the 16th Company of the Royal Sappers and Miners. He enlisted on 10 November 1826. During the late 1830s and early 1840s Trimble was stationed at Kilrush in County Clare. His primary duty on the Survey was examining work on the ground, and his daily rate of pay peaked at 1s 8d. This was reduced to 1 shilling during the summer of 1841, which suggests a disciplinary issue. He was later reduced to the rank of Private on 16 February 1842. He left the Irish Survey in early 1842.
12. *outside the Borough B.V.*: unknown abbreviation, but refers to the boundary of the borough.
13. *8*: unidentified nickname for an ally on the Ordnance Survey. Possibly Louisa Tyndall's mistype of 'B', Archibald McLachlan's pseudonym, although this is not certain; see letters 0235, 0244, 0259 and 0270 of this volume.
14. *Corpl. Bishop*: Thomas Bishop (*c.* 1802–?), a Corporal in the 16th Company of the Royal Sappers and Miners. Bishop enlisted to the army on 23 July 1825. He was stationed at Rathkeale in County Limerick, before moving to Killarney in County Kerry. His primary duties were perambulating and examining groundwork. His daily rate of pay was 1s 5d, rising to 1s 6d by the time he left the Survey. Bishop left the Survey during late 1841.
15. *they are were*: this phrase may reflect an error in transcription or an error in the original letter.
16. *L. Pool*: Liverpool.
17. *'Anti-Humbug'*: Louisa Tyndall annotation: '(Lieut. Hamley's answer:)'. These last two paragraphs appear to be an addition or enclosure in Ginty's letter.
18. *Smyth*: the same as the Smith referred to earlier in the letter; see n. 8.
19. *W.G.H.*: William George Hamley.

From Archibald McLachlan 9 September 1843 0233

Sunday[1] – 9 Sep–

I had scribbled the beginning of a reply to Veritas[2] – Lo! the postman gave up 2 letters & a N paper[3] – and now I am puzzled, I thought Veritas was a R.E.[4] or the letter was written by their authority, or permission – How to answer it, and not give too much importance to Veritas – If he really is as you say – it would be foolish to descend from your cloud to grapple with a worm– other worms will start up, this must not be – However, you could crush him sideways –

I think you & D[5] are fully equal to all that is wanted – I will therefore give you only hints – dont follow me I pray you – use my hints only as helps, not as dictations!!

To Sir R P [6]

I addressed you in the Mercury[7] of 1st Sp:[8] – on public grounds for public good – spoke of a great public work – and a great public injury. I attacked no man's private character, but public men, & public grievances &c. &c. To that letter a reply has by some party been attempted, and only attempted – I shall not stoop to notice the productions of inferior agents, men ignorant of the subject – grossly ignorant – I have animadverted[9] on the O.S.I.[10] & by consequence the professional qualifications &c. of the principle Directors of that work. And I now publicly call on Richard G.[11] C.E.[12] Dublin who conducted & conducts the valuation to contradict me and reprove me publicly if I am wrong – I call on the R.E. all or any of them to come forth and defend themselves – I do not want your name, sign yourself a R.E. and I am satisfied – If all has been done honourably & efficiently you will have no reason to fear from any encounter – the more the thing is discussed the more your valuable services to the public will be appreciated. Again I say the R.E. employed on the O.S.[13] come forth – a dignified silence will not be construed into a conscious feeling of rectitude on your part – Defend your conduct, your professional career in Ireland – in England – and do not allow any ignoramus to assume your position. An explanation from you is due to your professional character, to the B of Or:[14] to the Parliament & to the Country – I will not reply to inferiors – But lest you Sir R.P. should fall into the error of supposing the letter of Veritas in the Mercury of Sep. 8th is the effusion of some R.E. and thereby conceive a low estimate of their statistical!! knowledge on the points in question, I will go out of the line I had marked out for myself, and dispose of some of its assertions – but shall not give any promise of noticing such foolish production in future. –

I notice this letter because & only because you may fall into error, & suppose it is by some R.E. well acquainted with the subject, &c.
& 1st The estimated expense of the O.S.I. was 300,000 – & time 7 years – read the evidence given before the Com: of the Ho of Com[15] 1824 – this was the sum nearly – & time – nearly – less would not complete it – much more was not anticipated, and the estimate was as correct as estimates generally are, and made by parties accustomed to drawing estimates.
2nd This was the estimate for completing a 'Townland Survey'[16] – which survey was to form the basis of a land valuation, & for these, Acts of Par.[17] were passed.
3rd Did the R.E. understand the nature and description of the maps necessary for valuation of lands, houses &c. – If it be said the R.E. were deceived by the defective nature of their instructions, I ask them for a copy of their instructions – if they say the words 'Townland S.' are ambiguous, I ask when was this ambiguity discovered – I ask them who drew up the Code of regulations or instructions commonly called C.C's[18] blue book. This Blue book (The last time I saw it was in Clonmel,[19] with Cap. Boyle,[20] perhaps he has it yet – but I cant recollect the page – Has nobody in the Dis:[21] office a copy – the page is not so important – the fact is a true fact, it is there – I think about page 6) or book of instructions was written previous to the commencement of the actual operations, and demonstrated that the writer knew well the nature of the maps required – In this book it is stated 'that the plans are to be drawn by a scale of 6 inches to a mile with all the accuracy and minuteness of detail of which that scale admits' this is plain speaking, no word here of roads and boundaries, nor any distinction of relative accuracy of the detail – every thing was to be as accurate as the scale would admit – So the blunder cannot be attributed to any ignorance on the part of those who drew up the regulations or instructions. But if they still persist and say they were deceived, or had not proper orders &c – what follows? That a body of RE. – empowered by the British Par: commenced a grand national undertaking, which when concluded is to be creditable to the country & the scientific acquirement of the age, A Colonel 5 Cap: 25 Lts[22] commence and proceed – without any individual of that number knowing any thing of the matter, did not know whether they were acting right or wrong, and never enquired – however, they acted and continued to act till certain maps were sent to Mr Griffith CE. Dublin for the purpose of valuation – and he found them of no use –
4 When Mr G. after an attempt *[at]* a valuation, returned the maps as erroneous and useless, and a note of the errors existing in the maps. Here is the secret of the maps being abandoned. The R.E now found by experience that their maps were useless and dearly did the country pay for their experience – these maps were not condemned because of deficiency of detail, but because – Boundaries – Roads – houses – every thing was wrong. – even Trig heights & Distances were wrong –
5. Let Mr Griffith be examined before the Ho of Com: and a few simple plain questions will settle the point –

Sunday night – 10th Sep.

To day I have perpetrated a dreadful scribble. I had little time to do it in – and sent it to you 1/2 finished. I am half sorry that I did scribble it – for since I sent it to post, I have been engaged in conversation with one of the good ones, on various subjects, among the rest, Veritas. My Gent:[23] is of opinion that Veritas is either a R.E. or one writing at his dictation, he says a R.E. would hardly make such admissions and no person would speak so of Col C.[24] without permission 'that Col. C. very wisely and considerately thought it better to keep some' &c. My gent: says the expressions 'very wisely and considerately' are proof that it was written at the dictation of a R.E. – and that the whole tone of the letter bears marks of a R.E. having something to do with it; and also that Spec.[25] 'is very likely to pass by with a sneer of derision,' for 'Veritas deserves no other reply.'

I am half of this opinion myself – for if we take much notice of him, another Veritas may take the field and expect to be patronised by his R.E. – Address it to Sir R.P.[26] – Express a hope that he will not be misled into the belief that it comes from any person acquainted with the subject – you notice the letter merely for the purpose of expressing your contempt of his presumption and ignorance in introducing the name of Col. C. without authority & coupling the name of Col. C. with such absurd statements, for that his letter was received with shouts of laughter and derision by every man acquainted with the subject – give him 2 or 3 *[home]* blows and leave the creature –

The scribble I committed to day was done in haste to be in time for the post – I did not finish it; but on reflecting on what I have since heard, it would be better for Spec: not to enter into a lengthened reply, and to treat him as undeserving of such an honor!!! And reserve your heavy guns for a weightier opponent – now when I think on't, I am half ashamed of fighting such a spooney![27] – You and I may think otherwise – Kill him, you have an easy job! –

On my conscience when I think of it, it would be burning shame for Spec: to thunder on such worthless creature – Tho' not for the present, yet for future use, you may recollect the following – A field S.[28] was not by Mr G.[29] or any of his valuers considered necessary for their valuation; in certain cases general, or principal lines of fences, or where a leading large fence sub–divided a townland, such a fence would be found useful on the maps – a field S. was never dreamt by Col. C. nor Mr G. nor Parliament – nor were fences surveyed till after the revision had progressed considerably & until the R.E. had time to look about them a little – in Down, Antrim-Derry & some others, fences were not surveyed in the revision except now and then a leading fence.

The Blue Book – 'Accuracy & minuteness of detail' – I have said the R.E. fell into a mistake in surveying every field; now on comparing these together you get into a 'fix', well I must bring you and myself out of it. Now I maintain that those words 'accuracy & minuteness' are written by a scientific man on a scientific subject; I will take the words in an Engineering & Surveying signification, this is their plain obvious meaning, that every thing is to be done perfectly accurate; Everything is to be mapped, correctly mapped. – Now Sir! I demand of you or any man, what standard of minuteness & accuracy had the writer in view, when he wrote those words? had he the map of some gentleman's estate, which was done by some good practical Surveyor? had he a good Irish Baronial map in view? Such as were drawn for the Grand Jury, and adjusting Grand jury Ass.[30] such as that done by Armstrong[31] of the C°. Armagh?[32] – or was it a map drawn at the academy at Woolwich[33] by Cadets? – I maintain it was the latter he had in view; that he had no idea of 'Accuracy, or minuteness' beyond the accuracy of Woolwich Academy, and ignorant of the practical accuracy necessary for a valuation of lands, houses &c. &c.

If you reject the assertion of No. 2, as being inconsistent with the scientific character, and long experience of Col. C. – inconsistent with the idea Parliament & the B of Or entertained of him – Then you must adopt N°.1. I say you must adopt N°. 1. – Then what follows? Why, a 'Fix' – a Devil of a Fix – When was No. 1 written? it was written before, or at the beginning of the S. – When was it discovered, that it was found necessary! to abandon the old maps, and commence a detail S. – Why, 6 years after N° 1 was written – this is the real fix, and I leave Carbon[34] to take which he pleases – If he chooses N° 1. then I say, that the necessity of a detail survey was known by him from the very beginning, detail & accuracy, both were known to be essential. I am right then when I say, they or he fell into a mistake when he adopted a field survey – he was equally mistaken when he surveyed by actual admeasurement[35] small ponds – ½ acres of bog – & waste lands of 1 acre and many other trifles not considered necessary by Mr G. – if he says, no; I was not mistaken in 1832–3–4–5 &c then I reply you must have been mistaken in 1825–6–7 &c – your book proves it – your book – explain your meaning in that, and all is settled –

The apparent inconsistency of my words which place you in a 'fix' lies in this, that 'minuteness and accuracy' admits of two meanings – in my reasoning here I adopt the reasoning of N° 2 – yet in conclusion, I will retract my expressions in the letter on the OSI 'that they fell into a mistake of an opposite nature &c' because if you are thrown by those expressions into a fix, others might be so too – so– expunge the words – then I am out of the fix but Carbon is into it.

C.B.[36] has returned from London, black as thunder – It is reported that Acid[37] is to come with his Dt.[38] office – The Trig party[39] has gone to Preston and among them there are no trumps! – B.[40]

I do not speak mysteriously – I name a man – Mr G.C.E.[41] who was appointed to carry into execution the valuation – and I think that the R.E. should for the honour of their corps have the man examined.

Veritas says that experience had proved that a survey of mere roads and boundaries was insufficient – Who found out this insufficiency – experience proved it – it follows that inexperience of the R.E. was the cause of the Blunder – for had Mr G. himself surveyed the county his practical experience of the affairs of civil life would have saved him from such an egregious blunder – in fact the maps done by the R.E. were mere military skeleton maps, such as they make at Woolwich, they have no idea of what was wanted for land valuation. When it was discovered the maps were useless – the news of that fact came like a Thunderbolt, it spread panic and dismay – why all this if they were conscious of having acted up to their orders. The Col. blamed the Cap. Cap. blamed the L^{t}. the L^{t}. threw the odium on the men, the men retorted – & each party concealing the fact from the public why all this? Veritas would call this '<u>experience</u>'

The old method was abandoned: was there any new act of Par:? No – any new code of law or instructions? No – It was because Mr G. would not receive them

---

It is said that a detail S. would require thrice the time & money of a skeleton S. – Surely *[Nimms]*, Bald, Telford[42] &c. &c. such men could surely give a correct opinion on the subject – and their replies show that they and Parliament understood well the nature of the required survey – and cost –

Veritas asserts that the original, was a mere skeleton S. and that a detail S. occupies 3 times the time and 3 times the money of a skeleton S. – according to this the whole affair stands thus:

Original 7 years time misspent		——	7 – 200.000
Detail – 21 – –	= 900,000	——	21 – 900.000
			–28 –11.00.000[43]

I did not think the amount was so much, and I give this statement that the R.E. themselves may see the necessity of giving correct documentary proof of the facts – and not leaving it to such as Veritas.

As to the pay of the men – the same absence of facts is shown by Veritas; from 1830 up to the present moment there has been not much difference in respect to the pay of the men; it is certain that a <u>few</u> men receive good salaries – we should make a distinction between the <u>Boards men</u> – and Colonel Colbys men, also a great distinction between the Irish & English Surveys. On the Irish Survey the average amount of pay did not amount to 1/6. On the English Survey it does not amount to 2/6 – there are men who have been employed on the OSI 6–10–14–15 years who do not get 3/6 – take any party of surveyors in any part of the S. &

the pay does not amount to 3/. at least it was so lately. In 1842 when there was no deficiency of funds the men were as badly paid as at present.

There is no military men in Her Majesty's dominions endure so much fatigue as the Royal Sap: & M:[44] employed in the field work on the O.S.E.[45] & O.S.I. and there are no military men so shamefully treated &c &c in fact the men are and were paid not by any fixed standard but by the prices of Provisions in their locality – care being taken by the R.E. to discover the prices of meat, potatoes, milk &c. – & regulate the pay of the men according to that – I am borne out in this by every man who reads this letter on the O.S. – Some men after having toiled for 12 months, are recommended to the notice of their superiors as deserving men, & their pay is raised from 2/2 to 2/3!!!!! and as much formality about the action as if great public advantage were conferred –

Post time | more tomorrow.

RI MS JT/1/TYP/11/3794–3799
RI MS JT/8/1/4a

1. *Sunday*: September 9 was a Saturday; McLachlan appears to have made an error.
2. *Veritas*: a pseudonym under which one of Tyndall's critics wrote to the *Liverpool Mercury*.
3. *a N paper*: a newspaper.
4. *I thought Veritas was a R.E.*: a Royal Engineer.
5. *D*: probably a civil assistant in the 1st Division; see letter 0231, n. 6 of this volume.
6. *Sir R P* : Robert Peel.
7. *the Mercury*: the *Liverpool Mercury*.
8. *I addressed you in the Mercury of 1st Sp*: this refers to Tyndall's 1 September 1843 letter in the *Liverpool Mercury*, which was published under the pseudonym 'Spectator'. See letter 0228, *Correspondence of John Tyndall, Volume 1*.
9. *animadverted*: criticized, publically censured (*OED*).
10. *O.S.I.*: Ordnance Survey of Ireland.
11. *Richard G.*: probably Richard John Griffith.
12. *C.E.*: Civil Engineer.
13. *the O.S.*: Ordnance Survey.
14. *B of Or*: Board of Ordnance.
15. *Com: of the Ho of Com*: Committee of the House of Commons.
16. *a 'Townland Survey'*: a map of a town produced by the Ordnance Survey.
17. *Acts of Par.*: Acts of Parliament.
18. *C.C.'s*: unidentified abbreviation; possibly 'Colonel Colby's'.
19. *Clonmel*: a town in southeast Ireland.
20. *Cap. Boyle*: not identified.
21. *Dis*: District.
22. *Lts*: Lieutenants.
23. *My Gent*: My Gentleman, not identified.
24. *Col C.*: probably Thomas Frederick Colby.
25. *Spec.*: Spectator, Tyndall's *Liverpool Mercury* pseudonym.

26. *Sir R.P.*: Sir Robert Peel.
27. *a spooney*: a fool, a silly person (*OED*).
28. *A field S.*: a field Survey.
29. *M*r *G.*: Richard John Griffith.
30. *adjusting Grand jury Ass.*: 'Ass.' may be an abbreviation for 'assignments' or 'assessments'.
31. *Armstrong*: probably William Armstrong (*c.* 1781–1861), an Irish surveyor and engineer.
32. *C*o*. Armagh*: County of Armagh, a county in northern Ireland.
33. *The academy at Woolwich*: the Royal Military Academy at Woolwich, London.
34. *Carbon*: a nickname for Thomas Frederick Colby; see letter 0231, n. 9 of this volume.
35. *admeasurement*: a formal ascertainment of the dimensions or size of something (*OED*).
36. *C.B.*: Captain B., probably a nickname for Captain John Bailey of the Royal Engineers, who was stationed at Yorkshire; see letter 0195, n. 14, *Correspondence of John Tyndall, Volume 1*.
37. *Acid*: a nickname for an unidentified opponent of Tyndall's on the Ordnance Survey, possibly Henry Tucker.
38. *Dt.*: District.
39. *The Trig party*: a reference to a group of surveyors, a party of trigonometers.
40. *B.*: the pseudonym McLachlan used to avoid detection.
41. *Mr G.C.E.*: possibly an acronym for 'Griffith Civil Engineer', a reference to Richard John Griffith.
42. *Nimms, Bald, Telford*: not identified.
43. *11.00.000*: McLachlan error; the sum should be 11.000.000.
44. *<u>Royal Sap: & M</u>*: Royal Sappers and Miners; see letter 0231, n. 32 of this volume.
45. *O.S.E.*: Ordnance Survey of England.

To John Tyndall, Snr. 11 September 1843 0234

Preston | Monday morning[1]

Dear Father

Yours[2] was received yesterday morning. With respect to the additional expense I have been at on account of Emma's visit,[3] I think it quite unnecessary to enter into a detail of it. Whatever expenses I have incurred on her account will be liquidated in a couple of months after her departure, so that it is needless to speak to me further on that subject. All that Emma will require of you is money sufficient to bring her home when she is going. At present I'm unable to muster the sum. You ask for a <u>full</u> and <u>true</u> account of her health. I shall give you the latter at all events. Judging from the evidence afforded me, I would conclude that I never saw her in better health than she is at present. You say that my mind is set very much on America.[4] It most assuredly is. This frame of mind however is no momentary impulse. My convictions on this subject have been the result of mature deliberation. The thought of going to America has more or less occupied my mind for years, but within the last nine months it has become a settled determination on my part. As for the hope held out when you say: – 'You don't know, nor I either, but this may be the best America for you yet'.[5] It is rather illusive – rather <u>intangible</u> for me. I have learned the folly of trusting to such hopes, and every day's experience strengthens the conviction of their irrationality. Man to a certain extent is the child of circumstance, or at least circumstances operate powerfully on a person like me who stands in the world without prop or stay and has nothing to depend on but his own successful exertions. And judging from the circumstances in which I am placed, and from the evidence afforded me by my short experience, I can see nothing – no argument – which can successfully oppose itself to the step that I have resolved on. I think it quite useless to make any application to Somerset House[6] at present. Any emolument[7] which I would receive there would not weigh against my determination of crossing the Atlantic if good hopes were held out to me beyond. So it is best to wait until we hear from Wm Wright and Wm Tyndall, and after that we will be the better judges how to act. You may read my former letter to the Dean,[8] merely to show that I am alive to his kindness, but burn this, and keep its contents by yourself.

Your affectionate son | John.

RI MS JT/1/TYP/10/3297
LT Transcript Only

---

1. *Monday morning*: Louisa Tyndall annotation: 'Sep. 12th, 1843'. However, 12 September 1843 was a Tuesday. Since the letter was written Monday, the real date of the letter is probably 11 September. Louisa Tyndall frequently dated letters by postmarks, which were often stamped a day or two after a letter was written.
2. *Yours*: letter missing.
3. *Emma's visit*: John Tyndall's sister, Emma, was visiting him in Lancashire.
4. *You say that my mind is set very much on America*: see letters 0220 and 0230, *Correspondence of John Tyndall, Volume 1*, for Tyndall's recent correspondence about America with his father.
5. *'You don't know ... for you yet'*: not identified.
6. *Somerset House*: on the Strand in central London, the principal government offices in Great Britain; see letter 0230, *Correspondence of John Tyndall, Volume 1*, for further context.
7. *emolument*: advantage, benefit (*OED*).
8. *the Dean*: Richard Boyle Bernard, Dean of Leighlin.

From William Ginty 15 September 1843 0235

Liverpool Sunday 15th Sep. 1843. | Deceived! eh? | GIVE THE ENCLOSED TO GEO![1]

Dear Alpha[2]

> 'What nobler breast plate than a breast untainted
> Thrice is he armed who hath his quarrel just
> And he but naked tho' locked up in steel
> Whose conscience with injustice is corrupted'[3]

Where? tyrant is safety now
Thy long enjoyed security
A care-cloud settles on that brow
So lately lit with extasy

Say despot, didst thou fondly dream
Thy villainy and botchery
Would glide a subterranean stream
O'er which men trod with apathy

'Illusive hope had led thee on'
And lulled us in tranquillity
But, miscreant, now that hope is gone

We grapple with thy majesty
Or didst thou think a dastard soul
Filled every breast that cunningly
Had viewed thee climb the very goal
Of thy ambitious perfidy?[4]

Worst estimate of all this!
The Lion crouchant[5] flattered thee
To play thy gambols[6] – knowest thou this?
That Lion's left his lethargy!

Aye! well I ween thou know'st it now
In bitter sad reality!
Come brothers lay the despot low
He sprawls in dying agony!!!

Every day, every post, adds additional vigour to my courage courage! bah! Tis simple duty. There is only one thing about your incomparable letter to Tucker[7] that I regret – deeply regret – that is that my name is not at the tail of it. I would glory, I long to have one like it sent from Liverpool! But to us, to me, it is unnecessary! Hear my reasons; we are not situated as you are – there are many timid fools among us, if such a thing was done I question if even the 15 who did not sign would all agree to it and as for the remaining, 9 I verily believe it would cause these dirty devils to draw up a counter-request! One has done so already they sent for him, he got his name taken out – if Tucker had not the intention of forwarding it – if the spirit of one or two did not tell him he must forward it, he would never have done this – this is evident. This wretch had been sounded beforehand by that consummate scoundrel Scott.[8] He said to him he wished it withdrawn. Now if they had any notion that any of the rest would wish the same they would gladly comply with their request or I miscalculate the poor devils very much! In drawing it up we could not say it was at the request of all. Those every one who would not sign that request would be put down as opposed to it, questioned on its merits, and their names undoubtedly withdrawn. Now in the present state of affairs I think it will be sent off without this risk. Your letter will ensure this – but if all comes to all, those that will insist upon its being forwarded it will do so!!!

I seen O'Neill[9] last night, he was thunderstruck at hearing the memorial[10] was here yet. He calculated upon this memorial being — (when in the hands of the M.G.)[11] a corroboration of his statements. Off he started to Short,[12] and now they are discussing the merits of a letter to the M.G. calling on him to defer his decision until he sees our memorial. He will also say it was handed to Mr H.[13] on the 26th of ultimo.[14] Col. Colbey,[15] Col. Cunningham and Captain Tucker being then in Liverpool so that there might have been no delay. This I heartily concur in, O'Neill says he will fix them on that memorial. He insists 'it was to injure him it was kept back'. If O'Neill's determination comes to Tucker's ears (his next to no ears, curse them, I was hoarse yelling at him) I guess he will not lose much time, for his own sake. from this we have nothing to fear and it will rise the wind to a hurricane.

You need not boast of your 'fudgery'. I insist Yealand Conyers[16] will be first on the list of losses. Larry E.,[17] exd. it!!![18] It was computed here 4 or 5 times, and as many more times in Preston. There was great humbugging about it – comparing registers with computations. In short Yealand Conyers – the very name (here) goes in whispers – it contained only 1500 and some odd acres. They are at or have just commenced Caton.[19] This was surveyed by your party (I think).

Do you intend to touch Bobby T[20] on the rump for his letter opening? You should!

I will do all I can with 8!!![21]

Yes faith! a little more than 'common caution' is requisite I have a queer card to play. I laugh at all! and I'm blamed for all the 'mischief'.

And 'Hamley's' wrath and 'Tucker's' rage
Have swept o'er 'Hotspur'[22] yet he stands –
A firebrand in a patriot's hands
And 'Sappers'[23] threats and coward's shock
Have changed his nature to a
ROCK!

How is my brother-in-law[24] getting on eh?

Since I finished this this evening up to the present (9 o'clock) I have been cramming a sheet full to the Editor[25] 'I walked into him' I gave him a very fine report of the acting superintendent's speech and said his own reporter would not do it better. I told him of the circular desiring us not to give any information to publishers. I told him all about the memorial and sent him a copy of it for his perusal. I shewed him the necessity of issuing Spectator[26] into the world as soon as he could – next Friday if possible. I asked him to make an editorial comment said it would add considerably to its weight on the public mind and said I would call for the copy of the memorial early in the week and at the same time place the 3rd letter of Spectator[27] in his hands. Just read that last sentence again crazy skull. Your last letter to Tucker was 'the unkindest cut of all'. Why you told the poor man in very plain terms that you had no confidence in him or them – it smells very much of the 'serve you right' kind of argument. More power lay it into tight. During the late display some of the lads laughed outright at a fellow in the rear rank who was nearly dead with the fright. Holy St. Patk.,[28] if you only heard his teeth chattering, he would assuredly drop if he was in the front rank. Two devils kept plucking my coat and shoving me as my mouth was applied to one of the holes in Harry's head – they can't be called ears.

Hotspur.

Jack's[29] people have arrived in Dublin. She's[30] there et Alpha. What about the 'Northern Briton'[31] eh? Send me her name and address for God's sake. Oh you heartless dog – Would you call young &c. but hush I hear she is an ugly old devil – Oh you sublime rascal – 'Does she squint'?

Murder in Irish how is little Jim[32] during all this sport tell him if we had these chaps regulated I would write to him. Blast them they are putting us to a considerable deal of trouble and no thanks! Oh! Oh!

RI MS JT/1/TYP/11/3610
LT Transcript Only

---

1. *GIVE THE ENCLOSED TO GEO*: probably George Latimer.
2. *Alpha*: an occasional nickname for Tyndall.
3. *'What nobler breast plate ... with injustice is corrupted'*: W. Shakespeare, *Henry VI*, III. ii.233–6. The rest of the poem appears to be a Ginty original.
4. *perfidy*: untrustworthiness, deceitfulness (*OED*).
5. *crouchant*: crouching (*OED*).
6. *gambols*: an energetic leap (*OED*).
7. *incomparable letter to Tucker*: letter missing.
8. *Scott*: Corporal William Scott.
9. *O'Neill*: John O'Neill; see letter 0232, n. 6 of this volume.
10. *the memorial*: this may refer to an early draft of the letter of protest sent by the workers of the Ordnance Survey of England to George Murray, Master General of the Ordnance on 23 September 1843; see letter 0236 of this volume.
11. *the M.G.*: the Master General of the Ordnance Survey, George Murray.
12. *Short*: not identified.
13. *Mr. H*: possibly William George Hamley.
14. *ultimo*: last month.
15. *Col. Colbey*: Thomas Frederick Colby.
16. *Yealand Conyers*: a village in Lancashire, England.
17. *Larry E.*: probably Lawrence Eivers.
18. *exd. it*: i.e., executed the survey of Yealand Conyers.
19. *Caton*: a parish in Lancashire, England.
20. *Bobby T.*: not identified.

21. *8*: see letter 0232, n. 13 of this volume.
22. *Hotspur*: a nickname for William Ginty.
23. *'Sappers'*: see letter 0232, n. 10 of this volume.
24. *my brother-in-law*: not identified.
25. *the Editor*: John Smith (1792–1861), the editor of the *Liverpool Mercury*. Smith, according to an obituary in the *Liverpool Mercury*, was a 'warm and earnest advocate of the principles of civil and religious liberty, and he was one of the first to denounce the tax on the food of the people [i.e. the Corn Laws]' ('Death of Mr. John Smith', *Liverpool Mercury*, 7 January 1861, p. 2). After the apparent refusal of a number of other newspaper editors, it was Smith who published Tyndall's pseudonymous letters to Robert Peel in the *Liverpool Mercury*.
26. *Spectator*: Tyndall's *Liverpool Mercury* pseudonym.
27. *the third letter of Spectator*: Tyndall's third *Liverpool Mercury* letter under the Spectator pseudonym.
28. *Holy St. Patk.*: Holy Saint Patrick, the patron saint of Ireland.
29. *Jack's people*: possibly relatives of Jack Tidmarsh.
30. *She's*: not identified; presumably an old romantic interest of Tyndall's.
31. *the 'Northern Briton'*: in 1843 Tyndall was enamoured with a woman he called 'the Northern Briton', possibly a 'Miss Wilding' he mentioned in his journal; see letter 0325, n. 12 of this volume.
32. *little Jim*: not identified; possibly Phillip 'Jim' Evans.

To George Murray 23 September 1843 0236

TO THE MASTER GENERAL OF THE ORDNANCE.[1]

Sir,

We the Civil Assistants whose names are herewith sent humbly beg to draw your attention to the accompanying copy of a District Order, which requires the signature of each individual employed on the Ordnance Survey of England, in either the fourth or fifth column of a form attached.

After mature deliberation, during which the consequences which might result from our signing either column of the form were calmly considered, – we conclude, that though feelings of respect for the party from whom that order emanated might be an inducement to many of us to comply with its demands, still in justice to ourselves and in consideration of our own interests, we must most respectfully beg leave to decline signing either of the Columns alluded to.

The avowed cause of this Order's being circulated is that a reduction is to take place to bring the expenses within the amount of the Parliamentary grant of 1843 – this, Sir, is a system of reduction quite novel to us, our memories can furnish us with no precedent: if a reduction alone were contemplated is it not reasonable to suppose that the qualifications of the men employed would be taken into account? and that the services of those who might be least useful, would be dispensed with? This rational plan has ever been adhered to, and in the present case we are at a loss to conceive why it has been abandoned. We cannot think that a system like the present which hazards the loss of the most efficient men on the Survey, would ever have been resorted to, if it were supposed that these men would accept the proffered boon and return to their own country.

We are led to suppose from this order that those who <u>wish</u> to remain <u>may</u> remain; now will not their signatures in the pay list be sufficient testimony that they are still connected with the Survey: we have given this matter the coolest consideration, but we cannot conceive what the signatures of those who might wish to remain here have to do with a form which professedly affects those and those alone who may wish to go. The avowed object of the form could be effected with greater facility, and with less grounds for doubtful surmises, by each Division Officer obtaining the signatures of those who wish to return to Ireland: as these are the only persons who are to be affected by the contemplated reduction.

We earnestly and humbly request Sir that you will not interpret these remarks as originating from a spirit of opposition, or a wish to dictate to our superintendents – we solemnly disclaim any such motives, but as the thoughts from which these remarks have sprung have in some measure influenced us in our present proceeding, we deem it right to lay them before you honestly and distinctly.

Our grounds then Sir for declining to sign either column are briefly these. –

The fourth column is to contain 'the signatures of those who wish to remain on the Survey with no immediate prospect of an advance in their present rates of pay'. It is an easy inference that those who wish to remain under such circumstances are satisfied with their present condition. We cannot sign a document from which such an inference can be drawn, for the God of truth knows that we would lie in our hearts if we stated that we were satisfied with our present circumstances.

What surety have we to prevent our signatures from being made subservient to other ends than that stated in the District Order when the heading of this column would warrant inferences completely at variance with our real feelings? The expression 'who wish to remain on the Survey' as it stands here admits of being qualified to a great extent – We have no room allowed us to restrict it, and we might have no opportunity of doing so if in the event of our giving our signatures, a misapprehension should hereafter arise relative to our views of this expression.

When we say that we are not satisfied, we give utterance to a sentiment which pervades the bosom of every Civil Assistant employed on the Survey, with not half a dozen exceptions. And though fear or other causes may induce numbers to sign this column, yet are we convinced, that they feel as we feel, and would act as we have acted if they would follow the dictates of their inclinations.

Various considerations, unnecessary, and perhaps improper to recount induce us to decline placing our signatures in the fifth column: many of us actuated by the hope that we would be rewarded on the English Survey have given up other prospects which owing to neglect may not now exist. The statement 'if you are not satisfied with your present condition your passage will be paid to Ireland' may sound very fair. But Sir this is a sad alternative for men, many of whom have wasted their youth and energies in this employment, and few of whom, owing to their low rate of pay, have been able to realize a sum sufficient to allow them time to seek another situation.

It is said Sir that anonymous letters have been sent to you at various times complaining of the lowness of the pay, – We unhesitatingly record our censure of such a mode of proceeding, and publicly condemn it as both improper and impolitic – Permit us, Sir, to lay before you in an open, honest, and respectful manner a few statements in connexion with ourselves, fully aware that the safety of our situation depends on the accuracy of these statements, it is not our design to trouble you with a petition, you of course will draw your own conclusions from what we shall lay before you.

Our pay Sir is very low, so low as to preclude most of us from the enjoyment of many of the necessaries of life. On the subjoined list is stated the occupation and time of employment of each of us, together with our respective rates of pay: some of us have joined the Survey young men, full of hope and vigour, the prime of our days have been devoted to the duties of our situation, and we have ever been faithful and assiduous in the discharge of these duties; we now hang on the Survey liable to be discharged at a moment's notice, without superannuation or gratuity, or perhaps comforted by the assurance that if we wish to go to Ireland our passages shall be paid: on the Survey our youth and energies have been spent, and our bodies unfitted for the discharge of the duties of active life. What then, Sir, awaits us? A gloomy future indeed, a reference to our rates of pay will shew that we have been unable to make suitable provision for futurity. Many of us are young men who however feel the effects of confinement and application to our duties, prematurely stamping us with the evidences of age and infirmity, while our resources render us unable to administer to ourselves proper support. We have no allowance in case of sickness; our pay, small as it is, is then cut off; and in such cases numbers of persons connected with us have been compelled to depend on the aid of friends or to appeal publicly to the generosity of their fellow-workmen.

We are inclined to think that you are unacquainted with this state of things, or if informed of it, that it is from sources on which you can place no reliance, such as those alluded to in the District Order. We find that those persons who have been employed by the Board of Ordnance to assist in the execution of the Survey of England, are allowed the pay which their abilities deserve: if we select parties of equal talent from their numbers and ours, we find that their pay generally exceeds ours by twice its amount – we are totally at a loss to account for this disproportion.

The question may be asked – nay Sir it has been asked – 'What would satisfy these men' – We presume not to state the rate of wages which we conceive due to our abilities, but we beg leave to refer you to the accompanying copy of a 'Form of Recommendation' approved of and signed by Col. Colby in 1828. – The rates of pay awarded by this form to efficient draughtsmen and clerks are far from being high. The form was constructed when the operations of the Survey were going on in Ireland and in a part of Ireland remarkable for its cheapness. This Sir we deem a virtual Contract, a reference to our rates of pay will shew that not an individual amongst us receives the rate awarded to the higher classes of this form: the inference is that we are not eligible to fill those classes, or that the contract has not been kept: we can appeal to those who are capable of judging of our merits – we can appeal to our own work, and from both these sources we can prove that many of us are qualified to stand in the highest classes here stated, while the subjoined list will shew that we receive little more than half the rate of wages awarded.

Our discontent is not founded on any circumstance which could possibly result from a diminution of the grant of 1843. Far be it from us to murmur at the unavoidable results of such a diminution: our discontent arises from the consciousness that the remuneration which we receive for our services is not at all equivalent to the value of these services, and that there is not in Great Britain a situation similar in its requirements to the Ordnance Survey, wherein the persons employed are so badly rewarded as the Civil Assistants attached to that Survey.

In conclusions Sir we would express the hope, that you will pardon any irregularities which may exist in our mode of addressing you. If such there be they are the result of inexperience in such matters. The statement of our grievances and the sentiments of our hearts have been calmly, plainly, and truthfully, laid before you, and though our tale be an unvarnished one, still we trust that the feelings from which it has sprung will give it a title to your consideration.

Windsor, 16 September, 1843.[2]

District Order,

As it is necessary to make a further reduction of the number of persons employed on the detail Survey[3] to bring the expenses within the amount of the Parliamentary vote for 1843 – and as anonymous letters have been sent at different times to the Master General of the Ordnance complaining of the low rates of pay:

The Division officers are to make out nominal lists of their office and Field Parties according to the Form in Page___.

A copy of this Order is to be sent with the Nominal list to each party to be signed by the men of the respective parties in one column or the other according as the persons wish to remain on the Survey with no immediate prospect of any advance in their present rates of pay, or return to Ireland.

The passages will be paid for those who wish to return to Ireland.

The Non-Commissioned officers in charge of parties are to see the lists signed and to attach their signatures to them.

The Division officers are to transmit the lists to the District office with as little delay as possible.

(Signed)
Henry Tucker, Captain Royal Engineers.

To the Civil Assistants of the Division office party.

FORM

Nominal list of office Party of the ___ Division to be signed by each person either in the 4th or 5th column.

20th Sept. 1843.

Names	Rate of Pay.	Employment.	Signature of those who wish to remain on the Survey with no immediate prospect of an advance in their present rates of pay.	Signature of those who wish to return to Ireland.
Civil Assistants				
A.B.[4]	3/9	Drawing	A.B.	
C.D.	3/3	Plotting		
D.E.	2/0	Registering	D.E.	C.D.
Labourers				
A.E.	1/9	Chaining	A.B.	
C.D.	1/6	D°	C.D.	

Form of RECOMMENDATION for the PAY of CIVIL ASSISTANTS.

Sir, I recommend who is now receiving per diem in class as deserving of an increase of Survey Pay in that class, or of promotion to class.[5]

District Officer.

Class A. (Miscellaneous Duties) Rate of Pay not to exceed 2/– per diem.

Superintending Piling Parties and other duties of a responsible nature, requiring more intelligence than is necessary for the duties of a chainman.

Class B. (Clerks) Rate of Pay not to exceed 2/– per diem.

Copying and other duties which require persons who can write tolerably well.

Class C. (Clerks) Rate of Pay above 2/– and not exceeding 3/– per diem.

Duties which require superior qualifications in writing, considerable knowledge of Arithmetic, and skill in the particular Books and Accounts of the Survey.

Class D. (Clerks) Rate of Pay above 3/– and not exceeding 5/– per diem.

Great accuracy and rapidity in the execution of all those duties which are required to be performed in keeping the various books and Accounts of the Survey, and in checking Computations.

Class F. (Draughtsmen) Rate of Pay above 2/– and not exceeding 3/– per diem.

1. Road Surveying and Plotting neatly and accurately.

2. Penning in lines of all kinds, either for Roads or Boundaries neatly and accurately.

Class F. (Draughtsmen) Rate of Pay above 3/– and not exceeding 4/– per diem.

Qualified for Road Surveying and Plotting, and also capable of penning in lines of all kinds, either for Roads or Boundaries.

Class G. (Draughtsmen) Rate of Pay above 4/– and not to exceed 6/– per diem.

Qualified for Road Surveying, Sketching Gardens, and completing the drawing of fair Plans.

Class H. (Draughtsmen) 6/– per Diem, and upwards.

Qualified to perform the whole of their duties, both in surveying and drawing, in a rapid and efficient manner.

Classification of the Civil Assistants.

Class.

A	B	C	D	E		F	G	H
				1	2			

This Table is to be inserted in the Monthly Return, by the side of that showing the classification of the Royal Sappers and Miners,[6] and to be filled up in a similar manner.

In each Recommendation the Division Officer will fill up the Table with the members in the respective classes attached to his Division.

In recommending the Civilians for the respective classes, the District Officer will take care to state definitely the grounds on which he recommends the individual for the class, in order that I may not only put him in the class, but also assign to him the Rate of Pay in that class to which, on comparison with orders, he seems to be entitled.

Also when a Civilian has acquired higher qualifications, and contributes more to the progress of the work, than he did when I assigned his Rate of Pay, the District Officer is to send a fresh recommendation in his favour if he has not already attained the highest Pay of his Class.

(signed) | Thomas Colby, Lieut. Col. | Royal Engineers,
Ordnance Map Office, Tower. | April 28th, 1828.

RI MS JT/1/TYP/10/3895–3900
LT Transcript Only

---

1. *TO THE MASTER GENERAL OF THE ORDNANCE*: Louisa Tyndall annotation: 'This is the Memorial signed by the Preston men + handed to Cap Tucker on Sep 23/43'. See letter 0231 of this volume for Archibald McLachlan's first draft of this letter.
2. *Windsor, 16 September, 1843*: the following documents appear to be enclosures to the above letter.
3. *the detail Survey*: the branch of the Survey tasked with surveying and mapping topographical details, rather than boundaries.
4. *A.B. ... C.D.*: initials not identified. Since the letters of the initials occur in roughly alphabetical order, they are likely exemplary or pseudonymous.
5. *Sir, I ... promotion to class*: the gaps in this portion of the letter are intentional.
6. *Royal Sappers and Miners*: see letter 0231, n. 32 of this volume.

From William Wright 26 September 1843 0237

Cincinnati | Sept. 26th 1843.

My dear Jack

Yours of August[1] has come to hand 2 weeks ago and in endeavouring to commence a reply I am quite at a loss where to begin. Your letter contains so much that gave me joy and pleasure, and first your intention of visiting this country and so soon overcame all my other feelings and in anticipation of the future I calculated upon all the rambles we would take upon foot and horse back over the hills of old Kentucky and along the banks of the charming Ohio (they did not appear half so pretty before as since your announcement). Next came our native Shamrock[2] and although its pristine beauty has departed it met with a hearty welcome, every relic that I meet with of Ireland in this distant land are highly treasured by me and none more dearly prized than the idol of every true Irishman. Divided as we may be upon other subjects none will have their love questioned respecting the green immortal Shamrock. I would not have delayed answering your letter till now, but that I wished to be fully informed of what you might meet with were you to come out here. I am sorry that I am compelled to inform you that the occupation you now follow[3] would not be at all encouraging in this country. I have had it from the very best authority, the Surveyor General of the States of Ohio,[4] that there are 30 applicants for any vacancy that might take place in the department, the cause which he assigned for it was that the Internal Improvement system[5] was such a perfect mania some 7 or 8 years ago that the Civil Engineers had more to do than they were able and consequently received large compensations for their services. The goodly matrons seeing that to be a Civil Engineer was to be a coiner, made all their youths Civil Engineers, the tables have been turned, the improvement system has gone down and with it the employment of those young men in that line, but there is one satisfaction here, that if we cannot get employment at what we have been used to why we can at something equally as good. My uncle[6] has also made enquiry respecting it of several of the officers the result has nearly been the same with this exception that although there were a great many in the city acquainted with surveying the number of good draftsmen were very small. Now Jack[7] my candid opinion is that if you were in Cincinnati and could not find employment in your line that you could find employment at something else at which you could earn a respectable living. I was not too sanguine when I stated that you could even learn a profession it has been often done in this city and is doing to my certain knowledge this day by more than one. So that if you have got your parents'[8] consent, in God's name come along and my word for it you will not regret it, for there are a great many things here which tend to draw a man out that does not exist in the old country or if they do they are only for the favoured few. Many a towering genius has been crushed by the prejudice of our falsely called superiors, that in other climes and other circumstances would have been a pride to his countrymen. You can learn a regular system of book-keeping in 3 weeks and by that alone you can earn a decent living, the charges are very moderate for obtaining a knowledge of it. That would enable you to take hold of the most difficult set of books in this city (and close them with credit to yourself). A young active enterprising fellow like you need have no doubts about the propriety of emigrating for there are a number of ways by which a man can live comfortably in this country and which do not exist in Ireland. I have been doing business on my own acc/.[9] now for nearly 10 weeks and am doing a pretty fair business. I have a spare bed and apartment at your service and it is not necessary for me to say that you would be more welcome to them than any man in Ireland even my brother Thomas[10] would not receive as hearty a welcome. I have lately written to my sister Mary Ann[11] requesting her to come out but have not yet received an answer, you would much oblige me by sending her a line and asking if she would do it as I am anxious to hear from her about it and am not writing to any of the folks this time. Address her at No 12 Nichols Street Dublin. I am afraid this will not reach Boston before the steamer starts unless it goes by to night's mail and as it is now late and after a hard day's work I close this at all events, I would not cross it[12] as you would never be able to read it. I will write to you by the next steamer. Sarah[13] joins with me in the invitation to you to come out and she desires me to say that she will strive to make amends as much as she can for the dear friends you may leave behind. Many thanks for the poetry[14] you need not be ashamed of it.

Yours affectionately goodbye | William Wright.

There is much in your letter I would wish to note but time will not permit. We are all in good health but strange weather – the glass was up to and above 96° last week to day down to 50° oh I am so cold. If you would have an opportunity when writing to your father to mention if he would please to let my folks know you had a letter from me and all's well.

My kind regards to your father and mother.

RI MS JT/1/TYP/5/1811–1812
LT Transcript Only

---

1. *Yours of August*: letter missing.
2. *our native Shamrock*: it is possible that Tyndall enclosed a pressed shamrock in his letter to Wright; see letter 0190, *Correspondence of John Tyndall, Volume 1*, in which Tyndall asks his father to send him a sprig of shamrock so he can forward it to Wright.
3. *the occupation you now follow*: Tyndall was working for the English Ordnance Survey.
4. *the Surveyor General of the States of Ohio*: more officially, the Surveyor General of the Northwest Territories. A government official responsible for land surveys north of the Ohio River in the United States. Judge William Johnston held the office from 1841–5.
5. *the Internal Improvement system*: probably the Indiana Internal Improvement System, a large-scale project undertaken by the State of Indiana in 1836 to construct canals, railroads, and other transportation systems.
6. *My uncle*: not identified.
7. *Jack*: i.e. Tyndall. Tyndall himself rarely used the nickname 'Jack', but several of his Irish friends and family members used this nickname for him.
8. *your parents*: Sarah McAssey Tyndall and John Tyndall Senior.
9. *acc/.*: account.
10. *my brother Thomas*: Thomas Wright; biographical details unknown.
11. *my sister Mary Ann*: Mary Ann Wright.
12. *cross it*: Wright is referring to the practice of cross-hatching a letter, in which the author rotates the letter 90 degrees and writes over the top of the existing text.
13. *Sarah joins with me*: Sarah Wright, William's wife.
14. *thanks for the poetry*: not identified; see letter 0216, *Correspondence of John Tyndall, Volume 1*, for more of Wright's writing about Tyndall's poetry.

From William Ginty 1 October 1843 0238

Sunday 1st October 1843.

An account of the two days proceedings at Liverpool

Friday

The horrors of the 'Investigation'[1] having terminated, in proof of the line that 'satan finds some mischief still for idle hands to do',[2] the minions of monopoly cut out fresh work for themselves. 'Intimidate! intimidate! do every thing that ingenuity can invent to frustrate these fellows' designs' – seemed to be the order of the day. And as a beginning, that most consummate of all consummate rascals, Crpl. Scott,[3] ruled that no man was to speak above his breath in the office in future! (this as you will soon see was a grand step to keep men from talking 'mischief'). Between breakfast and dinner Hamley visited all hands – praised every man's work, which he never did before; called up some broken instrument and said he would certify it as having been broke by accident and so save a fellow's cash. By this and some other movements he created a good feeling, as may be expected – It was so infernally unusual to see the worthy man in a good temper! Thus he laid a foundation for his after movements! After this he went from room to room and discussed the 'form'[4] with the men, never once pretending he had seen the memorial.[5] (Read his speech,[6] and you will have all he said, nearly word for word, in our office, which was only a repetition of what he said in George St.,[7] in different rooms). You may depend on this report being authentic. Mark who he addressed himself to in the commencement – he said he granted the request, a lie – he never noticed it. Mark the threats at the ring-leaders – this of course was threatening all, for those that took the most active part were elected to do so by the rest. The rumour he spoke of was not 'circulated' – he lies. He merely said so for an excuse for his address. Mind how the rascal slipped the intention – Just what we all knew it was for – to gag us.

Saturday

He had a long discussion with 'Ye three gentlemen'. Taylor, Hansbrow,[8] and a fellow named Blakeney,[9] all in one room. 'If you don't like your pay' says he to Taylor, 'why don't you go elsewhere'. So I intend, when opportunity offers – replied T. After a great deal of talk, in which he got the worst of it, he said 'he did not come there to discuss the matter'. 'Corpl. Scott, take it round and ask them individually to sign'. Scott did. Scott silenced (veritas)[10] Hansbrow for saying to Blakeny – 'sign if you are satisfied'. Blakeney signed after a great deal of hesitation, adding that he was not satisfied and that it would not prevent him from remonstrating hereafter. Taylor and Hansbrow said they would sign if he allowed them to write a remark in connexion with their names – refused, and so they refused. Scott went to a room upstairs and gave two men to understand that Mr Hamley intended calling in several for recommendation that evening and immediately afterwards entered with the former. Frank Tidmarsh[11] refused to sign. Scott called him downstairs and assumed the office of guardian over him and forced him by saying many things of a threatening nature and winding it all up by saying mind he used no threats. He signed and says he was forced, and is half crazy for doing so. Thus Sir villainy[12] thinned our ranks of 10. 15 are left, thanks to their consciences. The 'God of truth' seemed to have little effect on the 10 who are now laughed at by all hands!

Hamley I understand was with Eivers – trying to persuade him to remain. Taylor got a thundering letter from Beatty[13] that fellow will be fixed I'll do his goose yet!

What of Spec.[14] Go ahead the sooner the better. Agitate! Agitate! Agitate! Tell them for God's sake about Liverpool borough.

Hamley when asked if the memorial was gone, said yes! And when referred to it another time said it was a private thing between them and the Master General[15] had nothing to do with it. Beatty says 'he hopes' Dhigan and Connolly[16] in Manchester 'will be dismissed'. I did not get the Sub-surveyorship. I am promised a clerkship at £50 a year here. Oh if I get it. There was *[150]* applications for them two berths.

No doubt when Hamley sends the news to Tucker he will try the same! in Preston. Be prepared!

Have we not enough to prove intimidation? I calculate we have. Their very act of presenting it to us was improper – and insult and a gross intimidation in itself when our relative position is considered. Oh! here's room for the lash and no mistake.

The Speech. NO INTIMIDATION!!!

After the disturbance created by the 'Gathering of the Clan' had subsided, Lt. W.G. Hamley R.E.,[17] proceeded to address the meeting. The honourable gentleman,[18] who was received with deafening cheers, looked unusually pale and during his address bent an intense gaze upon his mother earth with his hat pulled over his eyes. 'He (Mr H.) received a request signed by a number of Civil Assistants – many of whom he supposed were present, requesting time for deliberation with respect to the late circular[19] – he granted that request (cheers). The week would be up tomorrow and he would send the form round for signature (hear hear). He hoped they would all sign it (cheers and cries of 'we will') he was sure they had been led to think wrongly on this matter (hear). A cry of discontent and opposition had been raised by a few who induced others to back them – He had heard – he was certain a cabal of the kind had been got up by two or three who lent themselves to cause trouble, raise disturbance and create unnecessary expense, such as the late investigation (hear hear). Such men he would say were nuisances (cheers). They done considerable injury and if he (Mr H.) could find them out he would settle their hash[20] (tremendous cheers and cries of 'serve them right'). A rumour had been set afloat that all those who did not sign would be increased and that others would never obtain one – He begged to say it was false – that no such thing was ever contemplated. 'Pon my word and honour' exclaimed the Honble. Gent[21] with some warmth 'this thing was not got up to entrap or injure the Civil Assts'. A pause during which it was evident a reply was expected. None being given, he continued to say, that while he was speaking he might as well tell them all he knew of the matter (hear! hear! hear!). Letters[22] had been written to the Master General about low pay. Col. Colby had been accused of making promises and not keeping them; many men talked of bad treatment. They must be all aware of the trouble this would cause. The expense of searching over old documents to contradict such false statements. Now he (Mr H.) would let them into the secret – but it was not a secret. It was intended to put a stop to this kind of grumbling (loud cheers). If they did not like the remuneration they were getting, they were at liberty to go (hear hear). He would not order any of them to sign it, he had no authority for doing so – if they were military men they would be punished. Col. Colby would be very much displeased and would perhaps part with them readily – such conduct if carried much further would prove fatal to the Survey[23] (hear). He (Mr H.) would not be affected by such a result – he could return to his duty. So could the other officers. The sappers[24] could return to their duty – but the civilians would be thrown out of employment (applause). He begged them to remember he used no threat, he did not say there was much harm in not signing it (hear). If they took his advice they would sign it (cheers). He held out no threat to one side nor hope to the other. He came there merely to disabuse their minds and to request them to act on their own opinions tomorrow. It was of course optional with them. Particular men should state their particular grievances. They should take their chance of increases. A large grant[25] might be given next year, if so there would be a general increase (hear! hear! hear!). In conclusion, he had to request them to think and act upon their own discretion. The Honble. Gentleman raised his head amidst loud cheers to view the retiring multitude.

Serjt. Sinnett requested to be informed by the Honble. gent who had just stood up, if the order promulgated that evening at the Lower House[26] was to extend to this house, to the effect that the men were not to speak above their breath during the working hours.

(Mr Hamley replied) 'Speak', roared the Honble. Gent, 'if they don't speak 'mischief' here they will in some other place. They may speak as much as they like if they mind their work – if I find out those refactory[27] gentlemen that excited the rest, *[they]* and I will soon part company (cheers from all parts of the House).

Exit Hamley.

Several other gentlemen intended to address the meeting – prudence however said – 'let him alone – give him rope enough'.[28]

RI MS JT/1/TYP/11/3612–3613
LT Transcript Only

1. *the 'Investigation'*: see letter 0231 for discussion of the necessity of an investigation into the wages and productivity of the employees of the Ordnance Survey of England.
2. *'satan finds some mischief for idle hands to do'*: I. Watts, *Divine Songs Attempted in Easy Language for the Use of Children* (1715), 'Song XX: Against Idleness and Mischief', lines 11–12.
3. *Crpl. Scott*: Corporal William Scott.
4. *the 'form'*: probably the 'Form of Recommendation for the Pay of Civil Assistants'; see letter 0231 and letter 0236.
5. *the memorial*: the letter of protest sent by the workers of the Ordnance Survey of England to George Murray, Master General of the Ordnance on 23 September 1843; see letter 0236.
6. *his speech*: see below.
7. *George St.*: probably a reference to 25 Great George Street in London, the headquarters of the Institution of Civil Engineers.
8. *Hansbrow*: not identified.
9. *Blakeney*: not identified.
10. *veritas*: truth (Latin).
11. *Frank Tidmarsh*: see letter 0239, n. 20 of this volume.
12. *Sir villainy*: Ginty's nickname for William Scott.
13. *Beatty*: not identified.
14. *What of Spec.*: a reference to Spectator, Tyndall's pseudonym when writing to the *Liverpool Mercury* about Ordnance Survey affairs. See letter 0228, *Correspondence of John Tyndall, Volume 1*; and letters 0248, 0252, 0261 and 0287 of this volume.
15. *the Master General*: George Murray.
16. *Dhigan and Connolly*: not identified.
17. *R.E.*: Royal Engineer.
18. *The honourable gentleman*: the form of address used in Parliament's House of Commons.
19. *late circular*: recent written material intended for a large audience (*OED*); see letter 0236 of this volume.
20. *settle their hash*: i.e., resolve the dispute, create order (*OED*).
21. *Honble. Gent*: the honourable gentleman, i.e. Lieutenant William George Hamley.
22. *Letters*: see letters 0231 and 0236 of this volume.
23. *the Survey*: the Ordnance Survey of England.
24. *sappers*: see letter 0232, n. 10 of this volume.
25. *A large grant*: i.e., Parliamentary grants; see letter 0236 and letter 0246, n. 6 of this volume.
26. *the Lower House*: a reference to the House of Commons, Parliament.
27. *refactory*: this may be a misspelling of 'refractory', which means obstinate or stubborn (*OED*).
28. *give him rope enough*: an allusion to the proverb 'Give a *thief* rope enough, and he'll hang himself', J. Ray, *A Compleat Collection of English Proverbs* (London: J. Hughs, 1737), p. 161.

From William Ginty 4 October 1843 0239

Liverpool | Wednesday 4th Oct. 1843.

Dear Alpha[1]

I have just returned from the office. 'The bird had flown' before I got there, it matters not, a letter will do as well and that letter lies before me sealed and stamped for the gent.[2] I told him it would be in his hands by Saturday evening or Monday evening at the latest. I spoke of the meeting of C.E.'s[3] being rather premature at present – until Spectator's letter[4] had exposed the enormous errors of the work to an astonished public. For the life of me I can't make head or tail of the enclosed little document.[5] Who does S— t mean? Sinnett or Scott?[6] If Sinnett, he says we are fools! I tell him he dare not say otherwise! If Scott, he seems to be lowered very much in his own estimation. Of course his late conduct has raised him in ours! As for the other scrap, my comprehension herein fails me, or perhaps my memory. As for dates, faith my sweet fellow they are well impressed on my memory! I 'never can forget' – All happened together. As for persons – I know them all. But what of that, the ruffians here would swear to a thing here one moment and swear to the reverse the next. Bill Latimer heard Col. Colby say the Ph.[7] or Township at which they were looking (Downholland)[8] was doing for 4d per acre, to Col. Cunningham! This was on the last day of the investigation Thursday the 28th Sep. 43 (book this) as you will have more to book than me! Again our letter with 25 signatures[9] was left on Scott's table on the evening of <u>Tuesday the 26th</u> of Sep. by Wm. Latimer (no person seen him do this). Hamley's speeching[10] &c took place on <u>Friday the 29th</u> and the — form[11] was presented to those who had signed the remonstrance[12] to the Master General[13] on <u>Saturday the 30th (book all these mind that!)</u> As to Hamley's speech, I slipped a few sentences in that I did not hear him use. Here they are! (a2)[14] 'Col. C[15] would be displeased and perhaps part with them readily'. (a2) Again about the breaking up of the survey and it not affecting him or the sappers[16] but throwing the c's[17] out of employment'. These two trifling things <u>I</u> could not swear to, <u>but I can to all the rest of the speech as reported</u>. Latimer and Dick Hunter[18] and many others can swear to the extract marked (a2). I can't say who heard him say the other they were both used in George St.[19] on Friday the 29th. On Saturday *[Hamley]* said to Taylor, or in his presence, that 'he did not see much harm in signing the form <u>or even in not signing it</u>'. Now this may be important, in case we were reprimanded, if these are not the exact words it is the exact meaning of his words! This I say was encouragement to us not to sign, at least we can call it such, his words will bear us out! I will draw Frank Tid.[20] out when I see him. Monday night is <u>'our club' night – he will be there</u>. A club to meet once a fortnight we have established. It keeps us or rather <u>them</u> always <u>ripe for 'mischief'</u>. We sing a song get drunk and go home. That's all the harm we poor devils do! You ask me 'do I think the work in Liverpool exceeds the contract'? – Not a *[man]* in the office but would not swear it doubles it <u>on chance it is not open work</u>*! Would it not be very easy to obtain from the guardians of the respective townships what the contract was. Preston is a central point, a tête a tête[21] with a Preston 8[22] would *[do]* all, that is he would ask the question publicly and faith *[see]* they would answer. There are 3 copies of Downholland,[23] I guess there is *[sumat]* extra for these. Downholland is within 16 or 17 miles of Preston, it is joining Halsall[24] where Marquis was!

You must see I cannot manage to make a calculation even roughly. I have set two – Dick and Bill[25] – to draw some fellows out in the Lower Ho.[26] It may succeed. I have no access to journals nor anything else! I am sure Downholland will exceed the mark considerably, as those that have worked at it have high pay! They are very hard working men. That's a damn bad job!!! One in particular would do as much as three easily. A rumour is afloat today that the ring-leaders in Manchester[27] have been discharged 5 or 6 of them. I don't believe it. I hope tis false. I had two letters from there yesterday and nothing was said about it! *[Ya]*

Yours | A Rock.[28]

RI MS JT/1/TYP/11/3614
LT Transcript Only

---

1. *Alpha*: an occasional nickname for Tyndall.
2. *the gent.*: the gentleman, not identified.
3. *C.E.'s*: probably an abbreviation for Civil Engineers.
4. *Spectator's letter*: 'Spectator' was Tyndall's *Liverpool Mercury* pseudonym; see letter 0228 *Correspondence of John Tyndall, Volume 1*, and letters 0248, 0252, 0261 and 0287 of this volume.
5. *the enclosed little document*: not identified.
6. *Scott*: Corporal William Scott.
7. *the Ph.*: not identified, possibly an abbreviation for parish.
8. *Downholland*: a small parish in Lancashire, England.
9. *our letter with 25 signatures*: probably the letter of protest sent by the workers of the Ordnance Survey of England to George Murray, Master General of the Ordnance on 23 September 1843. See letter 0236 of this volume.
10. *Hamley's speeching*: see letter 0238 of this volume for Ginty's account of Hamley's speech.
11. *the — form*: possibly the 'Form of Recommendation for the Pay of Civil Assistants'. See letter 0231 of this volume for the text of this form and letter 0238 for further insight into Ginty's opinions of it. '—' may be an instance of Ginty or Louisa Tyndall editing out profanity, probably 'damned'.
12. *remonstrance*: see n. 9.
13. *the Master General*: George Murray.
14. *a2*: not identified.
15. *Col. C.*: probably Thomas Frederick Colby.
16. *sappers*: see letter 0232, n. 10 of this volume.
17. *the c's*: not identified, possibly the civil assistants.
18. *Dick Hunter*: a nickname for Richard Hunter.
19. *George St.*: this probably refers to the headquarters of the Institution of Civil Engineers; see letter 0238, n. 7 of this volume.
20. *Frank Tid*: probably Frank Tidmarsh, a relative of John Tidmarsh's who was working in the same branch of the Ordnance Survey as Ginty.
21. *tête a tête*: private conversation (French).
22. *a Preston 8*: not identified.
23. *3 copies of Downholland*: Ginty is suggesting that three separate survey maps have been made of Downholland.
24. *Halsall*: a small parish in West Lancashire, England.
25. *Bill*: William Latimer.

26. *Lower Ho.*: Lower House, a reference to the House of Commons, Parliament.
27. *the ring-leaders in Manchester*: Tyndall and Ginty's Survey allies in Manchester.
28. *A Rock*: apparently a pseudonym for William Ginty.

From Archibald McLachlan 6 October 1843 0240

Oct. 6 1843[1]

Notes[2]

(1) I would say – The Superintendence of the work was committed to the B of Or:[3] & the Irish Government – I am not sure that all was committed to the Board –

---

(2) – The determining, or Survey of Public Boundaries and the valuation was committed to Mr Griffith –

---

(3) Chosen, to carry into practical effect, the undertaking – or, to superintend actual operations – This would distinguish their duty from the General Superintendence of the Board –

---

(4) – 30 R.E.[4] (25 Lieuts & 5 Capts[5] = 30 total) – this is not strictly correct – there were 3 or 4 officers of Artillery – but let 30 R.E. stand as it is – the error is of no consequence –

Also there were a few officers on detached duty – such as Trig:[6] there were only a few – I dont remember the number – I would not mention them – or if you do don't say how many. I never could find out how many men or officers were employed on detached duty. I have a remark to this effect in the History of the O.S.I.[7] –
(5) – Any number say a great number the expense is the principal thing!! of civilians – or civil assistants – or a number of civilians as Assistants –
(6) 'This time' – 1825 or 1826 – mention the year –
(7) I would join this sentence to that part where the name of M^r^ G is first introduced, up at (2) – because this sentence (7) stands disconnected rather too much –

---

(8) This mention of additional officers on Trig. seems disjointed from what you say of 5 Cap. & 25 L^ts^.[8] I would say 5 Cap 25 L^ts^ & a few other officers 'were placed under the immediate control of the officer above alluded to' –

---

(9) These alterations were made – Your statement is correct.

---

(10) You are right here

---

(11) I remember one offset 13 chains

---

(12) I cannot certify the last sentence – in our district the officers merely signed their names – thus : A B C – L^{t}. R.E. 24 June 1828 – or A.B.C. Cap. R.E. 29 June 1828 *[&c]*[9]

General

All that relates immediately to Carbon,[10] would keep by itself – or, all that you say of him should lie in one long sentence.– All you say of Mr G.[11] should be kept by itself or, placed in one long sentence – At present Carbon is mentioned in 2 places, & the sense seems unconnected. Mr G. is mentioned in 2 places; the sense here would be clearer by saying all you intend to say of him, to say it in one place – What I mean is, that by mentioning these men twice the sense is not so clear as if all was said in one place.[12]

Perhaps it would be well to say more in your opening paragraph, of the great injury done by the R.E. in the 6 N. Counties,[13] to private enterprise, or the Civil Surveyor. – Also you might say the enormous expense of the S of 6 N[14] Counties and at this particular time he Sir RP[15] is not justified in continuing or, extending the operation until he looks for this other mode of distribution of the money. However D.[16] would give you a good hint on that point –

You embrace enough for one letter – and said quite enough to show practical men the nature of the thing, & the N^{o} of hands employed – & that is enough on the present subject. When you come to condemning the Maps – revision and Burning – then you can enter into Tarnation Eloquence –

Try to have it inserted in the first III[17] –

B.[18]

Tomorrow's post will bring you – a new point relating to London –

Friday – I promised in one of the other letters to send you the following tomorrow. But as I have time before post, I send it now. –

2 of our old hands are sent to Southampton – and another old hand to follow in a few weeks – so soon as these three old favourites are gone, then comes the work of destruction in this Division –

Carbon should have had the Bill passed last Session[19] – for London – he has delayed, the cause is his Estimate – but principally his book of instructions for officers in surveying large Towns. I got a peep at this Book – pamphlet 13½ pages

– it is not published yet – the copy I have is the proof sheet – notes & corrections in Pencil by Acid[20] – the errors are errors of the printer. The Book opens thus–[21]

Sanitary Improvement

Structural Means Within Legislative And Administrative Control.

App: AI. –

Draught Instructions for an Eng: Off:[22] for a S. and Report on the most advantageous Arrangements – for the Improvement of the General Sanitary Condition of the Population of + + + (there is a place left for the name of some town)

You are requested to make such a survey of the district in which the Town of — is included, as may serve to determine the necessary extent of area – which from the natural features of the Elevation – strata &c ought to be comprehended under our directory for drainage, for obtaining the best supply of water for the public as well as private use of the population, for the removal by sewerage from the houses & streets of the Town – & for its productive application, and for effectually & economically carrying out all such other structural arrangements as may hereafter be specified as requisite for the protection of health & Promotion of comfort and prosperity of the Population.

S-1. You will cause an accurate Horizontal Plan to be made on a scale of 60 inches to a mile – &c &c The Book is a clever manoeuver – an infernal humbug!! It will be laid before Parliamentary Committee &c &c And in the Appendix he quotes several C.E.[23] in Hamburg – Glasgow &c – On Waterworks, Sewerage. He quotes a Mr Thorn C.E. Greenock[24] – 'That he met with opposition from envy & private jealousy'!!!

Never mind, I will take good extracts – you will have them all –

B.

RI MS JT/1/TYP/11/3800–3802
RI MS JT/8/1/4a

---

1. *Oct. 6 1843*: date given by Louisa Tyndall.
2. *Notes*: this letter appears to be McLachlan's notes and corrections on the second Spectator letter, which Tyndall was preparing to submit to the *Liverpool Mercury*; see letter 0248 of this volume.
3. *the B of Or*: the Board of the Ordnance Survey.
4. *R.E.*: Royal Engineers.
5. *25 Lieuts & 5 Capts*: 25 Lieutenants and 5 Captains.
6. *such as Trig:*: possibly an abbreviation for 'trigonometers'.
7. *O.S.I.*: Ordnance Survey of Ireland.

8. *5 Cap. & 25 L^{ts}.*: annotation from unknown person, not in McLachlan or Tyndall's handwriting: 'better say 30 officers on the Survey besides those employed at the Engraving dept & primary observation'.
9. *29 June 1828 [&c]*: annotation from unknown person, not in McLachlan or Tyndall's handwriting: 'They were certified to be correct so their signature is quite sufficient'.
10. *Carbon*: a nickname for Thomas Frederick Colby; see letter 0231, n. 9.
11. *M^{r} G*: probably Richard John Griffith.
12. *if all was said in one place*: annotation from unknown person, not in McLachlan or Tyndall's handwriting: 'This depends entirely upon circumstance'.
13. *6 N. Counties*: six northern counties of England; see letter 0231, n. 16 of this volume.
14. *S of 6 N*: the survey of the six northern counties, see n. 13.
15. *Sir RP*: Sir Robert Peel.
16. *D.*: probably a civil assistant in the 1st Division; see letter 0231, n. 6 of this volume.
17. *the first III*: probably the first three letters written by Tyndall under the pseudonym 'Spectator' in the *Liverpool Mercury*.
18. *B.*: the pseudonym McLachlan used to avoid detection.
19. *Bill passed last Session*: i.e. the legislative bill should have been passed at the previous Parliamentary session.
20. *Acid*: a nickname for an unidentified opponent of Tyndall's on the Ordnance Survey, possibly Henry Tucker.
21. *opens thus*: The text 'Sanitary Improvement ... 60 inches to a mile – &c &c' appears to be an excerpt from an unidentified pamphlet by Thomas Frederick Colby.
22. *Eng: Off:*: English officer.
23. *C.E.*: Civil Engineers.
24. *Mr Thorn C.E. Greenock*: Mr Thorn, Civil Engineer for Greenock, a city in Scotland. No further information has been identified.

From Archibald McLachlan 8 October 1843 0241

Sunday 8th Oct – 1843 [York][1]

I have just received your last.[2] – I have read carefully all the pamphlet I spoke of:[3] it is a clever and comprehensive bit of work, and I must say that on the subject handled in it, I never saw so much excellent matter before; it appears all through, to be drawn up with great talent and address, and evidently the work of Carbon[4] – Sir R.P.[5] is to have a copy previous to next session, he will use the arguments and facts when he introduces the Bill[6] – In regard to a Town, there is no possible imaginable point, or question, which is not treated of in the pamphlet – And in making or drawing up the Report & Survey the R.E.[7] officer is told to look on the district as if it were the property of One Prudent proprietor, anxious for the Health, Prosperity, income, production capabilities – and future well being of his property & his tenants – or Population.

I told you some of our old hands were getting up an address, and a piece of plate,[8] to be presented to Cap. B.[9] – When 2 or 3 called on him to know when it would be convenient, they would (they the deputation) come and present the address and Plate – he, C.B. hesitated, and said Acid[10] is to dine with me today –

I will speak of it to him – 'You see I am a Soldier, doing my duty only – and my successor might expect a like address' – and – 'I am pleased to think we part in good feeling, & glad tho' other divisions are complaining, this Div[11] seems to be contented. It would not be prudent in me to receive your address, without first consulting my superior & successor' – !!! –

The consultation ended in this – 'Owing to certain Military regulations, it would not be prudent in me to receive it, I have spoken with Acid – he says it would be imprudent owing to my position, therefore I cannot receive it – but I am just as well pleased as if I did receive it – and feel highly pleased that we part with mutual good feeling' – !!!! –

Now you may see the hand of Acid here! –

C.B. would receive it, had it not been for Acid – Acid did not see the deep design of the Address, he did not see the design – but he foresaw how it might be turned against himself before long – and might cause an enquiry into the cause why one Div were congratulating and another Div complaining – besides he foresaw that Spec:[12] might get hold of this, and make it a 2 edged sword!! – Acid said to one of the deputation who is now going to Southampton[13] – 'that Cap B. could not receive the address &c owing to a military order – military men are subject to military regulations, but I am very glad that you are so well pleased with Captain B. I will mention the circumstances of such good feeling to Carbon'!! Och, The Liar –

I could not use the facts in the Pamphlet – it will not be <u>published</u> before next session – i.e. it will be printed and circulated amongst members of Par:[14] & no farther, at least I think it won't go farther than the members, until the Bill is past[15] – and then all parties may see it. I saw the <u>Proof sheet</u> – it was picked from other papers. I only saw it by stealth, no body in the Div: saw it but me & 2 or 3 of Windsor men. Should the facts be <u>too freely</u> used, he Acid, would at once know who the delinquents were – or he could point to the very individual through whom the information came!! – However we can use the facts in a negative kind of way, and escape suspicion – I am informed by some Windsor men!! that he heard Sir Delabash[16] say to Acid, words, from which it may fairly be inferred – that there exists a good feeling or understanding between Carbon & Sir R.P. – 'Carbon called on, or went to discourse with Sir R.P. on the Report' – This report is (I believe) the Pamphlet and perhaps the Estimate – – Of this I am almost sure that the Survey of London is 'Fore-ordained' and will come to pass –

B.[17]

NB. Tell Mr Marquis to send me his address.

Sunday

You should note down the dates of all important circumstances

---

Is the Master General's name[18] put on the Titles of your 3 chain plans?

I think it would be well if somebody or body of persons – or a certain number, when you are all together in one room, each to take part in the speech – say in the drawing room, when the principal parties are present – 2 or 3 – or 4 all together ask Stanley about the Memorial.[19] I dont think it would be wise for <u>one</u> person <u>alone</u> to do it, as he might afterwards be pointed at – You should see after it as soon as possible. Acid, <u>if he can</u>, will diddle you all of it. After hearing what Stanley says you can judge how to act –

B.

I would not like to advise – but I think you should see after it, the sooner the better – if Stanley is a good one, he could be spoken to in a friendly way – Acid is such a rogue, that no dependence whatever can be placed in him. Carbon was not so bad until enlightened by him. Carbon used to be a fair man to those on the 2 inch[20] – he sometimes gave a rise of 2/. per day – but those days are evaporated –

As you say; a slashing letter must appear. I will think on that point.

– B

<u>Cost of Plan</u> (6 inch)

	<u>Per Acre</u>	
Parish Trig Cal.	0.0.0.1/4	<u>or 6 pence per acre.</u>
Determining Boundaries	""""...1/2	
Surveying	"""..2.1/2	the calculation is based
Levelling	""""...1/4	upon the fact that on
Content Plot	""""...1/4	an average a man plots,
Plotting Plan	""""...1/4	Draws, Surveys &c
Tracing	""""...1/4	a certain number of
Examination on Grounds	"""..1.0	acres – and that this
Drawing Plan	""""...1/4	man on an average re-
Names & Name Books	""""...1/4	ceives 3/. per Diem –
Calculation of Areas	""""...1/4	
	0.0.6.0	

Now let us take a 6 inch plan[21] and from the Index table or margin of the plan The area and cost would be <u>strictly</u> as follows – 'Area 3891–' Acres

	Shillings	
Plotting the Plan	0.128.0	
Tracing " "	18.0	
Examining on the ground	108.0	or, one penny farthing per acre
Drawing	63.0	
Writing	18.0	
or shillings	3 35.0	

To which we must add Surveying – Plotting Plot – B^{d}.[22] remarks &c.

The cost would stand thus –		1 . ¼
and		4 . ½
	per acre –	5 . ¾

I have tried the calculations upwards of a dozen ways and I always come to nearly the same conclusion – or Six pence per acre of course this does not include botching, patching repatching of Boundaries – Name Banks – or Re-surveying – Officers allowances & &c &c. It only covers actual first outlay – or what the different duties would cost were they correctly done in the first instance –

Approximation

Ordnance Expenditure from 1841 to 1846 inclusive,

or 5 whole years at =	50.000 a year
	5
	£250.000
The Scotch Survey for the above period =	15.000
The Irish D°————————————D° =	30.000
	£45.000
	205.000 Leaving for the English Survey 205.000 £ for 5 years.

	Let us take it at 200.000 £	
	20	
	4000.000	
	12	
The pricing of one acre 6	48.000.000 pence	8.000.000
	8.000.000 –	acres furnished in England in 5 years at a cost of 6 pence per acre!!!! –

I cannot get the Ordnance Estimates. My friend called in three places in London and could not make it out – And in the absence of the Estimates or Grants I have to judge from the nature of the thing, and recollection of what I have heard from different parties – From what I know & from what I have heard I think the above guess, may be taken for want of a better –

I have no means of judging correctly of the actual number of men employed, but I could give a very near guess, as to the gross number

On the Trig:[23] –	50
Irish Contours –	50
Scotland –	50
Southampton –	70
In the N. of England	170
	390 or 400 –

Add to this about – 25 Engravers. on an average of 8/. a day –

Make allowance for instruments, Books &c. &c. And if you add the R.E. themselves you will find the annual grant vanishes like Jonah in the Whales Belly[24] –

NB. The above may include the Boardsmen but it does not include Labourers (Chainmen).

RI MS JT/1/TYP/11/3803–3806
RI MS JT/8/1/4a

---

1. *York*: Louisa Tyndall annotation.
2. *your last*: letter missing.
3. *the pamphlet I spoke of*: see letter 0240 of this volume; this appears to reference an unidentified pamphlet written by Thomas Frederick Colby.
4. *Carbon*: a nickname for Thomas Frederick Colby; see letter 0231, n. 9 of this volume.
5. *Sir R.P.*: Sir Robert Peel.
6. *the Bill*: a Parliamentary bill.
7. *R.E.*: Royal Engineers.
8. *an address, and a piece of plate*: McLachlan and some other allies apparently composed a letter of thanks to Captain B., along with a 'piece of plate', a metal plate frequently given as a gift or prize.
9. *Cap. B.*: Captain B., probably John Bailey; see letter 0231, n. 29 of this volume.
10. *Acid*: a nickname for an unidentified opponent of Tyndall's on the Ordnance Survey, possibly Henry Tucker.
11. *Div*: Division, i.e. of the Ordnance Survey.
12. *Spec:*: Spectator, Tyndall's *Liverpool Mercury* pseudonym.
13. *now going to Southampton*: the office of the Ordnance Survey moved to Southampton in 1841.
14. *Par:*: Parliament.
15. *until the Bill is past*: a bill that Colonel Colby was trying to pass through Parliament; see letter 0240 of this volume.
16. *Sir Delabash*: Sir Henry Thomas De la Beche (1796–1855) worked as the geologist under Colonel Thomas Colby on the Ordnance Survey (*ODNB*).
17. *B.*: the pseudonym McLachlan used to avoid detection.
18. *the Master General's name*: George Murray.
19. *the Memorial*: the letter of protest sent by the workers of the Ordnance Survey of England to George Murray, Master General of the Ordnance on 23 September 1843; see letter 0236 of this volume.
20. *2 inch*: maps that are drawn at a scale of two inches to one mile.
21. *6 inch plan*: i.e., maps that are drawn at a scale of six inches to one mile.
22. *B^d^*: Boundary.
23. *On the Trig*: i.e. number of men employed on the Ordnance Trigonometrical Survey.

24. *vanishes like Jonah in the Whales Belly*: Jonah 1:17: 'Now the LORD had prepared a great fish to swallow up Jonah. And Jonah was in the belly of the fish three days and three nights'.

From Archibald McLachlan 8 October 1843 0242

Sunday – England – N° 2

The expenses were obtained thus – 1114 men at 3/. – £52,000 – The number of men, laborers and all, prior to the Reduction was 1114 nearly – a few more came from Ireland after the calculation was made – but not many – I could not tell the pay and expense of Officers –Trig:[1] – Instruments – Boundaries – Stationary – Southampton &c &c – but raised the pay of the men up to 3/. and said that will clear all –

	1114
	3
	3342
for	311 – days
	3342
	3342
And speak in round numbers the expenses is about 60.000	10026
yearly – or was so lately –	137020
One of the gentlemen here told me, that about 8 months ago	
he knew from the Superintendent[2] that the quantity done	10020
in Yorkshire was – 600.000 acres, including the towns.	1039362
	51.9000

600,000 added to Lancashire = 1.100.000 = 1.700.000 – I conclude that Lancashire was then all surveyed – but did not know you did any of Yorkshire

I will suppose that Lancashire is all done	= 1-100.000
Our part of York	= 0-800.000
Your part of York	= 1.900.000
If the Manchester party has done any of York it has to be added	=

Because in my calculation the work done by B. & D.[3] was 600.000. I said 700.000 to make sure – since that time little has been done to affect the calculation, because of the reduction in the number of men – & the time spent on the towns – Now without knowing how much you have done of York – or how much the Manchester party has done of York – I would calculate in round numbers and say –

Lancashire————	1.100.000
York————	1.100.000
	2.200.000
Expenses————————	£140.000 = 2/. per acre!!

You will see by this whether I make any awful blunder such as a Million of Acres!! and 60.000 £ – !!

Now stating it as formerly————	Acres 1.800.000
Expenses = 120.000£ ————	2/. per acre –

A very good guess – and we cannot make anything but a close guess – Now I will state it my way to suit the alterations since the above–

	Acres 2.200.000
Expenses 140.000£ ————	2/ per acre ––

You will see that in both the results are nearly equal, 2/ per acre. And in speaking to the public I should say – 'The quantity done is in round numbers – 2,000.000 acres – expenses 130.000£ or = 2/. per acre – and if this statement is false, you, gentlemen, can set the public right by stating the true quantities'!!

Suppose I make some awful blunder and instead of the above

It should stand thus ————	Men = 1114 = 2/6 = £40.000
	2 years
	80000
	20
	1.600.000
and at the end of 2 years we have————	Acres = ———— 1.600.000

<u>Monday</u>

To day I have consulted a cool & knowing one – he says

<u>we</u> C.B.[4] division has done ————	500.000 acres
and will not be <u>finished</u> in less than 6 or 9 months –	
Annual cost of all the O.S.E.[5] – 50. or 60.000 – yearly.	
'It cant be less than 50 – nor more than 60' – a year. –	
'C. Dunford[6] has done ————	500.000 a[c]
and won't <u>finish</u> in shorter time than 12 months'	
– I say Acid[7] has done Lancashire ———	1.100.000
to be <u>finished</u> in 9 months – (<u>all</u> 6 inch work) –	2.000.000
So if you say 130.000£ – and 2.000.000 acres	

you come as near to the truth as the absence of official documents will permit – So you have here all that I can say on this point. – B.[8]

RI MS JT/1/TYP/11/3807–3808
RI MS JT/8/1/4a

1. *Trig*: this may refer to the men employed on the Ordnance Trigonometrical Survey.
2. *the Superintendent*: the Superintendent of the Ordnance Survey of Ireland, Colonel Thomas Frederick Colby.
3. *the work done by B. and D.*: this probably refers to the work done by the B and D divisions of the Ordnance Survey.
4. *C.B.*: Captain B, not identified.
5. *O.S.E.*: Ordnance Survey of England.
6. *C. Dunford*: not identified.
7. *Acid*: a nickname for an unidentified opponent of Tyndall's on the Ordnance Survey, possibly Henry Tucker.
8. *B.*: the pseudonym McLachlan used to avoid detection.

From Archibald McLachlan 9 October 1843 0243

Monday – N° 2 – read the other N° 2 first[1]

Awful rumours are afloat to day – of Acid's[2] being in Liverpool by order of the Mas: Gen.[3] to investigate into a row among the men, and certain representations to some Members of Par!!![4] He is to be here tomorrow when a number are to be discharged – and some reduced – Yolland[5] is to have a division, and to commence surveying in the South of England. Acid is to have Yolland's place in Southampton – and to be a Little Carbon.[6] All is now to be conducted by Acid, under the control of Carbon – Our Div:[7] is to be either broken up altogether or given to Yolland – in case of the Liverpool affair ending unsatisfactory, Hamley is to go off – Yolland gets his place and we are to be distributed among all parties – such is the Rumour – I have been consulting my knowing old gent:[8] – The plans of Derry – Tyrone – Monaghan – part of Armagh &c[9] were revised by him!! I asked what became of the originals – 'I don't know but I think they were burned, ask Mr –[10] he was there and knows more of the whole affair'.

So I will consult Mr – tomorrow–

I asked, did any of the original old plans pass revision, did you correct and revise any plans and how many – 'He tried a few of Derry – Do[11] of Tyrone and other counties, but very few of them would do – We plotted fresh plans from the old books and a few of these made good enough plans – but only a few of them, we found it took more time to make good plans from old books and revisions, than to make the plans from a complete fresh survey, so after a few trials in each county, we gave up patching & revising altogether & surveyed right out of a face' –

When did you begin to survey the lines – 'Think it was in 1832, but I dont recollect every thing we did – we plotted some parishes on new sheets, from old Books, we revised them by the line survey but found they would not do, some parts of the Revision we traversed, it was all no use' How many of the old plans did you revise – 'I dare say we could get one parish out of 20, 19, out of 20 of the old plans & books were not good, and we dropt using them altogether after some trial in each county'. – I asked particularly how much time was lost in the original survey – he said about 6 years. I said dont you think it was 7 years – he said no. From August 1825 to the beginning of, or middle of 1832 – not more than 6 years – So here we differ! perhaps he is right!!! – if he is right, I am one year wrong – I asked were other districts as bad as his – 'I don't know, I am not certain of it, but you may rely on it they were all as bad as us'–

I will repeat it 'The scientific labours of the first 7 years &c' – and dare them to disprove it – should they prove it was only 6 or 5 ½ years – I care not, the important point designed and insisted on is still correct – If they prove that all the old maps, plots &c were not burned, I care not, the grand fact stands correct – I will hazard the assertion that the order to Burn, extended to all the Districts, and to all the useless maps plots &c – The only possible mode of disproving is – producing the original maps plots, books – This I defy them to do – the more I look into it, the more confident I become – Tomorrow (Tuesday) I will consult Mr – it wont be before post time – but you shall soon hear – You will have an answer to the other points soon

B.[12]

Tuesday 1 o'clock – yours of Monday[13] came – I wont be in time for the post if I stay to say much – the reason for delay is that I would not be in time for Sunday's post – and wished to consult with the above mentioned –

This sheet may be read separate from the other remarks –

1st Mr G[14] - It is uncertain whether he was appointed by the Board or not – I cannot find any body who read his appointment or who saw it. He acted in 2 capacities, Bdy[15] Surveyor, and Valuer. The Bdy. S.[16] was paid by Government – Valuation is paid by the County – The County pays a part, if not the whole expenses of valuation. I am informed by knowing ones, that he was appointed partly by the Board, & partly by the Irish Government at the Castle, Dublin – The Irish Government is his principle employer as to the valuation – The Board is his employer as to Boundaries – this is all I am able to collect on the matter

---

2nd – Lt. Renwick[17] – Bickett[18] knows all about him – Think he went off the S.[19] about the end of 1828. I am pretty sure that was about the time.[20] The cause of his leaving was irregularity, and disobedience of orders – The great fault was, Surveying roads streams &c by lines – instead of by traverse. When he was reproved by Cap English,[21] he said 'By God I'm right, you will all do it that way yet' – He did not do all his parishes by lines only a few of the later ones – And in the revision his work was not so useless as the other officers.

3rd – £300,000 was for S. and not for Bdy or valuation. Griffith used to get a separate grant for Bdy ask D.[22] –

4th My calculation goes to say upwards of 200,000 was lost by the original Survey – Others who know the subject as well as myself say 250. or 270.000 – We are safe in saying 200.000 – In the Ho of Com:[23] 1829 Carbon stated the expense at 77.000!!! I understand his trick lies in not including military pay of man or officer!! 77.000!!! that sum is a bare faced prevarication – the mere pay of officers would be more than half that sum –

The First 7 years[24]

5th I would not repeat the word 'First'. I would rather say the First 6 years – the whole lies in the word 'First'. However you can state it as you will, here is my way of calculation –

	from August 1825 to 1830	= 5 ½ years,
1830 or 1831 revision commences	– Revision Experiment	= 1/2 year
	The First 6 years	= 6.0

From 1831 to 1835 or 6 – we lose another year in attempts to cobble up some parishes of certain counties. There was much more than one year lost – but say 1 year – Total 7 years. So we have 6 years lost, in succession, that is provable – and at least one year more at repeated attempts – If some old men are put on oath they would say, 'more than 7 years'. The only objection which could be found is with the word 'First' – which is not critically correct – let them find fault if they like – their error in correcting us will be soon detected.

Altitudes

6th The three alterations are notorious – Captain H.[25] deduced altitudes from one mountain, Captain L.[26] from another mountain – they differed when they came to close on each other. Both were right and no mistake!! – Neither would give in. Carbon was down there at dinner – the 3 parties disputed on the altitudes, Carbon agreed with Cap H. – Cap L. got into a rage (he was a violently passionate man) he jumped up from the table, and slashed his plate with all its contents against the door – and jumped, and tramped on the broken pieces – The heights were changed to agree with Cap H. –Afterward H was proved to be wrong. They were altered back again to the original – and still wrong – At the Revision the whole altitudes were given to, or done by, one officer – Portlock[27] – and he altered all that was done before. – B.

NB The above alterations were made in the principal County heights – some of the parish heights were altered ½ dozen times.

RI MS JT/1/TYP/11/3809–3812
RI MS JT/8/1/4a

---

1. *read the other N° 2 first*: i.e., letter 0242 of this volume.
2. *Acid*: a nickname for an unidentified opponent of Tyndall's on the Ordnance Survey, possibly Henry Tucker.
3. *the Mas: Gen.*: the Master General of the Ordnance Survey, George Murray.
4. *Members of Par*: Members of Parliament.
5. *Yolland*: William Yolland (1810–85), a senior officer on the Ordnance Survey who was responsible for many of the maps of Lancashire and Yorkshire (*ODNB*).
6. *Carbon*: a nickname for Thomas Frederick Colby; see letter 0231, n. 9 of this volume.
7. *Div*: Division.
8. *my knowing old gent*: not identified.
9. *Derry – Tyrone – Monaghan – part of Armagh &c.*: counties in Ireland.
10. *M^r^ –*: not identified; McLachlan chose to omit the name.
11. *D^o^*: Ditto.
12. *B.*: the pseudonym McLachlan used to avoid detection.
13. *yours of Monday*: letter missing.
14. *Mr G*: Richard John Griffith.
15. *Bdy*: Boundary.
16. *S.*: Survey.
17. *2nd Lt. Renwick*: William Turnbull Renwick (1802–90), a Royal Engineer who worked in the E District of the Ordnance Survey, under the command of Captain Frederick English. He joined the army aged 22 and became 2nd Lieutenant on 12 September 1825. He rose steadily through the ranks, achieving the rank of General on 27 February 1876. He served on the Irish Survey at his first post between September and November 1825. He came back to the Survey between October 1826 and March 1832. He then went to Jamaica. He later served in Ireland for three separate periods, as well as spending over eleven years in Canada and almost three years in the Cape of Good Hope. Renwick died in Ontario, Canada on 5 April 1890.
18. *Bickett*: Joseph Bickett, a civil assistant. He worked in the District Office in Captain Tucker's A District. Bickett joined the Survey on 26 December 1826. In December 1839 he was the highest paid civil assistant in the A District on a daily rate of 4/6. His primary duty at this point was examining plans. He was stationed at Waterford during the early 1840s. His pay rose to 4/9 in the autumn of 1840. He left the Irish Survey, presumably for England, between August and November 1841; the records for this period are missing.
19. *the S.*: the Survey.
20. *1828. I am pretty sure that was about the time*: Louisa Tyndall annotation: 'Wrong, it was in [18]31'.
21. *Cap English*: Captain Frederick English, a Royal Engineer. He worked on the Ordnance Survey of Ireland throughout the 1820s, and was commander of the E District. English served on the Iberian Peninsula and the Netherlands between 1808 and 1809, and in France between 1815 and 1818. He became 2nd Captain in 1813, and was promoted to Captain in 1825. He went on to become Lieutenant Colonel in 1837, and he died in

1849 at Woolwich Arsenal after an accident resulting from an artillery experiment.

22. *D.*: probably a civil assistant in the 1st Division; see letter 0231, n. 6 of this volume.
23. *Ho of Com.*: House of Commons; see letter 0252 for Tyndall's description of Colby's 1829 testimony before the House of Commons regarding the Ordnance Survey of Ireland. See letter 0211 in Volume 1 for an earlier reference to this testimony.
24. *The First 7 years*: McLachlan appears to be giving Tyndall notes on a draft of the third Spectator letter; see letter 0252.
25. *Captain H*: probably William George Hamley.
26. *Captain L*: probably Sir Thomas Aiskey Larcom (1801–79), a senior surveyor on the Ordnance Survey. Larcom had joined the Survey in 1824 and in 1828 was appointed Colby's assistant in the headquarters of the Irish survey at Mountjoy House, Phoenix Park, Dublin. Larcom was put in charge of the office for engraving maps, but he soon came to dominate the whole business of ordnance surveying. By early 1846 simmering tensions between Larcom and Colby came to a head. Colby was increasingly jealous of Larcom's independence and of his stature in Ireland and on the Survey. Colby successfully had Larcom dismissed in 1848. However, Larcom enjoyed a highly successful career after his dismissal. In January 1853 Larcom was appointed under-secretary for Ireland under Aberdeen's liberal-Peelite administration, a position he would hold until 1868, and was then created baronet and appointed to the Irish privy cabinet. He died in Heathfield, Hampshire in 1879.
27. *Portlock*: Captain Joseph Ellison Portlock (1794–1864), an officer of the Ordnance Survey (*ODNB*). He was born in Hampshire and came from a Church of England background. He joined the Royal Engineers at the age of 18. He advanced steadily through the ranks, becoming 2nd Lieutenant on 20 July 1813, 1st Captain on 2 September 1839, and eventually Major General on 25 November 1857, at which point he retired from duty. Portlock served in Canada between 1814 and 1822 and was involved in a successful military campaign in Upper Canada in 1814. He began on the Ordnance Survey of Ireland in July 1824. He was stationed in Corfu after leaving the Ordnance Survey in April 1843. He returned to Ireland in 1847, and was married in Cork in December 1849. Portlock died at Blackrock, near Dublin on 14 February 1864.

From William Ginty 13 October 1843 0244

Friday 13th Oct. 1843

Dear Romulus[1]

Your attempt at an excuse for your want of punctuality adds another proof to what indeed was proved to my satisfaction long ago – id est,[2] your skull is a 'crazy skull'. I am quite certain that my abbreviation for 'morning' and 'evening' could never be confounded by any but a being touched in the upper story. You admitted you were under the impression that Monday evening was the promised time. I believe you were! This however does not interfere with what I have asserted and you have proved. 'But let that pass'. In faith I pity you. Love is a sorry customer to creep into a man's cranium.[3] Love thoughts and sensible thoughts should not be huddled together. They invariably confuse each other if not kept apart – That's fact! The preface to your just remarks[4] on my /b/ulls is another proof – there was not the slightest occasion for spouting about the spirit that actuated you. Yes! I know you well! Your studied apology was indeed unnecessary, but the proof of insanity herein lies; first you apologize to all intents and purposes and wind up by saying 'I feel Bill[5] that apology is unnecessary' (we felt alike) if so – why use it? Or vice versa! This I call incontrovertible testimony against the sanity of your mental materials. Now for the remarks themselves. My answer to the first is that I plead habitual ignorance, if however any person had hinted at an error in the sentence I could and would have detected it. Your criticism is just, well-timed, and duly appreciated (if anything coming from such a chaotic mass of heterogeneous thoughts and actions could be valued). Second 'I seen'[6] – Just so. 'I seen the man that saw the man' &c. A most beautiful Irishism, and one that I fear I will never rid myself of, without being in perpetual remembrance of it, nor you either!!!

'I – saw – seen (copy)
I see – I saw – I have seen'

Now this is a correct copy of your illustration, look close at it! At the time you wrote this you were for the instant off your guard and the Irishism slipped – You then thought of it and corrected it – this was no fault of the pen. Oh no! I have laughed heartily at this, all your powers of argument won't do. Try sophism. I anticipate all you can say, perhaps wisdom would silence you! I quote 'I shall expect a similar mode of treatment from you' – Knowing that an opportunity for complying with your expectation will seldom if ever offer. I am most happy to return the compliment in the present case. But seriously, when ever you see an error tell me of it. With respect to the memorial[7] you write; 'my own opinion is &c'. In the name of common sense could your opinion be other than your own. It is only superfluous, perhaps justifiable for emphasis's sake. You can now judge that I am looking for bulls. Enough!

I most heartily concur in your own opinion my sweet fellow! Acid[8] has some regard for his own safety none but a madman would stop it. With regard to the 'nominal' capacity of the H.O.[9] I have to tell you that the more you pitch into these *[ruffians]* in your last letter the better. By the Tithe Plans[10] and the announcement from the M.G.[11] to the surveyors, the case stands thus – Some one of the three, the man that ordered that heading to be written – 'The Master General[12] or Mr Patten[13] is a liar'.[14] Mark now that heading, when laid before such a numerous and respectable community as the Board of Guardians &c was calculated to deceive them, it is neither more nor less than the publication of a libel on the Master General. Yes you can prove, this is proof enough that they 'exceed their orders' and undertake matters 'on their own authority' and more, that they publish manuscript libels on the Master General amongst a respectable and numerous class of the gentry and professional men!

With respect to Downholland,[15] I have learned since that the calculation was not made accurate enough. If it – the expense exceeds the estimate or contract price it does no more, that is beyond controversy, but to go to the exact letter, as if before a court of justice, expenses could be insisted upon which indeed would throw the balance in favour of — your argument.

There was very little of the contract work done here. Downholland 3395 acres and Yealand Conyers[16] – acres.[17] This latter exceeds the estimate with a vengeance, I could swear the computation cost about 4d an acre. They were young in the trade at that time. And now you may refer to this with confidence. I believe it will be the best you can refer to on the survey. I will send you all about this in my next. There was only two done here and in each case the same as yours with respect to copies &c. A call to the lower House! The Captain[18] is there, he wants those only who signed the memorial. Hurrah! Now Ginty – firm resolve! The answer! The answer! Just as I half expected another trial at the intimidation. No go! old boy. You often called me a Bladder but I bladdered today to the unutterable consternation of Tucker and Hamley. He said he addressed our 'Brethern' in Preston this morning. Just so, my dash 'thank him for the term' (as Dan[19] says) – 'our Brethern' ought to be proud of us! We were all 25 of us assembled in our room – present Hamley, Tucker, and Scott.[20] He had our petition there in his fist! (Are they justified in delaying it so long?) He made a long speech and at last asked some questions which led to a long and even animated discussion, on my part and on his. It was animated enough he had previously picked out a half dozen or so and among those were the youngest and oldest hands and among the latter was your humble servant. He asked and I answered and I asked and he answered a nice little volume of questions. I will give you a few of them. 'Do any of you wish to have a number of men discharged and the remainder increased'? Some few said 'they did not wish it'. I said 'It is hard Sir that the majority of the men should be kept in perpetual misery to support a number of inefficient men'. 'Yes' says he, 'but are you to judge of efficiency or inefficiency'? 'I don't pretend to do it, it has nothing to do with our memorial; one thing is certain Sir – We are not all alike in qualification'. 'No' says he, 'but we don't bind ourselves to discharge the useless or useful, we can discharge which we like?' I nodded + repeated 'I had nothing to do with that, and that had nothing to do with the memorial' – 'Now', says he making a sign of division of those present. 'Suppose you were discharged and you kept, would that be just treatment'. Taylor replied 'if there is no alternative, you may adopt it if you think proper'. Ginty – 'we have nothing to do with that, it is not for us to decide'. 'But you have tied Col. Colby's hands behind his back, he can't do anything' Ginty 'if his hands are tied, there is nothing in them. The Master General may loose them and get an extra grant put in them'. (a titter behind) – T.[21] 'You seem to think that an extra allowance can be got – it is impossible. See the difficulty last year in obtaining even a low grant' – Ginty – 'perfectly possible Sir – not without precedent, and the revenue is increased 1½ million this quarter'! (Hamley laughed) and I repeated it in most emphatic terms and said we might as well be in work houses &c &c &c. (I heard or rather I saw that announcement about the revenue in the Mercury[22] this morning and it was fresh in my memory). He spoke in long terms about the evil results of anonymous epistles and alluded to Spec.[23] He spoke also of an article in the 'Architects and Engineers Magazine' saying the Board of Ordnance ought to be relieved of their office with respect to the survey, and said if we supplied or wrote such letters – I interrupted him – 'Sir we will not allow ourselves to be identified with anonymous writers – we condemn it. And if for nothing else that memorial (pointing to it) is perfectly necessary and justifiable, as it records our censure and protest against conduct of which we have been accused'. – T.[24] 'Oh! I don't accuse you or any of you of that' – 'Yes Sir you suspect us' – 'No No!' – 'Yes, or why state it in and send round that district order'. Need I say this was a settler – Hamley stood up and walked about. Willy[25] and Taylor assailed Tucker again and I pitched in a few more clinchers, and in the middle I may say of the confusion, he said 'Go to your works' (X).[26] So we did, leaving him in a precious stew. By Jupiter they are quaking. Oh */how/* we lacerated him! I am all anxiety to hear how 'our Brethren' got on. Now he thought to get us to withdraw it. He acted foolishly in taking me into the Deputation to answer for the rest. I spread a spirit among these men that their host of tyrannical petty foggers could never crush! Taylor and Willy spoke like men. Willy was raging and frightened me with the yells he gave at bothered Harry.[27] This I suppose was the grand and dernier[28] resort – He failed – We conquered. 'Fraught with this fine intention and well-fenced'. 'In mail of Proof, our purity of soul' layed low the insidious attempt of the old pot-headed dog! Previous to the summons to the bar of the 'Lower House', an order came round saying 25 men were to be discharged from this division. I suppose the fool that issued this thought at least some of us would be afraid, we the 25 malcontents would be hopped off – a trick of intimidation worthy of an engineer ignoramus. He dwelt a long time on the evil results of anonymous writing. Oh! fire, Spec – They dread it – they tremble, they see, they know, what effect an exposure will have on the public. They will lose their lucrative situations and their credit. This news only adds a momentum extraordinary to the fall of humbug. A multiplied velocity to the descent of base trickery and long practised tyranny and injustice. Crush the wretch! War to the goose quill Ink! Ink! Ink! By God, I think they have no right to delay a letter – a public letter so long. They got our memorial on the 26th ultimo, and it is now the 13th proximo. This to me appears a piece of business they had no right to keep back so long. He says the writer or informer of Spec is in Preston. He speaks of some figures being in the letter which were known only to a few, and was only used in private correspondence. Just so, by dash, have they got a clue to Spec. What do you say? Can they trace the matter out? – I will search out that magazine if in Liverpool. I will refer 8[29] to it.

He repeated many times that we had a right to petition Colby and not the M.G. But I suppose all this was merely the re-echoing of his Preston Demonstration. He looked misery itself when he ordered us to our duty – no doubt swearing vengeance internally against poor

Omega[30]

I would not wonder if Col. Tom[31] would try us next – he will fail. He made no distinction between those that signed the form and us that signed the memorial alone. The fellows behind me kept pricking me most unmercifully, and so shoving me – Oh if I dare to let the stopper of vial of bitters I could have made him writhe fifty times worse than he did.

Hamley seemed to be on a broiling iron all the time.

Enough!

Tullus Hostillius[32]

Excuse the grammar of this for pity's sake, I am not Ginty just now. I'm a personification of smiles and contempt &c.

Capt. Tucker's speech &c in Liverpool.[33]

Will you answer me these simple questions?

How many men have been dismissed since we came to England – Some hundreds! How is it that we have not a superintendent less – A *[word]* from Spec on this will get an officer stripped and that officer Tucker. If we had only enough then, we have too much now – Handle this! What is Tucker's capacity! Why he does the work of a silly doting and expensive old Col. His pay would take 20 or 30 men out of misery!

Dear John

Morning has beamed upon me. Now I tell you what I want you to do – You have a good many of my Rhymes or verses[34] – You have them all but I have not one, nor do I recollect one. Send me the best of them – there is a host of them not worth sending and would only make me curse the hand that wrote them were I to see them now I don't mean to insinuate that I am improving in a poetical way, only I am less an ass than I have been. Nor do I intend to *[twin]* my own 'Dear Murray'.[35] Only I would like to have them. A hoary head may doat upon them in after years! I want no fair copies only pencil out one of them roughly when you write. I want to copy them into a little book where a dozen or two of yours is already.

W.G.

RI MS JT/1/TYP/11/3616–3617
LT Transcript Only

1. *Romulus*: a new nickname for Tyndall. In Roman mythology, Romulus was the son of the god Mars and the founder of Rome.
2. *id est*: that is to say; i.e. (Latin).
3. *Love is a sorry customer to creep into a man's cranium*: Ginty was implying that Tyndall was enamoured with an unidentified woman.
4. *your just remarks*: letter missing.
5. *Bill*: Ginty's nickname.
6. *I seen*: Tyndall appears to have been teasing Ginty about his use of 'seen'; see, for example, letter 0235 of this volume, where Ginty writes 'I seen O'Neill last night'.
7. *the memorial*: the letter of protest sent by the workers of the Ordnance Survey of England to George Murray, Master General of the Ordnance on 23 September 1843; see letter 0236 of this volume.
8. *Acid*: a nickname for an unidentified opponent of Tyndall's on the Ordnance Survey, possibly Henry Tucker.
9. *the H.O.*: Home Office.
10. *the Tithe Plans*: see letter 0231, n. 19 of this volume.
11. *the M.G.*: Master General, i.e. George Murray.
12. *The Master General*: George Murray.
13. *Mr Patten*: John Wilson-Patten (1802–92), Baron Winmarleigh and Member of Parliament for North Lancashire. Born John Wilson; his father Thomas Wilson had changed his name from Patten to honour a relative whose fortune he inherited. The family changed their surname to Wilson-Patten in 1823. Patten matriculated at Oxford in 1821 and entered the Commons in 1830; he represented Lancashire there until 1874, when he was created Baron Winmarleigh. He was primarily a Tory although he allied himself with Robert Peel's Whigs for a time. During the controversy between the civil assistants and the Ordnance Survey leadership in Lancashire, Patten investigated the civil assistants' claims and declared that the Master General of the Ordnance, George Murray, had not authorized the Survey leadership to undertake Tithe maps.
14. *The Master General or Mr Patten is a* liar: Louisa Tyndall annotation: 'This is worked in to 'Spec[tator]' letter 4'. See letter 0261 of this volume.
15. *Downholland*: a parish in Lancashire, England.
16. *Yealand Conyers*: a parish in Lancashire, England.
17. *– acres*: this number has been left out of the transcript.
18. *The Captain*: possibly Captain Henry Tucker.
19. *Dan*: not identified.
20. *Scott*: Corporal William Scott.
21. *T.*: Taylor.
22. *the Mercury*: the *Liverpool Mercury*.
23. *Spec.*: Spectator, Tyndall's *Liverpool Mercury* pseudonym.
24. *T.*: Henry Tucker.
25. *Willy*: William Hunter.
26. *(X)*: not identified.
27. *Harry*: possibly a derisive nickname for Henry Tucker.
28. *dernier*: last (French).
29. *8*: see letter 0232, n. 13 of this volume.
30. *Omega*: an occasional nickname of Ginty's.

31. *Col. Tom*: Colonel Thomas Frederick Colby.
32. *Tullus Hostillius*: the Third King of Rome, who ruled from 672–641 BC and was renowned for his military victories.
33. *Capt. Tucker's speech &c in Liverpool*: Louisa Tyndall annotation: 'This – I think, is only a label, – written – across the *[*letter*]*.'
34. *my Rhymes or verses*: Ginty's most recent extant poem to Tyndall can be found in letter 0235.
35. *my own 'Dear Murray'*: this is probably a reference to Lord Byron's relationship with his publisher, John Murray (1808–92) of the Murray publishing family. In *Don Juan* (first published in 1809), Byron mentioned his editor in C1.2: '(Plain truth, dear Murray, needs few flowers of speech)'.

To Captain Henry Tucker 14 October 1843 0245

Saturday, | October 14th, 1843

Sir,

A meeting having been held on the evening of Friday the 13th of October,[1] at which the information and arguments adduced by you on the morning of the same day, relative to a memorial to the Master General,[2] forwarded to you on Saturday the 23rd of September,[3] received a calm consideration; we, the undersigned individuals, being called upon by the unanimous voice of those present, respectfully beg leave to lay before you their sentiments as regards the document alluded to.

We have to state distinctly that in their late proceedings Memorialists[4] were actuated by no momentary impulse, no sudden ebullition of feeling caused by the unexpected appearance of your district order;[5] the course which they have adopted towards that order is the result of calm convictions which remain to the present moment unimpaired.

With respect to their applying direct to the Master General, we would briefly remark that it is through his interference alone that Memoralists hope to have their grievances in due time redressed; in reference to that part of the Memorial which relates to the insufficiency of the rates of pay, we are to state that your known aversion to applications or complaints on this subject coupled with your relation to Col Colby force on us the opinion, that any reference to the latter would not be attended with success. Another part of the Memorial speaks of no allowance being made in case of sickness. You know, Sir, that Col Colby has been applied to in such cases and his answers shew that we have no hope of redress of this grievance at his hands. In reference to the grant,[6] we would respectfully state, that the feelings which we have lately given expression are not of recent birth. They had an existence for years before the grant was produced. If Memorialists saw on the part of their Superintendents a disposition evinced to ameliorate their condition they would never have engaged in the present line of action.

In compliance then Sir with the wishes of those concerned we earnestly beg that you will have the goodness to forward the Memorial in its present form to the Master General.

This letter will explain to the M.G. the reason why Memorialists obtrude themselves directly upon his notice.

RI MS JT/1/TYP/10/3298
LT Transcript Only

1. *A meeting having been held on the evening of Friday the 13th of October*: i.e. a meeting of the surveyors who felt aggrieved; see letter 0244 of this volume.
2. *the Master General*: George Murray, Master General of the Ordnance Survey.
3. *a memorial to the Master General, forwarded to you on Saturday the 23rd of September*: see letter 0236 of this volume.
4. *Memorialists*: the authors and signatories of the letter of protest sent by the workers of the Ordnance Survey of England to George Murray on 23 September 1843; see letter 0236 of this volume.
5. *district order*: see the 'District Order' in letter 0236 of this volume, dated 16 September 1843.
6. *the grant*: the Parliamentary grants of 1843, meant to fund the operation of the Ordnance Survey; see letters 0236, 0238 and 0246, n. 6 of this volume.

To George Murray 14 October 1843 0246

To the Master General of the Ordnance.

Sir,

Having been assured from various sources, that you have ever evinced a willingness to listen to the voice of complaint, when that voice was raised in a proper spirit, and had sprung from a just cause: we the Civil Assistants of the Ordnance Survey of England whose names accompany this, feel emboldened to come forward and crave your kind attention to the following remarks in connexion with our situation.

We have been employed on the Survey of Ireland for many years, and we would humbly state that our assiduous exertions contributed materially towards the furtherance of that great work, our pay in Ireland was so low as to render as in many places where we have been located 'a proverb and a bye-word'[1] among the inhabitants: we however were induced to hope for an improvement in our condition when the Survey of England would be undertaken; this we imagined would offer a fair field for the exercise of talent. We do not say that our hopes in relation to this were grounded on any official statements made to us, these hopes were induced by many corroborative circumstances: we conceived, that the interest which it was natural to suppose our superintendents[2] took in those employed under them would induce them to rescue us from that state of poverty which had so long chilled our energies in Ireland, and which must form a striking contrast with the state of persons in similar situations in England, that we, filling situations in a highly important scientific undertaking would receive at least sufficient pay to enable us to support that appearance of respectability which would be expected from persons engaged in such an undertaking. We had seen men drafted from the Survey of Ireland to that of England who immediately after their arrival in this country had their wages considerably increased, (these instances occurred previous to the embarkation of the main body of the Irish Ordnance Surveyors). Considering these as precedents, we imagined that our treatment would be similar. – Our calculations on this matter proved false: during the past year we were all drafted to this country, while no individual whose pay in Ireland had amounted to twelve shillings per week received on his arrival here one farthing increase: Our hopes, though shattered by this circumstance were not altogether destroyed, we still anticipated better days, when the reproach under which our poverty compelled us to labour in a strange country would be taken away. The appearance of the accompanying District Order[3] at once struck at the root of our anticipations, and reduced us to the necessity of either signing a document from which inferences may be drawn contrary to our sentiments, or else adopting a course which our present circumstances pronounce inimical to our interests.

We humbly beg to draw your particular attention to this Order, the fourth column requires the signatures of those who wish to remain on the Survey with no immediate prospect of an advance in their present rates of pay. Those, Sir, who now address you have placed their signatures in this column; we were induced to act thus from a fear of incurring the displeasure of the party from whom the order emanated;[4] reflection however convinces us that this step was ill-advised, that we have incurred the risk of having our feelings misrepresented, by signing a document which would justify the inference that we were satisfied with our present wages. It is with considerable reluctance that we come forward to qualify those signatures, and were we not supported by an unwavering conviction of the righteousness of our cause we would not come forward at all.

The heading of the fourth and fifth columns taken conjointly may be resolved into this question, 'Is it your wish to go to Ireland or to remain on the Survey without any immediate prospect of advance in your present rates of pay?' We have chosen the latter, though for what reason we have been called upon to certify that choice we know not, inasmuch as our remaining on the Survey could be attested by our signatures in the pay-list. Many of us left our homes years ago to follow the fortunes of this employment; time, sir, affects materially the relation subsisting between a man and his home, it has placed us in such circumstances as render our continuance in our present situation preferable to returning to Ireland. But, sir, our signatures prove only the existence of such circumstances; we would by no means have these signatures regarded as proofs of contentment on our part, if this form afford arguments to support our being kept on our present rates of pay, we respectfully assert that these arguments are drawn immediately from our private circumstances and not from our merits as workmen – which latter we conceive to be the only just standard whereby our pay can be regulated.

You sir may have a desire to know what is the pay of the men who thus complain. For your satisfaction on this point we forward you a list containing our respective occupations, our time of service, and rates of pay. A glance, sir, will convince you that we have reason to complain, we wish not to descant on our own merits, we have no desire to fix the rate of wages due to our exertions, but sir we would appeal to our work; let that work be tested by competent judges, both as to its quality and as to the reward due to the producer. By the decision of such judges we will most willingly abide, being perfectly assured that it will remove every ground of complaint and dissatisfaction.

Our hours of working are from __ a.m. to __ p.m. with __ for breakfast, and __ for dinner,[5] we receive pay for six days in the week provided sickness or some other untoward circumstance does not prevent us from working the due hours; when laid upon the bed of sickness we have no resources, our pay is then kept back; we wish not to enter into details, but sir we could unfold instances of distress resulting from this cause which we doubt not would call forth your warmest sympathies and which would prove the necessity of amending a system which subjects meritorious and talented men to such poverty and humiliation.

The whole tenor of this letter proves that our complaints are not referable to the decrease in the grants for 1843[6] – the same grounds of complaint existed in 1842 and for many years previous – we knew not how to make known these causes of complaint without incurring the resentment of our superintendents, and for this reason we remained silent. At the present moment the example afforded us by other divisions of the Survey who have openly and respectfully laid before you a statement of their grievances, stimulates us to act in a similar manner – we do so sir praying that you will pardon whatever improprieties may exist in the mode we have chosen to lay before you the circumstances of our situation.[7]

RI MS JT/1/TYP/10/3893–3894
LT Transcript Only

1. *'a proverb and a bye-word'*: Deuteronomy 28:37.
2. *our superintendents*: including Thomas Frederick Colby.
3. *the accompanying District Order*: see the 'District Order' in letter 0236 of this volume, dated 16 September 1843.
4. *the party from whom the order emanated*: the District Order is signed from Captain Henry Tucker; see letter 0236 of this volume.
5. *__ a.m. to __ p.m. with __ for breakfast, and __ for dinner*: Louisa Tyndall's transcript has omitted these numbers.

6. *the decrease in the grants for 1843*: the Parliamentary grants of 1843, meant to fund the operation of the Ordnance Survey; see letters 0236 and 0238 of this volume.
7. *we have chosen to lay before you the circumstances of our situation*: see letter 0236, which was re-sent along with this letter. Tyndall and the other 'Preston Men' had attempted to forward letter 0236 to George Murray via Henry Tucker in September, but Tucker had returned the letter; see letter 0245 of this volume.

From Archibald McLachlan 18 October 1843 0247

Wednesday 18th. Oct. 1843.[1]

Your last[2] came to day – Mr Marquis wrote to me on Sunday morning, told me of C.T.[3] – the meeting and of the determination of sending the mem:[4] as originally intended – the remarks I made were grounded on the above – and before your account came – I never thought he opposed you – only that he demurred in signing – did he oppose? that's queer!! I don't think he will speak of Spec:[5]

I don't know what change has taken place since. I don't think any changes have taken place – I think the Commissioners[6] originally appointed had their appointment still – here they are as they stood in 1837 – William Blamire Esqre
Captain T. Wentworth Buller R.N.[7]
Reverend Richard Jones.

I can't make out whether there are any more Commissioners. You observe the 3 distinctions – Blamire Esqr – he is for the land owners – Buller R.N is for the Admiralty – and Jones is for the Church – each party is to watch and preserve the interests of his Caste – They all act for the general good, but each has a peculiar charge – for instance Jones, looks into and sees that the Church has not been robbed, by mistake, or Bribery &c &c – There are a number of assistant Tithe[8] Commissioners – for instance Dawson[9] is one – I think there are only 3 Commissioners, all the rest are assistants – 1 Commissioner for the Church – 1 for the people – 1 for the shipping interest –

Griffith | Of Mr G. I can only make this much – sometime in 1829 – or towards end of 1829 – he was progressing with the valuation – how long the experiment lasted I can't make out – I am trying to get all about it from one in Griffith's office Dublin – if an answer comes, I am afraid it will be too late for next weeks III[10]

Clew Bay[11] | In 1836, when I wrote to Hume[12] – Clew Bay was the point taken – thus – by Hodger's[13] prospectus, so many counties published – so many in the Engravers hands – so many in the Surveyors office – so many in progress of surveying – Joining all the knowledge we had – the conclusion then was, about 14 counties; I never saw Clew Bay in any document. I took that point as being a conspicuous place, easily found on the map – and as a tolerable fair point from which to draw a line through Dublin – It showed pretty nearly the quantity done – you may shift the line further North if you please, I should say 1/3 of the Island was done – that is not over stating it –

Veritas[14] | I once thought he was a R.E.[15] – but the thing is so lamely done, that <I> could hardly believe any of them could do such a botch – yet I would not like to swear that an R.E. did not do it!!!

Colonel's Ex. | I can't recollect the date of the Colonel's[16] examination of Parishes – but on thinking over and over in my bed, and calling over a number of old occurrences – I am inclined to think it was the latter part of 1832 – or Early in 1833–, but I wish you would say 1832 – because one of the parishes was Hollywood[17] which I drew myself – and it was the first parish our Division ever did by lines – the date is doubtful – But I would rather (after long reflection) say 1832, – However the fact is a true fact – A small error in date is of no great consequence.

Engravings –> They were burned or otherwise destroyed – if not destroyed, where are they?

Renwick[18] | I would leave Renwick out altogether.

Maps destroyed – Those destroyed were to those saved as 3 to 1 – some say 10 to 1 after Cº Down[19] was finished (or before), Portlock[20] got the Trig:[21] new triangulation was made, few of the old points used – Consequently old plans could not be used – if a few in Cº Dublin escaped destruction it was a mercy – that is in our district. They were just in the same scrape all over the Survey – 3 to 1 – !!!? – taking the whole survey they were 20 to 1 –!! – There were but a few saved in each District.

200.000£ – You know that before 1827 no C.A.[22] were employed – and for a long time after, some districts had not many – Taking this into account – and not forgetting the Phenix park[23] – I would say 200.000 was not far off the sum – I have heard it stated to be 200.000 – 150,000 – and so on, – 200.000 is not a bad guess!! I don't at present recollect what I told Hume - it was somewhere about 200.000.

I will look more closely into our levelling – I know the Trig party when they were here made awful work – levelled some lines of main road 3 times!! –

Extracts from the Evidence taken before the Select Committee of the Ho: of Com:[24] 1837 – 'On the Survey of Parishes under the Tithe Commutation Act'

As I shall refer to these, you may as well peruse them beforehand–

Question 118 –Did you not contemplate at the time you wrote to Spring Rice,[25] the necessity for a general survey? – Answer (by W. Balmire)[26] – We so far contemplated it, that we thought it would be very imprudent and injudicious in the country not to have it. We have perhaps gone farther, in calling attention to the matter, than we ought to have done by the circulation of Mr Dawson's reports; but we thought the country ought to be quite aware of the situation in which they stood in that respect.

121 – Did you not propose to the executive Government the having a general S.[27] made? – Answer. No doubt we thought it our duty to call the attention of the Government to what we considered might be done advantageously.

123 – Do you then recommend a general S. of all England &c – Answer – Undoubtedly &c &c

142 – Do you think the Land-owners of England are prepared to have the B^dy[28] of their Estates, for ever superseding all other evidence, determined by a S. taken behind their backs, that S. being taken, well or ill, as it may turn out, an S. performed by young men, & by contract? Answer I am quite aware that it is a most important consideration &c &c

194 – Did the correspondence with Mr Dawson precede your communication, to Government, or did it follow it? Answer – It preceded it – after being appointed as Commissioners we had communication with Government, which led us to appoint Mr Dawson as an assistant. Upon his appointment, we wished him to take the whole matter into consideration, & he, in consequence of our so doing, sent us those two reports. (those reports contained instructions for the Land-surveyors, entered into statistics – hinted at a general survey and much good scientific stuff all I believe dictated by Carbon)[29] –

255 – Had you not contemplated in great detail, the prospective advantages of a general S. – Answer, Certainly we had.

Answers by – Revd R. Jones

349 – If the point of difference be, whether the Proprietor, the Tithe owners, or the State should bear the expense, do you not think, as a National question, it would be very desirable that a S. should be made even though at the National

expense? Answer, Certainly, much more desirable if at the National expense than if at the expense of the land-owners.

391 – Are you not taking the question of general policy distinct from the Tithe act? Answer. I separate the subject into 2 parts; I have already stated that which was necessary for the immediate purposes of the Tithe Act. The Committee have now led me to state what I consider politic, independent of the purposes of the Act.

408 – Your views are founded on L[t] Dawson's report? – Answer Yes, all other Surveying work will cease; all the good Surveyors in the country will be at our command the moment this operation commences – Bravo!!

437 – Supposing the S. to proceed, you cannot tell at what time it will be completed throughout all England & Wales? – I think it may be carried on with great rapidity, if we have an unlimited fund to go to – Bravo!!!

499 – Are you to be understood to say that your object in calling the attention of Government to this was a double one; first, as there was a necessity for mapping 2/3 of the kingdom, namely 2/3 of the Titheable Lands, you thought advisable to take the opportunity of getting a general Survey at the public expense of the remaining third? Answer – Certainly!!

500 – Does the last proposition appear in your letter to M[r] Spring Rice? – No, I have already stated that we did not think it decorous, when calling the attention of Government to the subject, to chalk out the particular measures, we thought that they would come more properly from this Committee – Good – cunning – very deep– –!!

Answers by Dawson

645 – What is your idea of centralization? – Answer It would be better, perhaps, if I am to express my opinion to the Committee, that it should be entirely in a central office in London, or, if in the country, that the whole should be divided into 3 Districts, there might be an office established in each district, in which the plans should be received as sent in from the district Surveyors & by the contractors, &c &c –

659 – Would it be necessary still to have some central controlling authority in London? – I think there must be a central controlling authority in London.

660 – Some such person as Colonel Colby in Ireland? – Yes, the orders must proceed from one office

The following question relates to Surveyors of his proposed Districts – or his district, principal managing Surveyors –

678 – Do you expect you should find any difficulty in procuring surveyors competent to undertake that business? – No, a sufficient number might readily be found, such as are employed in the Ordnance Survey. Young men, well educated in the business of surveying, active and intelligent –
696 – That Patronage you propose to rest in the central controlling Board in London? Answer The responsibility would be theirs, & they should have the Selection of the Parties.

The above are a few of the questions I shall refer to – Some difficulty was encountered by the Commissioners in certifying the maps, Dawson advised a Committee and try to get up a Government Survey – and this was the Principal Cause of the Committee being appointed – Sir James Graham gives them all a horrible cross examination. The whole manoeuver was easily seen through Carbon was at the bottom of it – however it was no go – ¼ past 1 o'clock time to go to bed – B[30]

RI MS JT/1/TYP/11/3813–3817
RI MS JT/8/1/4a

---

1. *Wednesday 18th. Oct. 1843*: Louisa Tyndall annotation: 'Heading by Tyndall: "Clew Bay explanations"'.
2. *Your last*: letter missing.
3. *C.T.*: likely Captain Henry Tucker.
4. *mem*: the letter of protest sent by the workers of the Ordnance Survey of England to George Murray, Master General of the Ordnance on 23 September 1843; see letter 0236 of this volume.
5. *Spec*: Spectator, Tyndall's *Liverpool Mercury* pseudonym.
6. *the Commissioners*: Tithe Commissioners, civil servants charged with mapping parishes for the collecting of tithes. The tithe commissioners were Thomas Wentworth Buller, William Blamire, and Richard Jones; see letter 0231, n. 19 of this volume for further information on the Tithe.
7. *R.N.*: Royal Navy.
8. *Tithe*: see letter 0231, n. 19 of this volume.
9. *Dawson*: likely Robert Kearsley Dawson (1798–1861). Dawson joined the Royal Engineers in 1818. In 1836 Dawson began working in London as Assistant Commissioner to the Tithe Commutation Commission, based at Somerset House, in which role he determined that Tithe Maps should be made at a scale of 1 inch to 3 chains, the equivalent of 26⅔ inches to 1 mile (*ODNB*).
10. *next week's III*: i.e., Spectator's third letter.
11. *Clew Bay*: an ocean bay on the East coast of Ireland.
12. *Hume*: possibly Joseph Hume (1777–1855), radical MP for the Montrose Burghs and inveterate campaigner against extravagance in public finances, 'whose criticisms of Ordnance

spending were practically an annual Commons ritual', R. Oliver, 'The Ordnance Survey in Great Britain 1835–1870' (PhD dissertation, University of Sussex, 1985), p. 113.
13. *Hodger*: not identified.
14. *Veritas*: a pseudonym under which one of Tyndall's critics wrote to the *Liverpool Mercury*.
15. *R.E.*: Royal Engineer.
16. *The Colonel*: not identified; possibly Colonel Thomas Frederick Colby.
17. *Hollywood*: more commonly spelled Holywood; a parish in Ireland.
18. *Renwick*: William Turnbull Renwick; see letter 0243, n. 17 of this volume.
19. *C^o Down*: County Down, a county in Northern Ireland.
20. *Portlock*: Captain Joseph Ellison Portlock; see letter 0243, n. 27 of this volume.
21. *Trig*: Ordnance Trigonometrical Survey.
22. *C.A.*: Civil Assistant.
23. *Phenix park*: Phoenix Park near Dublin, Ireland.
24. *Ho: of Com*: House of Commons.
25. *Spring Rice*: Thomas Spring Rice (1790–1866), later Lord Monteagle, former Chancellor of the Exchequer in the Whig administration, of Anglo-Irish ancestry (*ODNB*).
26. *W. Balmire*: William Blamire.
27. *general S.*: general Survey.
28. *B^dy*: Boundary.
29. *Carbon*: a nickname for Thomas Frederick Colby; see letter 0231, n. 9 of this volume.
30. *B*: the pseudonym McLachlan used to avoid detection.

To Sir Robert Peel via The Liverpool Mercury 20 October 1843 0248

LIVERPOOL MERCURY, October 20th., 1843 | Letter II.

To the Right Honourable Sir Robert Peel, Bart, &c, &c.

Sir – a few weeks ago I addressed you[1] on the subject of the proposed survey of London; my letter sprung from the conviction that the claims of a certain class of individuals to public patronage were of a very questionable character. I used what seemed the best means of opposing, with success, the spirit of greedy monopoly, which, to the prejudice of civil surveyors, has characterized the proceedings of these individuals since the commencement of the survey of the six northern counties of England. Besides, Sir, I considered that the outlay of public money should be a subject of deep interest to the head of the British Ministry. Your ears are no strangers to the murmurs of discontent arising from heavy taxation, and if the Ordnance survey afford a specimen of the manner in which public works are in general conducted, I hesitate not to trace one great cause of the burdens of the people to the waste of the national funds, caused by the ignorance of those to whom such works are entrusted. I addressed you anonymously, because my statements require not the adventitious aid[2] of a name to invest them with the garb of credibility. They challenge public inquiry. To support them I can adduce the testimony of living witnesses, who are at present employed on the English survey. This, with other proofs of a different nature will, I feel assured, bear my statements, unshaken, through the ordeal of investigation.

In the letter referred to I have averted to many circumstances in proof of the position I assumed. The necessary brevity of a letter, and the extensive range I have taken, compelled me to pass over, in a very cursory manner, many points which are deserving of more attention. This consideration induces me to resume the pen, and devote a little time to the development of some facts connected with each particular subject to which I have referred.

The survey of Ireland occupies the first place.

Previous to the year 1824 numerous complaints were made of the unequal allotment of local taxation in Ireland. To remove the cause of these, the Government decided that a survey and valuation of the island should be made. The mutual relations of these departments stood briefly thus: – 1st. A survey was to be made, and maps constructed, to serve as the basis of the valuation; 2nd. A valuation of lands houses &c., based on the data furnished by the previous survey, and, of course, liable to be affected by the correctness or incorrectness of that data. The superintendence of the work was committed to the Board of Ordnance.

Two gentlemen of high professional standing[3] were chosen to carry into practical effect the designs of the legislature. The survey department was confided to an officer of the Royal Engineers; the valuation to Richard Griffith, Esq., Civil Engineer. The determining and sketching of public boundaries were also entrusted to the latter gentleman.

A force of five captains and twenty-five lieutenants of the corps of Royal Engineers (this was the main force; a few others were on detached duty) and three companies of Royal Sappers and miners,[4] raised expressly for this work, was placed under the immediate control of the officer alluded to. He had also liberty to employ any number of civilians or civil assistants he thought proper. This force commenced operations in the north of Ireland in 1825. Each captain had charge of a district numbering five lieutenants; each lieutenant had a number of assistants and sappers under his superintendence. The <u>practical</u> knowledge of the officers was at this time very limited, and to remedy this defect about one year was devoted to personal practice; each officer was furnished with a book of instructions for his guidance; from the abstract merits of this book I have no wish to detract; it proved however practically useless.

The first duty which engaged the attention of the Royal Engineers was the fixing of points throughout the country by trigonometrical observations: parties were engaged in observing the most prominent features of the country, mountains, hills, &c; from these points the officers of districts determined others, and the division officers or lieutenants (who were furnished with a list of these) observed any additional points which might be required for the survey of the parishes entrusted to them. There were thus three series of observations; and from these the altitudes were determined. But, as no definite arrangements had been made respecting a proper datum line from low water mark,[5] the altitudes were completely involved in error; owing to which the levels of some districts were altered three different times. This affords one of the first evidences of a want of forethought, & even later improvements in this branch of the duty cannot be regarded as the result of prescient skill but of accident. Each parish was divided into triangles by connecting the points previously observed; the sides of these triangles varied in length from one to two miles; the detail surveyed consisted of the parish and townland boundaries, roads, lanes, houses, and principle rivers; the boundaries were surveyed by triangulation;[6] the roads were traversed by a theodolite;[7] and from the line of traverse others branched off to take in the houses and other objects of detail. These branch lines were, in general, left hanging unchecked. Thus the boundaries and other details were surveyed independently of each other, and, owing to this, their relative position was, in most cases, ridiculously erroneous. The vertices of the subsidiary triangles used for the boundaries were seldom checked, and no care was taken to avoid obtuse angled triangles,[8] the frequent use of which every professional must condemn. The offsets or perpendiculars taken to the boundaries were of incredible length, measuring in some instances upwards of one thousand links. These were taken by the eye alone; and almost at random; the aid of an instrument in such cases was not at all deemed necessary. Such were the material from which plans were to be constructed for the purposes of valuation! Some of these records of scientific skill were stamped with the signet of durability by being engraved. I will not anticipate events by letting you know, at present, the fate of these engravings.

I shall close this letter by remarking, that when the plans of each division were drawn, the officer took them to the ground which they represented, to test their accuracy; he made whatever alterations he deemed necessary, and previous to forwarding the plans to his district officer, he certified the correctness of each, by attaching to it his signature.

(Spectator)[9]

RI MS JT/1/TYP/11/3910–3911
LT Transcript Only

1. *a few weeks ago I addressed you*: letter 0228, *Correspondence of John Tyndall, Volume 1.*
2. *adventitious aid*: external, additional aid (*OED*).
3. *Two gentlemen of high professional standing*: probably Thomas Colby and Richard Griffith.
4. *Royal Sappers and miners*: see letter 0231, n. 32 of this volume.
5. *a proper datum line from low water mark*: a datum is a base elevation used to calculate elevations of the surrounding land. The phrase here implies that the surveyors were meant to take the nearest low water mark as their datum line (*OED*).
6. *the boundaries were surveyed by triangulation*: triangulation, a technique frequently used by surveyors, enables the surveyor to calculate the location of a distant point by measuring angles to it from two known points. The location of the unknown point can then be determined from the two known angles and the distance between the two known points

(*OED*).

7. *the roads were traversed by a theodolite*: when surveyors traverse a road or landscape, they place survey stations along their path of travel and then use these stations as a base for observing their next point. Measurements between points would be made using a theodolite, a specialized instrument for precision measuring of horizontal and vertical angles.
8. *obtuse angled triangles*: triangles containing an angle greater than 90^0. Considered problematic for triangulations.
9. *Spectator*: Tyndall's *Liverpool Mercury* pseudonym.

From William Ginty 21 October 1843 0249

Saturday.[1]

Dear Alpha[2]

I expected 'Thou' by yesterday's post and by this evening's post – disappointed – how's that? I hope it will be there tomorrow as I promised S[3] to let him have it by Monday evening at the latest. I suppose you have seen the III.[4] I infer from the notice that there is more than one in the field, it reads: 'Some important communications'. Perhaps some C.E.S[5] about the meeting we recommended the week previous. Did you con the pages of the III? The death of – Wood Esq. M.P.[6] in the Manchester Literary and Philosophical Institution. Awfully sudden. The deceased entered eternity with the 'Ordnance Survey of the six northern counties'[7] on his lips. You can infer from this that the thing is talked about among the nobility! I hear that they have got 20.000 acres more of Tithe work![8] Is it so? If I don't mistake, something appeared in a Preston paper about this matter, to the effect that the Board had been remonstrated with and that the C.E.[9] officers were ordered to do no more after the old contracts had been fulfilled! Was this a fact? If so – that about the 20.000 acres must all be talk. A fact came to my ears yesterday which astonished me and at the same time pleased me. I only regret I did not know it sooner, perhaps tis time enough? Not content with monopolizing the professional rights of Civil Engineers in the Tithe commutation Surveys they carry the rascality further and contract for the surveying of farms on estates!!! Such is the fact. A short time after I left George Street[10] office a map of an estate or farm (a small one) was drawn in the office, and it is said it was done for 10 shillings although the cost of its execution was nearly if not all out £3. This happened in last February or March. I never heard it till last night. I have put chaps on the scent to find out the particulars. This I call important! Do you? (!!!) But just hold hard yet awhile on this matter! Have you ever heard of Hornby Castle?[11] if not, I have, and seen it too. It is situated on the river Lune in the northern extremity of this county. Well! Hornby Castle is a very fine castle. I have seen the crimson tints of a departing sun shed a glorious halo on its domes and its minarets. To me in fancy its mouldering towers and ivied-turrets have hymned a lonely requiem to 'the deeds that were done in their time'[12] I have seen its splintered and now useless flag-staff totter in the winds of heaven! I have seen the tumultuous Lime roll its restless flood around the hallowed precincts of its firm base, that seemed to 'mock the lapse of ages'[13] and 'bid defiance to all-conquering time'![14] I have gazed upon that venerable superstructure enraptured! I have turned from that 'sad relic of departed worth'[15] to the surrounding scene in its rich, gay, loveliness of sublimity. The dark wild forest and wide-spread woodland – the emerald valley and burnished lake and said 'These! all these, I wish to heaven they were mine!!'[16] Within the portals of this time-hallowed domicile our recreant taskmaster takes up his abode when he goes to that part of Her Majesty's realms to examine. That old Baronial Hall has degenerated into the hands of serfs and menials. Oh shades of the Baron and the Chieftain, dream ye now who tread the halls that bore ye. I ween not, or your hoary phantoms would return to dispel the dastard plebeians 'who revel o'er the ruin they have wrought'.[17] Infamous pollution! 'The long-drawn aisles and fretted vaults'[18] of that proud tenement of other days – is indeed outraged! 'Hornby thy castle is a Holy Place'[19] – Thy Hall a Palace, for 'twas trod by 'Chiefs and ladies bright'[20] – by noble hearts and warrior arms, but now alas, the starched-up exquisite – the scented puppy, and the petty tyrant strut in pampered churlishness thro' thy portals, emblems, sad emblems of 'thy tale of ruin'.[21]

'Thus generations in their turn decay
So flourish these (alas!) when those (oh God!) are passed away'.[22]

The present occupier of this castle and Hamley are intimate and yesterday I heard that Hamley is going to get this place drawn on a large scale. He to pay just what the men's time will amount to, and all this for friendship's sake. Give him rope enough[23] say I – He will put his foot in it, this is worse than any charge against him yet. Not a word but whispers, I'll watch his movements – The other job was done certainly. How gloriously these will harmonise with the Tithe robbery!

I heard last night (I heard a devil of a lot last night! did I not?) for the first time that Jim Newton deserted last spring left the wife (Cuddy's[24] sister) who is now at home. Jim has not been heard of since, I understand. Not a word from Manchester or Skipton yet! All damn lies about the Discharging. I knew it. The wish was parent to the lie with the rascals!

Yours | Comracht Moran.[25]

By the time you get this I hope to have yours. Saturday 4 o'clock P.M.

RI MS JT/1/TYP/11/3820
LT Transcript Only

1. *21 October 1843*: Louisa Tyndall annotation: '*[*October*]* 23'. However, 23 October was a Monday, and the letter was written on a Saturday. Louisa Tyndall probably derived her date from a postmark stamped two days after the letter was composed.
2. *Alpha*: an occasional nickname for Tyndall.
3. *S*: not identified.

4. *the III*: i.e., Spectator's third letter. Spectator is Tyndall's *Liverpool Mercury* pseudonym.
5. *C.E.S*: not identified, possibly Civil Engineers.
6. *Wood Esq. M.P.*: George William Wood (1781–1843), Member of Parliament from 1832–43. He died of a stroke during a meeting of the Manchester Literary and Philosophical Society, of which he was vice-president. He was also a Fellow of both the Linnaean Society and the Geological Society.
7. *the six northern counties*: see letter 0231, n. 16 of this volume.
8. *Tithe work*: see letter 0231, n. 19 of this volume.
9. *C.E.*: Civil Engineers.
10. *George Street*: probably a reference to the headquarters of the Institution of Civil Engineers; see letter 0238, n. 7 of this volume.
11. *Hornby Castle*: a country home in Lancashire developed from a medieval castle. In 1843 the castle was owned by Pudsey Dawson (1778–1859), who was later High Sheriff of Lancashire.
12. *'the deeds that were done in their time'*: unidentified reference.
13. *'mock the lapse of ages'*: F. Schiller, *Fiesco,* II.xiii.
14. *'bid defiance to all-conquering time'*: unidentified reference.
15. *'sad relic of departed worth'*: Lord Byron, *Childe Harold's Pilgrimage,* II.lxxiii.1: 'Fair Greece! sad relic of departed worth!'
16. *'These! all these, I wish to heaven they were mine!!'*: unidentified reference.
17. *'who revel o'er the ruin they have wrought'*: unidentified reference.
18. *'The long-drawn aisles and fretted vaults'*: a common description of Gothic cathedrals.
19. *'Hornby thy castle is a Holy Place'*: possibly a reference to Lord Byron, 'On the Castle of Chillon', line 9: 'Chillon! thy prison is a holy place'.
20. *'Chiefs and ladies bright'*: T. Moore, 'The Harp that Once Through Tara's Halls', line 9: 'No more to chiefs and ladies bright / The harp of Tara swells'. On Moore, see Biographical Register.
21. *'thy tale of ruin'*: unidentified reference.
22. *'Thus generations in their turn decay | So flourish these (alas!) when those (oh God!) are passed away'*: Homer, *The Iliad,* book VI, ll. 185–6: 'So generations in their course decay, / So flourish these, when those are past away'. The interjections are Ginty's.
23. *Give him rope enough*: this alludes to the proverb: 'Give a *thief* rope enough, and he'll hang himself', J. Ray, *A Compleat Collection of English Proverbs* (London: J. Hughs, 1737), p. 161.
24. *Cuddy's sister*: Martin Cuddy; sister's name unidentified.
25. *Comracht Moran*: a joking pseudonym for William Ginty; meaning and origin not identified.

From Archibald McLachlan 22 October 1843 0250

Sunday night

The following is a correct copy – (for I copied it myself) – It is a copy of Instructions to the Boundary Surveyors, or those who perambulated the Boundaries in the North – Note, the perambulators[1] are all RSM[2] – I copied it at Preston in 1842 – it was used in 1841 & 1842 and may still be used for anything I know to the contrary. The thing is too good to pass over in silence and if well used might touch John Bull's[3] Bumps of Respectability[4] – The Perambulators are to pay those who show the Boundaries, or those whom we call the Meresmen[5] – Those persons who are best qualified to show the Boundaries of Parishes Townships &c are pitched upon by the Magistrates & the parties are Bound to show the Boundaries when called upon by the Ordnance Surveyor – The Magistrates do not know how much the parties receive for the work they perform, the remuneration therefore of Meresmen is left to the Honourable B.O.[6] i.e. Acid[7] – and he drew up the following scale. I copied it from his own handwriting!!!!!!

1st	Common Labourers –	2/. per diem
2^{d}	Farmers appearing a little more respectable than a labourer –	2/6 – "
3^{d}	Farmers that work on their own farms – Assistant overseers of Townships, & Game Keepers –	3/. – "
4th	Farmers who work on their own farms but employ labourers, Bailiffs of farms &c –	3/6 – "
5th	Yeomen Farmers – Stewards of estates – wheelwrights and millers &c &c –	4/. – "
6th	Those who farm but do not follow it as a business; Attornies &c. who live more by mental than manual labour –	5/. – "

There! – The highest rate allowed to an Attorney at Law, for trudging over hedge & ditch from 8 A.M to 5 ½ P.M is only 5/. & he must do it if nobody else can be found.

In Article 2^{d} you will see that a Farmer appearing a little more respectable than a laborer is to gain six pence per diem by his respectable appearance, and who is to be judge of this English Farmer's Respectability? Why, none other than a private in Her Majesty's Corps of Royal Sappers & Miners[8] – Should a farmer rich in this world's good come before this Royal Warrior with old trousers & an old hat, a dress fit for the nature of his day's occupation, he must be scanned from top to heel, & should he not appear Respectable he loses Sixpence!! Again observe the nice shades & gradations of character and pursuit, and each increase of Respectability has its increase of sixpence!! – Oh: this table of wages could be made the subject of cutting sarcasm

I think you should insist on the unfairness of 200.000£ being spent in surveying & perhaps 500,000 more spent in England and not one English Surveyor got a farthing of it however bad the times may be, as they were in 1842 – Touch John Bulls pocket & you make him growl. Show the impropriety & in fact the injustice of a Great National Survey being carried on in England & yet no English Surveyor being employed on it; that would touch John Bulls Pride

---

I believe that it is determined on that Carbon[9] is to leave the O.S.[10] and a Colonel Hall[11] is appointed in his place, the change is to take place February or March –

I forgot to say yesterday that if the R.E.[12] depended in any degree on the progress of the work, if their Salary was in proportion to work done, they would find it their interest to prosecute the work vigorously – and would exercise forethought and not Survey a Town 3 times – & encourage the men who have the practical duties to perform –

Your Sinly[13] | B.[14]

RI MS JT/1/TYP/11/3818–3819
RI MS JT/8/1/4a

---

1. *the perambulators*: i.e. the surveyors (*OED*).
2. *RSM*: Royal Sappers and Miners; see letter 0231, n. 32 of this volume.
3. *John Bull*: a national personification of England, or sometimes the United Kingdom.
4. *Bumps of Respectability*: referring to the phrenological belief that the size, shape, and bumps of each person's skull correlate with their character.
5. *Meresmen*: people who are employed to determine the boundaries of a parish (*OED*).
6. *B.O.*: Board of Ordnance.
7. *Acid*: a nickname for an unidentified opponent of Tyndall's on the Ordnance Survey, possibly Henry Tucker.
8. *Her Majesty's Corps of Royal Sappers and Miners*: see letter 0231, n. 32 of this volume.
9. *Carbon*: a nickname for Thomas Frederick Colby; see letter 0231, n. 9 of this volume.
10. *O.S.*: Ordnance Survey.
11. *Colonel Hall*: Colonel Lewis Alexander Hall (1793–1868), a Royal Engineer who replaced Colby on the Ordnance Survey; however, Colonel Hall would replace Colby in 1847, not in 1844 as McLachlan anticipated. Hall was born in Quebec and came from a Church of England background. It is unclear when he joined the Royal Engineers but he became 2nd Lieutenant on 21 July 1810 and rose steadily through the ranks, achieving a final rank of Lieutenant General on 3 August 1863. He saw action in the Netherlands and France between 1815 and 1818. Hall served in Ireland between 1821 and 1825. He then went to the West Indies, and married in Barbados in 1827. He also served in Corfu and Gibraltar. He worked on the Survey of Great Britain between April 1847 and July 1854. Upon retirement he was awarded the rank of Colonel Commander, and he died at Southampton on 16 March 1868.
12. *R.E.*: Royal Engineers.
13. *Sinly*: Sincerely.
14. *B.*: the pseudonym McLachlan used to avoid detection.

To John Tyndall, Snr. 23 October 1843 0251

Preston, Oct. 23rd, 1843.

My dear Father

Immediately on the receipt of your last letter[1] I wrote to two London Journals for the names of the Commissioners.[2] The Weekly Chronicle was one. On account of the great space taken up by Irish news, they were unable to answer any of their last week's correspondents. I have however obtained the names of the Tithe Commissioners as they stood in 1837, and my informant tells me that he is of opinion the same individuals occupy the situation at present – William Blanmire Esqr.[3]

Capt T. Wentworth Buller R.N.[4]

Reverend Richard Jones

My informant tells me that he cannot say whether there are any more than three – he thinks not, each of these has a distinct charge. Mr Blanmire is engaged on the part of the landowners, Capt Buller on the part of the Admiralty, and Mr Jones on the part of the Church. They act in concert for the general good. Perhaps it would be as well to wait till Saturday. We may get more explicit information on this subject from the editors to whom I have written. I received yours with Wm Tyndall's[5] enclosed this morning. He is a good writer and I believe a man of feeling and generous sentiments. Well the ray of hope which shone over Philadelphia has been extinguished. I however am no way disheartened at his account of things. The man seems to think that my profession is to me what its niche is to a statue, that I cannot stir out of it, why there are many things that I could turn my hand to in America. I'm no way fastidious. I am prepared for anything – I have often looked to the workings of my own heart, and I think from the knowledge I have of myself that a change of fortune, either for the better or for the worse, would not affect me much. I'm ready for anything – from the situation of a cowboy to a Seat in Congress. I have licked up a good deal of information here and there and I'm morally certain that I could turn some of it to account in America even taking the picture at its darkest side. There is no hope in this country – too much competition. The first readings of political economy, nay the plainest common sense will tell you that the value of labour is depreciated by the increase of competitors. They, (the Surveyors) have gone on increasing in England for years, without a corresponding increase in the work to be produced, and the country is now overstocked with them. I shall now give you a few extracts which I think will stand fair against Wm Tyndall's account. That account is confined to a certain locality, and I have not a doubt of its accuracy. The following is from a friend of mine at present[6] in York– a cool reflective Scotchman, I have previously acquainted him with my intention of going to America. – I quote – 'So you are about to vanish as many have done before you, I wish I could vanish too. My energies are completely withered in this employment. Let me know when you intend to go, I have two friends who left the Survey of Ireland in 1829 and went to America. They I'm sure will do all in their power to forward you. I wrote to one of them on chance three months ago. He received my letter and has replied to it. He tells me he will give me everything but Scotch whiskey if I go out to him. He has realized a handsome fortune, and can now return to his own country an independent gentleman' – Mark, his independence is owing to his success as a Surveyor. That's a cheering fact, is it not? I shall now make one or two quotations from a letter received[7] last week by a young fellow who works in the office beside me,[8] from a brother of his who left the Survey for America about two years ago. I may remark that the letter bears internal evidence that it was written by a man of sense and reflection. It is dated Erie[9] Sept 4th 1843 – 'If my uncles and aunts were here they could with very little labour procure all the necessaries and many of the luxuries of life. This country differs in many respects from Great Britain and Ireland. There a man by recommendations from gentlemen can get into many public situations, but here a stranger coming from a foreign country must go into the first that offers and trust to his talents for preferment. With ordinary abilities and strict integrity he must succeed, unless he is more than commonly stupid'. Again – 'Energy and self-dependence is what a man requires when he come to this country'. Again – 'I could not promise you that the moment you would land there would be a situation for you'.

You expressed a wish to know who the East India letter was from. It was from an officer in the 78th Highlanders who are stationed within 140 miles of Hyderabod.[10] He was my constant companion, messmate and bedfellow for many months in the South of Ireland. His uncle[11] is Captain in the same regiment, and thro' his interest young Higginson[12] obtained a commission. He is a most affectionate young fellow. We were very much attached to each other. He has given me a lengthened detail of his adventures since he left this country. I cannot imagine the reason why his letter should be returned. Was it not directed properly? I owe Mr. Conwill a letter. The arguments contained in his last[13] would be most unjust premises whence to draw conclusions relative to the prospects of a general Surveyor in America at present. I give him credit for the best intentions though I could not at all grant that his reasoning is valid.

Emma[14] is very well

Yours & c | John Tyndall

Would you send me a fair copy of the letter you wrote to Wm Tyndall?

Send me all my scraps of poetry.

RI MS JT/1/TYP/10/3299–3300
LT Transcript Only

1. *your last letter*: letter missing.
2. *the Commissioners*: see letter 0247, n. 6 of this volume.
3. *William Blanmire Esqr.*: William Blamire.

4. *R.N.*: Royal Navy.
5. *Wm Tyndall's*: William Tyndall. See letter 0234 of this volume. See also William Tyndall to John Tyndall, Snr,, 24 September 1843, RI MS JT TYP/10/3312–3313, in which William warned Tyndall, Snr, that he feared Tyndall, Jnr, would not find employment in America: 'there are already a great many of the profession [surveying] in this country now out of employment who already have much influence and many influential friends and yet they find it impossible to get employment'.
6. *a friend of mine at present*: not identified.
7. *letter received*: letter missing.
8. *a young fellow who works in the office beside me*: not identified.
9. *Erie*: a city in northwest Pennsylvania, USA.
10. *Hyderabod*: probably a misspelling of Hyderabad, a city in India.
11. *His uncle*: not identified.
12. *young Higginson*: Thomas Charles Higginson (*c.* 1821–98), a civil assistant, originally from Dublin, in the 5th Division, C District of the Irish Ordnance Survey, having joined on 8 May 1839. His duties were primarily surveying in the field party under Corporal James Mulligan, working in Counties Kilkenny, Tipperary and Cork. He was transferred to the English Survey in September 1841, where he worked in the 1st Division, stationed near Skipton in north Yorkshire. In the spring of the following year, Higginson left the Survey and evidently enlisted in the Royal Artillery's 78th Regiment of Foot on 10 April 1842.
13. *his last*: letter missing.
14. *Emma*: Emma Tyndall.

To Sir Robert Peel via The Liverpool Mercury 24 October 1843 0252

LIVERPOOL MERCURY October 27, 1843. | Letter No. III.

To the Right Honourable Sir Robert Peel, Bart,, &c.

Sir – Without any prefatory remarks I shall at once come to that stage of the Ordnance survey of Ireland, in 1829, when the evidence of the superintending officer[1] was given before a Committee of the House of Commons; the substance of this evidence was, that the survey was progressing in a vigorous and satisfactory manner. In the course of this or the following year, some plans were forwarded to Mr Griffith, who sent competent persons to commence the valuation of the district which they represented; their utter unfitness for the required purposes was proved, and consequently they were rejected by Mr. Griffith.

Now, either Mr. Griffith had, at the commencement of the survey been consulted relative to the nature of the plans required, or he had not. If he had, then his rejection of the plans shows that no attention had been paid to his demands; if he had not, does it not prove a degree of imprudence almost incredible on the part of the Royal Engineers, to persist so long without any reference to the Superintendent of the valuation? The wisdom, of submitting at first, the maps of a few parishes for his inspection, appears so evident, that it is hardly possible to believe that it could be overlooked by a number of individuals, specially selected on account of their superior scientific attainments. The Royal Engineers, however, had been over one-third of the Island before this was discovered. How can this indiscretion be accounted for? Only by assuming the existence of a degree of unwarranted self-sufficiency, which impelled these individuals onward, till Mr. Griffith arrested them in their blind career, and exposed the absurdity of their operations.

An attempt at a reply to my first letter has been made by some evidently unauthorized person, who styles himself 'Veritas'.[2] This individual ascribed the rejection of the plans to a deficiency in the quantity of detail. I merely notice this single remark of his crude production, because I have reason to believe that the opinion expressed by him is entertained by many. Granting that the plans were rejected, purely because of a defect in the quantity of detail, and assuming for argument's sake the accuracy of what was done, one point was proved – that the Royal Engineers were ignorant of the quantity of detail necessary, and, therefore, should have applied to Mr. Griffith for information on this head. I, however, assert, that the plans were not only defective in point of quantity, but that they were grossly incorrect also. Mr. Griffith returned the maps, with a list of their defects and errors; if that list to be in existence, to it, Sir, I fearlessly appeal in support of my assertion. That this list comprised letters of commission as well as omission; it referred to errors in the areas, boundaries, roads, houses, and all the other detail surveyed. The superintendent of the survey took several plans to the ground, and tested them by personal examination. Had this been done sooner, it might have averted many subsequent disasters. The roads and boundaries occupied his special attention; they were erroneous, the maps which he examined were condemned, and the country which they caricatured was re-surveyed. If this be not proof of the incompetency of the individuals who had previously certified the accuracy of these plans, I labour under sad delusion.

The mortification of the Royal Engineers, at this time, is more easily conceived than described; they saw themselves subjected to the animadversions[3] of scientific men, and they trembled at the idea of their proceedings being made known to the public. In 1831 some parties had advanced as far as Dublin and Athlone,[4] all were now ordered back, and an ineffectual effort was made to revise the work, and render the old plans serviceable. The vain attempt was in most cases abandoned, and a re-survey resorted to. In these cases the original plans and other documents were reduced to ashes or otherwise destroyed. The Superintendent will doubtless remember a portion of an order which appeared in 1833 – 'The plans are to be burned in the presence of an officer'; and the plans referred to were burned, while an engineer officer, with cocked hat and waving plume stood by, a pensive witness of this memorable scene, till the last dearly purchased smoked-wreath had mingled with the atmosphere. Between 1831 and 1836 some maps were patched up and rendered passable, but it were well if this had never been attempted; the cost of examining, re-examining, scraping, altering, and shifting, exceeded what would be required for a new survey. Even at the present moment, so much do many of these maps and their field notes disagree, that were the former destroyed, nothing similar could be compiled from the latter.

Will it be said that I exaggerate? One fact will prove the credibility of all my statements – a sum of £200,000 was wasted in the production of this 'Comedy of Errors'.[5] Judge, then, Sir, the extent of that failure, which incurred such an enormous expense. Am I not justified in stating that if all our public works be executed in a similar manner, to them may be referred one great cause of the present enormous taxation?

How necessary, then, Sir, it is that persons engaged in these important undertakings should have some interest at stake depending on the success of their exertions. If this work had been entrusted to civil engineers, I feel assured that such disgraceful blundering would never have taken place. The jealous regard which a civil engineer must entertain for his professional character, compels him to think before he acts, and would prove an insuperable barrier against such a course of errors as I have detailed.

The ordnance survey of Ireland is now, I believe, completed. It is a highly useful work, though by no means matchless. But to whom, I ask, is the country indebted for its present usefulness? Without question to Mr. Griffith. Had he been deficient in either professional knowledge or moral courage, when he discovered the true state of things in 1829 and 1830, this public work would have proved a discredit to the age – a record of national incapacity, unparalleled in the history of similar undertakings.

Fearing that my statements, if used without proper restriction, might injure the character of deserving men, I would, in conclusion, distinctly say, that I unhesitatingly acquit both assistants and sappers[6] of all blame; they acted according to order. In those days, as in the present, these classes comprised many persons of superior abilities, whose usefulness would have been apparent, had not their exertions been paralyzed by a system of discipline incompatible with their vocation.

SPECTATOR[7] | Dublin, October 24, 1843.

RI MS JT/1/TYP/11/3912–3914
LT Transcript Only

---

1. *the superintending officer*: Thomas Frederick Colby.
2. *some evidently unauthorized person, who styles himself 'Veritas'*: a pseudonym under which one of Tyndall's critics wrote to the *Liverpool Mercury*.
3. *animadversions*: criticisms, public censorship (*OED*).
4. *Athlone*: a city in central Ireland.
5. *'Comedy of Errors'*: a reference to W. Shakespeare, *The Comedy of Errors*.
6. *sappers*: see letter 0232, n. 10 of this volume.

7. *SPECTATOR*: Tyndall's *Liverpool Mercury* pseudonym.

To John Tyndall, Snr. 28 October 1843 0253

Saturday night | Stormy and rainy!

Dear Father

This week's Chronicle[1] contains a reply to my question relative to the Tithe Commissioners[2] they are William Blanmire[3] Esq Thos Wentworth Bullen[4] Esque and the Rev Richard Jones.

My former informant made a slight mistake in the name of the second gentleman – it is Bullen not Bulwer or Buller[5] – he is not styled 'Captain'[6] by the Chronicle.

Yours &c | John

Kindest remembrances to Mr Conwill. He will scarcely pardon my long silence. If he knew how I am situated he would judge leniently of me.

J.T.

RI MS JT/1/TYP/10/3301
LT Transcript Only

---

1. *this week's Chronicle*: probably the *Preston Chronicle.*
2. *Tithe Commissioners*: see letter 0247, n. 6 of this volume.
3. *William Blanmire*: William Blamire.
4. *Thos Wentworth Bullen*: Thomas Wentworth Buller.
5. *it is Bullen not Bulwer or Buller*: Tyndall was mistaken; the correct spelling was in fact 'Buller'.
6. *he is not styled 'Captain'*: 'Captain' was Buller's appropriate form of address, as he was a Captain in the Royal Navy.

To John Tyndall, Snr. 30 October 1843 0254

Preston | Monday evening | October 30th, 1843.

My dear Father

I forward you a letter from Wm Wright.[1] His account respecting Surveyors nearly agrees with that of Wm Tyndall.[2] But he takes a wider view of the case, and does not limit his remarks to my present line of life. He of course can form a good idea of what I am able to do, and on that knowledge he founds the very cheering encouragement which he holds out to me. I however have some hope even in my present business. You may remark what he says respecting good draughtsmen. Their number is very few. Now drawing is my forte. I am able to produce as respectable a plan as many of my compatriots and this consideration holds out a hope of success in this line. However, as I stated in a former letter, I'm ready for anything.

Your affectionate son | John.

RI MS JT/1/TYP/10/3302
LT Transcript Only

1. *a letter from Wm Wright*: almost certainly letter 0237 of this volume.
2. *Wm Tyndall*: William Tyndall.

From Martin Cuddy 31 October 1843 0255

Oct. 31st 1843[1]

Dear Dear Tyndall

With fond pleasure I heard from my friend Holmes that you were well – and wished to receive a letter from me. Now what put such a thought into your head. If you had any yearning towards me why not write me, as well as I you – I was thinking not to write you but when the light of days that's past flash'd across my vacant mind and laid bare to my view the many boxing bouts we enjoyed together, this and this alone was enough to kindle a dying flame, and rouse my dormant faculties – in the fond hopes of receiving a few lines from Snooks.[2] – Well what do you think of late agitation I dare not say ours, true there was a few that would 'brave the lion in its den', but treason flourished around in all its appalling forms – slaves crouch'd beneath the tyrannizing and fiendish gaze of a beggarly Sapper,[3] Now for Repeal[4] – 'I speak not of men's creeds – they rest between Man and his maker – but of things allowed Averred and known – and daily, hourly seen.

'The yoke that is upon us doubly bow'd
And the intent of a tyranny avow'd'[5] –

Well Tyndall are you a repealer?[6] I would say not, but perhaps I do you an injustice, by my supposition – I hope I do – 'Tis my opinion that every Irishman should be one, 'tis a fact that no sophistry can get over that our country is a land of slaves, by English oppression, are we not looked upon by bigbellied Bulls[7] with every sort of contempt? But I sincerely hope that the grovelling sway of our taskmasters will soon be abolished, and forever crushed. Even now the voices of raised up millions is shouting Liberty Liberty. Away with the hydra headed monster that still would eat up our country's remaining wealth, by striving to bring to punishment the <*1 word excised*>[8] of the day.

I could say much upon the above subject – but I will defer it till I hear your judgement, I shall conclude this radical epistle with a few verses of mine own – mad brain's concocting – and hoping they may meet the approbation of so very able a critic as Snooks – –

Oh! Erin,[9] My Country awake from thy slumber
The trumpet of Liberty re-echoes the land
Shake off the vile shackles that does thee encumber

And crush'd thee for years, with a death dealing <u>hand</u>

Arise then brave Erin of freedom benighted
And hail the lov'd morning of liberty's dawn
Remember the curse by which you are blighted
Arise – then and vanquish the Tiger and fawn

Men of Erin unite, round the common cause rally.
Let your voices go forth on the wings of the wind.
Let repeal be the word through mountain and valley.
And starvation to the demon that would us dare bind.

So long you have suffr'd by demon oppressors
And your country laid desolate by the fire and sword
But vengeance ere long will o'ertake the oppressor
For heaven and justice their aid will afford.

The Sun – Its brilliant beam on Tara[10] is shedding
The spirits of our forefathers look smilingly on
And the music that once enlivened that dwelling
Will burst with fresh ardour when our freedom is won.

I have not taken any pains in the writing of the above – when you answer this I will send you other verses, but in a different strain. I would have written to Jim[11] long since – but me behaving so badly towards him I could not bring myself to do it – but I will write him yet – what could I <u>do</u> I know <u>not</u> it not being in my power to act <u>otherwise</u>. No doubt he deems me a <u>villain</u>, surely my finances are villainous – but my heart's not. What about the fair ones you once were the delight of the <u>ladies</u>, are you not so yet – you shall have a <u>love</u> ditty in my next – I am almost devoured by the flame.

Farewell my dear Tyndall

Yours Cuddy.

N.B. Direct: | Richmond Place | Pendleton | Manchester

RI MS JT/1/TYP/11/3755–3756
LT Transcript Only

---

1. *Oct. 31st 1843*: date given by Louisa Tyndall.
2. *Snooks*: Walter Snooks, a pseudonym under which Tyndall published poetry. See 'Introduction', *Correspondence of John Tyndall, Volume 1*, pp. xxxix–liii, at p. xlix; see also letter 0109, *Correspondence of John Tyndall, Volume 1*.
3. *Sapper*: see letter 0232, n. 10 of this volume.
4. *Now for Repeal*: i.e., the proposed repeal of the Act of Union of 1801 that united Ireland with Great Britain.
5. *'I speak not of men's creeds ... a tyranny avow'd'*: Lord Byron, *Childe Harold's Pilgrimage,* IV.xcv.1–5.

6. *are you a repealer?*: a supporter of the Repeal Association, which campaigned for the repeal of the Act of Union between Ireland and Britain.
7. *bigbellied Bulls*: i.e., Englishmen. A reference to John Bull, the often-portly personification of England.
8. <*1 word excised*>: Louisa Tyndall annotation: '?' It appears the word has been excised because Louisa Tyndall could not decipher it.
9. *Erin*: a modern derivative of Éirinn, the Irish-language name for Ireland.
10. *Tara*: the former coronation place of the Kings of Ireland.
11. *Jim*: probably Phillip 'Jim' Evans.

From William Ginty [October] 1843 0256

'Twas Even – I said– a lovelier one
On summer roses never shone!
Blue calm and clear – October's sky
Seemed borrowing hours from bright July!
Round red and large – the setting sun
R<*1 word excised*> o'er the mighty landscape flung
In mingled gold and crimson rays
The lustre of his distant blaze –
Pale in the eastern heaven the moon
As dreading she had come too soon
To visit there – has sought apace
Behind a cloud a hiding place –
The leaflets as they earthward roll'd
Were all that of a winter told!
An Eve like this – in this drear clime
In such a place – at such a time
Is worth a dozen when the rays
Of hundreds glow around our ways
It looks so lone and yet so sweet
Last of the lovely band to greet
Our wearied eyes – like one bright flower
Which still remains to grace the bower
With hues and perfume – only one
When all the rest are wither'd – gone!
By Heaven, twas sweet! A brighter scene
Ne'er graced an Houri[1] in a dream!
The Enchanter's wand ne'er mystic drew
A fairer vision to the view!
Before me spread like one vast chart
The Landscape mocks the painter's art!
Far as the eye can reach I scan
The deep blue ocean fading wan!
Beneath me madd'ning as they go
The Lune's swift waters foaming flow
The Herald of many a tale of woe! (1)
High on the north – from out the clouds

'The Fells'[2] peer forth in snowy shrouds!
Beyond me – bursting on my sight
O'er crag and fell and burnished bright
Like livid gold which lightning mocks
In amphitheatre of rocks
And leafless trunks of oak – appear
The crystal 'lakes of Windermere'![3] –
Oh! for some pure one at that time
To fold – to clasp – to call her mine!
To gaze with her on that sweet scene
Such love as that were no 'fond dream'!
And looking towards the sun and wave
I thought that one might see him bathe! (2)
Far westward! – in a lovely isle
Were such are even seen to smile
And mutt'ring one deep loving prayer
I sighed and wished that I was there
Then turned to see bright Cynthia[4] steal
A peep from 'neath her golden veil
She rose chill cold and pearly bright
The Queen of shadows! now the night
Was gathering round me – and I threw
One glance towards Erin[5] – all was blue
The sun had set – the *[cloudlets]* perched
Along the fells seemed fixed for rest
Then clambering down – I cried 'Farewell –
'To thy bright summit Berkenfell!'

(1) The rapidity of this river has caused many accidents in the upsetting of Ferry boats – About 18 months since a man committed suicide by throwing himself over the bridge of this town – Commonly called the 'Devil's Bridge'.[6]

(2) This will bring an idea of Byron's to you,
'And oft I thought at C<*1 word excised*>[7] noon'
'While gazing o'er the Egian wave'
'How Thyria gazes on that moon'[8]

Tis better confess my guilt than be detected in the theft.

RI MS JT/1/TYP/11/3594
LT Transcript Only

---

1. *an Houri*: a beautiful young woman (*OED*).
2. *'The Fells'*: the famous fells of the Lake District in Cumbria County, England.
3. *'lakes of Windermere'*: a reference to the Lake District. Lake Windermere is the largest of the Lake District lakes.
4. *Cynthia*: an archaic reference to the moon.

5. *Once glance towards Erin*: i.e., towards Ireland. Erin is a modern derivative of Éirinn, the Irish-language name for Ireland.
6. *the 'Devil's Bridge'*: an infamous bridge in the Lake District town of Kirkby Lonsdale, Cumbria. In legend, the bridge was built by the Devil to allow an impoverished widow to cross the river and reach her runaway cow. In the story, the woman tricked the Devil into accepting her small dog instead of her soul in exchange for the bridge.
7. *C<1 word excised>*: this word should be 'Cynthia', but it does not appear in the transcript. See n. 8 for details regarding the source of this stanza.
8. *'And I oft thought ... gazes on that moon'*: Lord Byron, 'One Struggle More, and I Am Free', 4.v–vii.

From Archibald McLachlan [October] 1843 0257

Sunday

Dr Alpha[1]

Something must be done and done quickly, you know that at the present moment upwards of 20 of our men are now employed in London, almost every man feels resentment towards the O.S – [2]

Very few among them have the necessary knowledge of Facts – and still fewer have the Temper to do anything calmly – The letter in the Daily News[3] had been from one of the 20 – (at least I think so)

Last night, I was told by one of our own gentlemen at the O.M.O.[4] – 'That it was a poor thing that nobody would expose the O.S. in good style like Spectator,[5] stop a little, you will see quare[6] work before long – there is a confederation & commotion among the Boys in London, and perhaps they are clubbing their money to have the expenses of letters paid in the Newspapers, I wonder what they intend to do' – Now, if Steel-nib[7] is first in the field it might deter the above gentlemen from executing their design – Among them they could furnish much information, but who is to arrange and compose? I dont think they have a man among the 20 able to do justice to the Subject. Therefore let Steelnib be first in the market –

Do you mean that I should make a list giving each mans name and his rate and length of service? – or will something of this sort do

No of men	rates	Years employed
6	6/.	18
15	4/.	10
30	3/.	10
14	3/.	7
	&c———	&c———

I think perhaps the number at so much per day and length of service and duty employed on would be enough – would that do? –

Are you aware that the Excise – or Customs (I don't recollect which) have all got an increase of £50 a year added to their salary – This has been done owing to the repeated Custom House frauds &c &c – This increase is intended to place the men beyond temptation, and to make the Situation worth holding –

I am completely posed, I cannot see how you are to commence – I cannot imagine what will be the drift of your First Letter[8] – I imagine that each letter will be devoted to some <u>one point</u> – that you will take every[9]

RI MS JT/1/TYP/11/3775
RI MS JT/8/1/4a

1. *Alpha*: an occasional nickname for Tyndall.
2. *the O.S.* : the Ordnance Survey.
3. *the letter in the Daily News*: not identified.
4. *the O.M.O.*: unidentified acronym, possibly Ordnance Mapping Office.
5. *Spectator*: Tyndall's *Liverpool Mercury* pseudonym.
6. *quare*: an Irish regional pronunciation of 'queer'.
7. *Steel-nib*: unidentified nickname.
8. *your First Letter*: Tyndall may have been contemplating another series of newspaper letters about the Ordnance Survey.
9. *that you will take every*: letter ends here. This letter appears to be a fragment.

To Archibald McLachlan [October] 1843 0258

My Dear B,[1]
This moment within 10 minutes of post time I have received your letter[2] – I forward N° 5[3] without delay – do you evince the same promptness in sending it back – I have many reasons for not wishing the late movements to be mentioned – they will do very well for a corroborative <u>afterclap</u> – if however you think the mention of them of importance <u>they shall be mentioned</u> – <u>Omega</u>[4] <u>is in Connaught</u>[5] – he complains of our tardiness in bringing N° 5 forward – make haste! And get as much information as possible on the present state of things – we are disunited but we are still able to make them shiver – we'll see the thing out –
as of old | Alpha![6]

RI MS JT/8/1/4a

1. *B*: the pseudonym McLachlan used to avoid detection.
2. *your letter*: possibly letter 0257 of this volume.
3. *N° 5*: the fifth letter by 'Spectator', Tyndall's *Liverpool Mercury* pseudonym; see letter 0261 of this volume.
4. *Omega*: a nickname for William Ginty.
5. *Connaught*: a province in western Ireland.
6. *Alpha*: a nickname for Tyndall.

From William Ginty 2 November 1843 0259

O.S.O.[1] Thursday 2nd Nov. 1843.

Dear *[1 word illeg]*

The first in my mind shall be first on the paper 'don't curse my delay, Bill'[2] but I have cursed you aye, in no mean terms either. You should have had it ready for this week's. I'm afraid it will be no go in the next, as this holy 2nd of Nov., is the 'first day of term' on which that illustrious countryman of ours[3] is to appear at the bar of the Court of Queen's Bench to answer the charge of 'conspiring and misdemeanour'. Our worthy rulers are mad. Even if Dan was imprisoned during secula secularum,[4] it would not stop Paddy's[5] wrath. Lord Byron says 'persecutions or coercion will never crush a popular opinion'[6] but stop – this is no address to the people of Ireland. Now for Chadwick's[7] letter, it is commenced 'Dear Hunter' (Dick),[8] Ginty &c' – There is a brother of his[9] here, he had a letter from him lately in which he requested us to write to him – so we did – And oh tell it not in Gath[10] he answered it! I shall quote some of it. 'Hurrah three cheers for you my old shots' 'faith I give you credit for bearding the old chief constable of the lobsters' ——— Again 'go it ye cripples' ——— 'Ginty inspired me with such dread by adjuring me in such a solemn manner to look after Miss McCarthy[11] that I immediately shut shop and ran like blazes down to the Market Place. They were all well and unchanged I think they are more extensive in business than they were, as I saw a great quantity of penny loaves, butter, onions bacon &c &c. I was advising her to get into the fruit trade. Well I saw Miss McC. When I mentioned Ginty's name she blushed like a posy and fell in a faint under the counter I leaped over and caught her in my arms and after administering a pint of heavy wet[12] which stood by fortunately, she was able to answer all enquiries. I showed her the letter and she watered the well known characters with her tears, after the manner of Amanda in the Children of the Abbey.[13] I stopped up her countenance with what the bad times had left me of an ould wipe and she became composed. She is a fine girl and as there is no marrying men in this town she will be likely to remain single until Ginty returns to claim her for his Bride (loikly).[14] She desired me to send her fondest love tenderest regard and all the rest, and to believe her to be Ginty's 'truly devoted well wisher and chosen Houri'.[15] This is peerless – rich beyond comparison, it outshines all modern pretenders to the comic. The more I read it the more I roar – Mind the polite scandal of the fellow about the extensive business, and his advice, then her blushing like a posy, then the reference to the Children of the Abbey, in such beautiful keeping with the fainting, tears, and blushes, and then 'the ould wipe' – Its worth all Boz[16] wrote these two years! One quotation more and you have the cream of his inimitable letter. 'By heavens nothing could give me greater satisfaction than knowing that the bloody damned upstart ignorant haughty high coclelorum[17] officers of the Royal little goods should be taught their level and that by the worms they trampled on in their pride – the said worms having now thrown aside their chrysalis of dirt and meanness and emerged in the wing state which will allow them to soar to that height of respectability which their high and talented members in many cases have a claim to'. Again 'Prepare to welcome me next summer as one whom "Cubs delighteth to honour"'.[18] he says in another place that John Abbot wants to emigrate with him to America and he is likely to agree to the proposal! Of course your name is used by him in very affectionate terms! The letter is too long to copy and too heavy to send by post – you have here at all events its choicest parts! 'Coclelorum' – there's a word for you! He says he is passing his life in a sober monotonous obfuscation and like a bloody fine river gliding along decently to the say of futurity'. He is father of two daughters.

I am no advocate for scurrilety[19] father John. I am of the same opinion as your absent friend.[20] The sooner the Tithe humbug[21] is unfolded to public gaze the better, and hurry it for pity's sake. He told me long ago 'he would like to get them in after other' and now post your affair on Sunday evening so that Monday evening will see it in his clutches. I am very certain the old 8 chap[22] will chuckle over the exposure of the tithe work and the botchering of the survey in general. He published the short march on the strength of what is to come. Of course past and present are very different – but be quick. Damn your quadrilles[23] and French and love affairs. They are things of this world when compared to the grand project of Spectatorisms.[24]

I am very certain you may gain a very fine proficiency in phonography[25] without attending to lectures. It would cost me 14s, aye £1 to attend the lectures here. I am however 3 weeks aye 3 months before some that have attended them. I will send you the two books I have when I have done – they are very complete in themselves. 'The Phonographic Classbook' and 'The People's Edition of Phonography or writing by sound a natural method of writing all languages by one alphabet composed of signs that represent the sounds of the human voice'.[26] As for the lectures they are only catch-pennys,[27] and no service to any except fellows remarkably stupid. The Liverpool lecturer is just come from Manchester – he is a Rev. Benjamin Pitman,[28] brother to the inventor Mr Isaac Pitman.[29]

So my poor fellow 'they are all completely Irishized' or you are Englishized – If I don't mistake the latter version is the correct one. You are a proselyte,[30] your anti-Saxon soul is bamboozled into a belief that these Englishers – these semi-barbarians sentimentalists are even equal to the Houris of the Emerald Isle.[31] Verily you are in a pitiable plight – Poor unsuspecting Paddy – did Miss Bull[32] blarney[33] you into a belief that her charms were charms indeed. How damned fortunate you are 'preengaged' – you are not out 'in danger'. Well you are in love and ergo out of danger of being fancy struck with another. This is consolation with a vengeance – Talk of your philosophy now, Tyndall – pon my conscience you have made the most contemptible display in the love trade of any yet – Miss —[34] the 'Southern Briton'[35] is I calculate some ugly skinny romantic – rawbone – long-legged devil like yourself. Why don't you send me her address and if I don't do you justice I'm a fish that's all. Send me her name anyway until I make an <u>acrostic</u>[36] for you. I hear she is the wrong side of 30 – Oh damn your taste. I never fell in love with an ugly girl that's my shine. Come rake up the memories of the past and see Ginty's enamoratas will strut before you in the characters of Haidees – Dudus – Nourmahals. Golatis and Katinkas.[37] Bright Houris of bygone days to the devil I bob you all. I am a <u>man</u>. Is Jim[38] in love? I suppose not unless he dropped into the wake of the Bessie species.[39] Jim is very fond of pigmy beauties – it is very fortunate that Bessie and him did not 'strike up' – To such combinations may be attributed the present degenerate and anti-behemoth state of the human family. Poor little devil, celibacy will be the end of him.

I have been thinking – we have been thinking that you ought to come to Liverpool at Xmas – Faith will give you enough to eat anyway – <u>Such as it is</u>. Jack[40] and I will invite you and Jim and Bill Latimer. Geo and you may bring the full of a carriage if you like. Say what you think of it. Plenty of time yet – If you do not scout the idea at once we will draw up a proper instrument of indictment for ye all. You could come Saturday evening. Then Sunday, Monday for 'bladdering' &c and on Tuesday morning you could one and all 'go to hell out of this' – 'I wonder would they come'. – 'We ought to ask them'. – 'It's no use I'm afraid'. 'It's so expensive'. – 'They could walk part'. – 'They can come at the contract price'.– 'Since they would eat the <u>fare</u> anyway in Preston'. And so on, such are our cogitations on the matter. 'Come! dwell with me'. Come dwell with me.

> And you soon shall see and you soon shall see
> That our house is bright with the image bright
> Of Irish laughter mirth and glee
> Yours Pastorine.[41]

Come! Come! Come! By heavens man it will make you a year or two younger. Lord how I will pummel you and abuse you about the romantic looking caricature you have fell in love with. You have two months yet to ponder on it.

Be sure to make up your mind at once and none of Jim Evans's Cork expedition hesitatings nor no repetition from Jim – Tell Geo.[42] he will hear a great sermon. Tell him he will get a big plate for himself.

Hip hip hurrah – old Tucker sent my letter to Hamley who I have just been with – A memorandum accompanied it which Hamley was desired to read for me. After saying he had received a letter from me, he says to Hamley tell him that the words 'if there were fewer men on the survey then the men would be better off', are not easily misunderstood – The old dog. I never made use of such words he is a damned liar – he also says tell him he ought to have forwarded it through you. Well Sir says I (without a comment on the note) if I have any further remarks to make on the subject I will send them through you! And thus ended the interview. Now what's to be done? Is it worth further answer I am greatly inclined to treat the old humbug with silent contempt – Either that or tell him in a roundabout way that he is a liar. What is your opinion? We had Danefield[43] – that personification of Chesterfield's principles of politeness[44] here all the week – When Sinnett found out who he was, he was looking for him to eject him – but lo the bird was flown. I hear he has got 30 shillings a week as commercial or travelling clerk! Every man in the house has been with me during the last 10 minutes to learn the purport of my interview with Hamley. I have told them with the utmost gravity that Hamley read an order for me to hold myself in readiness to join St. Mc.Kerlie's[45] surveying party, for Scotland, on the 1st of the ensuing month! They all believe it – What damn fools!

I can hear two fellows this moment at the fire muttering. 'Damn them, I knew they would hail Ginty, oh man Tucker had it in for him, I knew it 'he is sent away from here because he is a spirited chap!' &c &c &c

I will expect a letter from you by the first post, if you approve of my further writing to the old bothered fellow. I'll walk into him.

Yours faithfully | Ivanhoe[46]

This letter is what Lt. Hamley called our memorial.[47] 'A damned long yarn'. 'Did you forward the memorial Sir?' 'Yes and a damned long yarn it was!'

After dinner

8 sent me the copy of the memorial I sent him for his perusal – On the envelope appears the following:– 'I. S.[48] is sorry he has been so frequently out of late when Mr G[49] called'. This I take to be a hint to call on him, for I was not there this week at all. I long to see his honour and I don't want to go with my finger in my mouth now that he has been disappointed by you. Bad luck to your quadrilles and balls and such flimsey gew-gaw trash occupies too much of your time.

Yours | 'The Attorney General'.[50]

RI MS JT/1/TYP/11/3618–3619
LT Transcript Only

1. *O.S.O.*: Ordnance Survey Office.
2. *Bill*: Ginty's nickname.
3. *that illustrious countryman of ours*: Daniel O'Connell (1775–1847).
4. *secula secularum*: saecula saeculorum, forever and ever (Latin).
5. *Paddy*: a common personification of Ireland.
6. *'persecutions or coercion will never crush a popular opinion'*: quotation not identified; may have been attributed to Byron in error.
7. *Chadwick*: John C. Chadwick.
8. *Dick*: a nickname for Richard Hunter.
9. *a brother of his*: this could refer either to an unidentified brother of John C. Chadwick or to Richard Hunter's brother, William Hunter.
10. *tell it not in Gath*: 2 Samuel 1:20.
11. *Miss McCarthy*: an unidentified love interest of Ginty's, possibly a relative of Michael McCarthy or the daughter of Mrs McCarthy, Tyndall's former landlady in County Cork. According to Louisa Tyndall's transcript, Ginty drew rays emanating from her name, suggesting particular radiance or beauty.
12. *heavy wet*: Louisa Tyndall annotation: '?' It appears that Louisa Tyndall had some difficulty deciphering the exact phrase Ginty wrote.
13. *Amanda in the Children of the Abbey*: Amanda is the main character in R. M. Roche, *The Children of the Abbey* (1796).
14. *loikly*: not identified.
15. *Houri*: a beautiful young woman (*OED*).
16. *Boz*: a nickname for Charles Dickens (1812–70), celebrated English novelist and social critic. The nickname comes from Dickens's 1836 book *Sketches by Boz.*
17. *coclelorum*: a variant spelling of 'cockleorum', a nursery word for jokes and jokers.
18. *'Cubs delighteth to honor'*: Esther 6:7, 'For the man whom the king delighteth to honour'.
19. *scurrilety*: scurrilous, i.e. coarseness or indecency (*OED*).
20. *your absent friend*: not identified.
21. *Tithe humbug*: see letter 0231, n. 19 of this volume on the Tithe.
22. *old 8 chap*: see letter 0232, n. 13.
23. *quadrilles*: a dance popular in the nineteenth century.
24. *Spectatorisms*: a reference to Spectator, Tyndall's *Liverpool Mercury* pseudonym.
25. *phonography*: Louisa Tyndall annotation: 'i.e. shorthand'.
26. *'The People's Edition ... the human voice'*: I. Pitman, *Phonography; or Writing by Sound: A Natural Method of Writing All Languages by One Alphabet, Composed of Signs that Represent the Sounds of the Human Voice: Adapted Also to the English Language as a Complete System of Short Hand etc* (London: Samuel Bagster and Sons, 1842).
27. *catch-pennys*: publications of little value or merit (*OED*).
28. *Rev. Benjamin Pitman*: Benjamin Pitman (1822–1910), an English author; he is known for popularizing the Pitman shorthand, see n. 29.
29. *Mr Isaac Pitman*: Isaac Pitman (1813–97), inventor of the most widely used form of English shorthand.
30. *proselyte*: a person who has converted to Judaism (*OED*).
31. *Emerald Isle*: a poetic nickname for Ireland.
32. *Miss Bull*: not identified; it may be a nickname meant to suggest that the young woman is English.
33. *blarney*: flattering talk (*OED*).

34. *Miss —*: it is unclear whether Ginty omitted the name himself or if Louisa Tyndall excised it.
35. *'the Southern Briton'*: unidentified person. In 1843 Tyndall was enamoured with a woman he called 'the Northern Briton'; see letter 0235, n. 31 and letter 0325, n. 12 of this volume.
36. *an acrostic*: a form of poetry in which the first portions of each line combine to form a word or phrase.
37. *'Haidees – Dudus – Nourmahals. Golatis and Katinkas'*: this appears to be a reference to legendary beauties.
38. *Jim*: probably Phillip 'Jim' Evans.
39. *Bessie species*: not identified; evidently a young woman named Bessie whom Ginty thought might attract Evans's attention.
40. *Jack*: John Tidmarsh.
41. *As you soon ... Yours Pastorine*: unidentified poem, possibly written by Ginty.
42. *Geo*: possibly George Holmes or George Latimer.
43. *Danefield*: not identified.
44. *Chesterfield's principles of politeness*: Lord P. Chesterfield, *Principles of Politeness and of Knowing the World* (1786).
45. *St. McKerlie*: John Graham McKerlie (*c.* 1814–1900), a member of the Royal Engineers.
46. *Ivanhoe*: a reference to W. Scott, *Ivanhoe* (London: Archibald Constable, 1820).
47. *our memorial*: the letter of protest sent by the workers of the Preston Ordnance Survey to George Murray, Master General of the Ordnance on 23 September 1843; see letter 0236 of this volume.
48. *I.S.*: not identified.
49. *Mr G*: not identified, likely Richard John Griffith.
50. *'The Attorney General'*: a nickname for William Ginty.

To John Tyndall, Snr. 3 November 1843 0260

Preston | Thursday night, very foggy | Nov. 3rd, 1843.

Dear Father

I sent you a letter which I received from Wm Wright[1] a few days ago, as I wish to reply to it I'd be glad if you would return it as soon as possible. When you have an opportunity, let Mrs Wright[2] know that I have heard from him. You mentioned my uncle Thomas[3] in one of your letters. Where do you think he is located? Is it in Nottingham? If so I will write to the post master there and make enquiries about him. Wm Tyndall after near thirty years silence has at length favoured you with an epistle. Who knows but we might squeeze one from the pen of my uncle Thomas, if he has not winged his way 'to that undiscovered country from whose bourne no traveller returns'.[4]

I am pressed for time | Goodbye | Your affectionate son | John.

RI MS JT/1/TYP/10/3303
LT Transcript Only

---

1. *a letter which I received from Wm. Wright*: probably letter 0237 of this volume.
2. *Mrs Wright*: William Wright's mother, first name not identified.

3. *my uncle Thomas*: unidentified relative of Tyndall's, possibly on his mother's side.
4. *'to that undiscovered country from whose bourne no traveller returns'*: W. Shakespeare, *Hamlet*, III.i.79–80.

To Sir Robert Peel via The Liverpool Mercury 4 November 1843 0261

LIVERPOOL MERCURY November 24, 1843 | Letter IV.

To the Right Hon. Sir Robert Peel, Bart &c., &c.

Sir – My two last letters[1] referred exclusively to the Ordnance survey of Ireland and its abuses; I now essay to address you on a subject of more immediate interest – the competition of the Royal Engineers with the civil surveyors of England, in offering contracts for the mapping of townships for the purposes of tithe commutation.[2]

The longing eyes of the Royal Engineers appeared to have been fixed on a survey for these purposes long before the competition alluded to took place. In 1837 a select Committee of the House of Commons was appointed to consider the best means of effecting the survey of parishes under the Tithe Act.[3] The parties examined were tithe commissioners.[4] Their opinions were strongly in favour of a general survey of England, and these opinions were founded on the report of an engineer officer.[5] The latter advocated a system of centralization, similar to that pursued on the Ordnance survey of Ireland. Is it not a fair presumption that this officer had the advice of others of his class? It was expected that the Irish survey would be completed about 1838; how desirable, then, would it be for the officers employed on the latter to move across, and engage without delay, in an undertaking which promised both emolument[6] and fame. This idea, however, was abandoned, and in 1841 the survey of the six northern counties, on a scale of six inches to a mile, was commenced. In the latter part of this year certain townships in Lancashire, to be surveyed for tithe commutation, were advertised in the usual manner. Then, for the first time, were the civil surveyors compelled to shrink before the prowess of a government competitor.[7] An inkling to nibble at the cheese, the monopoly of which was previously attempted, was evinced by Captain Tucker, Royal Engineers. He put in a tender for the townships advertised, and owing to the unparalleled lowness of his estimate, he was successful. This extraordinary proceeding drew from the Lancashire surveyors a memorial[8] to Sir J. Graham; it was forwarded through J. W. Patten,[9] Esq., M.P., and the reply of the latter contains the following:–

'The Master-General[10] and the Board[11] have inquired into the circumstances adverted to in the memorial, and have made such communication to the officer in superintendence of the Ordnance survey as will obviate the recurrence of any further ground of complaint of a like nature'.

'The proceedings which led to that contained in the memorial were unauthorized by the Master-General and Board of Ordnance'.

Before this disclaimer appeared the maps of several townships had been executed by the Royal Engineers, and upon each, in broad, legible characters, was written: 'Drawn from the Ordnance Survey of Great Britain, by order of the Master-General of the Ordnance'.

Now, Sir, which are we to believe? Can we suppose that Mr. Patten would, under the circumstances, state a deliberate falsehood? Assuredly not. On the other hand, it is difficult to imagine that an officer and a gentleman, holding a highly responsible public situation, would lend himself to imposition. There is, however, no alternative; the statement of Mr. Patten and that contained on the tithe plans are at direct variance. If one be true the other must be false: if Mr. Patten speaks truth the Royal Engineers must have acted on their own unauthorized caprice, and the superintendence of the Board, is in this instance a mere mockery.

Again, Sir, it is stated that 'a recurrence of any further ground of complaint would be obviated'. What is the meaning of these words? Is it not that the competition should cease? But the competition has not ceased. The Royal Engineers, since that statement was made, have contracted for half a dozen townships. Must we, then, believe the communication of the Master-general to be a mere 'tinkling cymbal?'[12] It may be said that there is at present no public competition by tender. This is an unworthy ruse. What care the land surveyors of Lancashire how the competition is carried on, whether openly or clandestinely, when its effects in wresting from them their professional employment are precisely the same? How can it be said that the grounds of complaint are obviated, when a change of means only, in the accomplishment of the ends complained of, has taken place? The answer which the surveyors received has thrown dust in their eyes, and they have slumbered in fancied security ever since, while their insidious adversary[13] with untiring perseverance, has continued to strike secretly at their interests. By the exercise of the most polite condescension in giving instructions to landowners, as to how they should apply for tithe maps, he has enlisted these powerful auxiliaries in his unfairly privileged crusade against the civil profession.

The Act for the survey of the six northern counties contemplates that they will be finished in 1846–7; by a former calculation I showed that they would not be completed till 1854. Does this not prove an utter disregard for the conditions of the Act, as well as of the demands of the Chamber of Commerce at Manchester who were instrumental in getting it passed? Does it not defeat the designs of the Legislature – to employ the men on an undertaking quite foreign to the subject for which they were appointed? And this is done, without any reference to the Parliament who sanctioned their appointment!

The Ordnance survey is conducted under one Act; the tithe survey under another; the men employed under the former are paid out of funds voted specially for the purpose of carrying out the designs of that Act. A question here arises: – Out of what fund are those men paid who are withdrawn from the Ordnance survey and employed at the tithe maps? Would it not be an unwarranted stretch of power to pay them out of funds voted by Parliament for another purpose? How do these men stand in the accounts of the Royal Engineers? Are they not entered as being employed on the Ordnance survey? If so, are not these accounts inaccurate? Do these accounts show whether the public gain or lose by such undertakings? If the former, do they show the transfer of that gain to the public treasury? If the latter, do they show how the deficiency is made good? These, Sir, are questions to which I, as a subject taxed by the British Government, have a right to demand replies.

I would here notice an argument which has been used in justification of the competition of the Royal Engineers. It has been said, that, as a survey for the six-inch scale was, at all events, to be made, and as the expense of the plotting and drawing only was to be covered by the contract price, the Royal Engineers would, therefore, be able to furnish tithe maps at a cheaper rate than any civil surveyors; and as the ratepayers had borne a portion of the expense of the original survey, they, in justice, should possess whatever advantages might accrue from it. This reasoning is specious; but, Sir, it is false as 'dead sea fruit'.[14] I here repeat, in stronger terms an assertion contained in my first letter, that, notwithstanding these advantages, and the wretched pittance awarded to the men who drawn them, the actual expense of the tithe maps executed by the Royal Engineers, are, in many cases, more than double the price charged for them. This, Sir, is no secret matter: it is a fact notorious in the Ordnance survey. A Royal Engineer contracts for a township at 4d. an acre – the true cost of that township is perhaps a shilling an acre; whence comes the 8d difference? Is it not, after all, from the public purse? If so, I as a citizen, pay my proportionate part of that 8d as well as the landowner who reaps the benefit of it, and I protest against the justice of taxing me for the sole benefit of another class of the community. If the contract price be insufficient to cover the expenses, the contractor should be the sufferer. The Royal Engineers have had ample proof of their inability to furnish plans at their contract prices, why then do they not raise their estimates or else relinquish the business altogether?

Will an endeavour be made to bring my statements into disrepute by pronouncing them the offspring of private jealousy? Miserable evasion! The question, Sir, is not whence my statements spring, but are these statements true? The gauntlet is before them; I publicly challenge the concentrated sagacity of the Royal Engineers to disprove what I have adduced. Are the civil profession, then, Sir, to be trampled on to gratify the unwholesome ambition of a griping monopolist, while I can furnish them with the means of opposing him? Never, Sir! – never shall my pen repose till I see these individuals stand forth divested of the mysterious veil which has hitherto enveloped their proceedings, and the voice of an enlightened public pronounce between them and

Spectator[15] | Dublin, November 4th, 1843.

RI MS JT/1/TYP/11/3914–3916
LT Transcript Only

1. *My last two letters*: letters 0246 and 0252 of this volume.
2. *tithe commutation*: see letter 0231, n. 19 of this volume.
3. *the Tithe Act*: see letter 0231, n. 19 of this volume.
4. *tithe commissioners*: see letter 0247, n. 6 of this volume.
5. *the report of an engineer officer*: possibly Richard Griffith.
6. *emolument*: salary, benefit (*OED*).
7. *a government competitor*: i.e., the Royal Engineers.
8. *a memorial*: the letter of protest sent by the workers of the Ordnance Survey of England to George Murray, Master General of the Ordnance on 23 September 1843; see letter 0236 of this volume.
9. *J.W. Patten*: see letter 0244, n. 13 of this volume.
10. *The Master-General*: George Murray.
11. *the Board*: i.e., the Board of Ordnance.
12. *'tinkling cymbal'*: 1 Corinthians 13:1.
13. *their insidious adversary*: i.e., Henry Tucker.
14. *'dead sea fruit'*: an idiom meaning that something appears promising but in reality is disappointing or false.
15. *Spectator*: Tyndall's *Liverpool Mercury* pseudonym.

From Henry Tucker 7 November 1843 0262

O.S.[1] York, 7th Nov., 1843.

Notice was issued to the Detail Survey[2] on 7th June to warn the Civil Assistants and chainmen that it would be necessary to make further reductions to lessen the expenses. In obedience to orders I have received to effect a reduction, the following persons are to be discharged on the 21st November inst. from your Division: –

Lt. Stanley,
Royal Engineers.

Civil Assistants	Labourers.
William Branter[3]	Patrick Donnelly[4]
James Butler[5]	Matthew Donohoe[6]
Peter Carroll[7]	George Doyle[8]
John Conner[9]	Matthew Doyle[10]
Jeremiah Griffin[11]	John Howard[12]
Philip D. Evans	Michael McCarthy[13]
John Holland	Michael Murphy[14]
John Kelly (1st)[15]	Daniel Mylot[16]
George Latimer	James O'Neill
James S. Lindsay	Daniel O'Sullivan[17]
Dennis Malone[18]	James Smyth

Robert Martin
John McGowan
Patrick McCord[19]
Henry McIntyre[21]
Oliver O'Railly.[23]
Edmond B. Sayers
John Tyndall.
Patrick Walsh.[24]

Charles G. Stuart
John Thornton
Robert Westhead[20]
John Tole[22]

Nineteen Civil Assistants
Fifteen Labourers.

(signed) | H. Tucker | Capt., R.E.[25]

A true copy.

RI MS JT/1/TYP/10/3902
LT Transcript Only

1. *O.S.*: Ordnance Survey.
2. *the Detail Survey*: the branch of the Survey tasked with surveying and mapping topographical details, rather than boundaries.
3. *William Branter*: joined the Irish Ordnance Survey on 1 October 1837. Branter remained in Ireland until the winter of 1841, when he left for the English Survey.
4. *Patrick Donnelly*: not identified.
5. *James Butler*: joined the Irish Ordnance Survey on 1 October 1840. Butler left for the English Survey at some point during the autumn of 1841.
6. *Matthew Donohoe*: possibly Matthew Donoghue; Donoghue joined the Irish Ordnance Survey on 1 May 1839. Donoghue was discharged from the Irish Survey on 27 May 1841; if he is the person dismissed in this letter, he probably accepted a demotion to labourer in order to join the English Survey.
7. *Carroll*: Peter Carroll; joined the Irish Ordnance Survey on 1 May 1837. He appears to have left the Irish Survey for the English Survey during March 1842.
8. *George Doyle*: not identified.
9. *John Conner*: possibly John Connor. However, Connor was discharged from the Irish Ordnance Survey on 27 May 1841, making it somewhat unlikely that he would have been hired on the English Survey; this may be a different individual.
10. *Matthew Doyle*: not identified.
11. *Jeremiah Griffin*: joined the Irish Ordnance Survey on 1 June 1838. He later joined the English Survey.
12. *John Howard*: joined the Irish Ordnance Survey in June 1838, but had left by January 1839. It is likely that he left for the English Tithe Survey, which offered better pay, before rejoining the Ordnance Survey in England.
13. *Michael McCarthy*: not identified; see letter 0288, n. 26 of this volume.
14. *Michael Murphy*: not identified.
15. *John Kelly (1st)*: not identified.

16. *Daniel Mylot*: joined the Irish Ordnance Survey on 1 October 1840. Mylot left the Irish Survey during the autumn of 1841 and later joined the English Survey.
17. *James O'Neill | Daniel O'Sullivan*: not identified.
18. *Dennis Malone*: not identified.
19. *Patrick McCord*: joined the Irish Ordnance Survey on 1 September 1840. McCord left the Irish Survey during the autumn of 1841 and later joined the English Survey.
20. *James Smyth | Charles G. Stuart | John Thornton | Robert Westhead*: not identified.
21. *Henry McIntyre*: joined the Irish Ordnance Survey on 1 May 1834. He remained on the Irish Survey relatively late, departing sometime after August 1842 for the English Survey.
22. *John Tole*: joined the Irish Ordnance Survey on 1 December 1839. It is likely that he left for the English Tithe Survey, which offered better pay, and later made his way back to the Ordnance Survey in England.
23. *Oliver O'Railly*: probably Oliver O'Reilly; he joined the Irish Ordnance Survey on 1 November 1840. O'Reilly left the Irish Survey during the autumn of 1841 and later joined the English Survey.
24. *Patrick Walsh*: appears to have joined the Irish Ordnance Survey in late 1839 or early 1840. He left the Irish Survey during the autumn of 1841 and later joined the English Survey.
25. *R.E.*: Royal Engineers.

From William Ginty 8 November 1843 0263

Wednesday evening | Nov. 8th 1843.

Dear Jack

No man knows what a day may bring forth – hurrah for the glorious 21st.[1] I have received notice to quit[2] and deliver up all that and those that part of the Ordnance Survey office in my possession for the last 7 years. Holy Nelly I'm classed among the inefficient C.A's[3] and also Taylor and James Livingston[4] – This worthy soul was one of the leading nuisances in the memorialists,[5] he is 3rd or 4th best man in this division 3 and something per day – and of 7 years standing. His dismissal surprizes me more than my own. Now – I expect there is a fellow named Tyndall on the list of inefficients too and Allen, and Evans, and Latimer[6] eh? Speed! Postman speed! Now what's to be done – We must rise a row in some way and as you are Master Devil I submit, if you tell me to burn the office it shall blaze, aye in graceful columns it shall curl towards heaven as the smoke of their torments. Grogan[7] is dismissed – this is the fellow that got the 9d for not signing the petition, he was recommended to be mulct of some of his pay[8] last month for some bad surveying – the Captain[9] said it was no time to be reducing men and so recommended him to be dismissed. Johnson[10] he that lived with Taylor goes too, but he is entitled to it to some extent. I can't say more tonight – This is unfortunate at the present time. I might have been better prepared. If there is a job to get in Liverpool I'll fork it out

Yours | The, Attorney General[11] | defunct.

RI MS JT/1/TYP/11/3620
LT Transcript Only

1. *the glorious 21st.*: the date on which the dismissed surveyors would depart; see letter 0262 of this volume.
2. *I have received notice to quit*: see letter 0262 of this volume for a copy of this notice.
3. *C.A's*: Civil Assistants.
4. *James Livingston*: unidentified member of the Ordnance Survey.
5. *the memorialists*: the authors and signatories of the letter of protest sent by the workers of the Ordnance Survey of England to George Murray, Master General of the Ordnance on 23 September 1843; see letter 0236 of this volume.
6. *Latimer*: George Latimer.
7. *Grogan*: unidentified member of the Ordnance Survey.
8. *he was recommended to be mulct of some of his pay*: i.e., to have some of his pay withheld or taken (*OED*).
9. *the Captain*: probably Henry Tucker.
10. *Johnson*: unidentified member of the Ordnance Survey.
11. *Attorney General*: one of Ginty's nicknames.

To John Tyndall, Snr. 8 November 1843 0264

Wednesday night[1]

My dear Father

Well, the thunderer[2] has at length flashed and I have been seared, not shivered by the bolt. I am dismissed – dismissed forever from the legion of Ordnance Surveyors[3] – Thank heaven however I leave it with a character unstained. I suffer in a righteous cause, and suffer in common with some of the best and oldest hands in the employment. I gave you a hint before that some danger was impending. I had no notion however that vengeance was so near. I did not anticipate so sudden a blow, coming as it did without previous signal. It might be thought sufficient to unman me this however is not the case. My convictions would prompt me to a reacting of the scenes which led to my dismissal, even if I were aware of the consequences. My last letter[4] to Mr Conwill goes more into the detail of our proceedings than any I wrote to you. This however is but a sketch. I shall shortly lay before you a full account of the business which caused our Superintendent[5] to take this step. At 10 o'clock this day we heard the order[6] for the dismissal of <u>nineteen</u> read. Among the number are – John Tyndall, Philip D. Evans, and about ten more of the best draftsmen on the Survey. The rest were useless hands. We are to start on the 21st inst. Think not however that we intend to bend passively beneath this stroke. Our case shall be laid before Sir Robert Peel. – It's not improbable that your worthy son will be sent as a deputy to this illustrious man. It behooves me to look about me. I must be stirring. – One thing I want you to do for me – While we were in Leighlin Bridge Mr John Alexander expressed a wish to have his estate surveyed. I know it was not done at that time. It may remain undone to the present moment. Will you ascertain this? I would willingly engage in this job, if it were Mr Alexander's wish to have it done. I venture to say I will furnish him with as respectable a map and as correct contents as any Surveyor in Lienster.[7] If it be his wish I shall furnish him with some small specimens of map drawing in various styles. Send me a reply to this as quickly as possible. Let not my dismissal trouble you or my mother. The situation was a paltry one. My salary might answer a settled person pretty well, but for one who had to pay for furnished lodgings &c it was very poor. It is now a little better than £50 a year.

I'm neither grieved nor cast down by this circumstance. Brighter hours will come; one unwavering conviction bears me up, that I suffer by the hand of arbitrary power, not by the hand of justice, that the proceedings which led to my dismissal[8] had their origin in uprightness of intention, and that they were such as would receive the approbation of every honest man.

Your affec. son | John

RI MS JT/1/TYP/10/3306
LT Transcript Only

---

1. *Wednesday Night*: Louisa Tyndall annotation: 'Nov. 8th, 1843'. From Tyndall's journal, 8 November 1843: 'An eventful day – I have been dismissed from the Ordnance Survey which I joined on the 1st of April 1839. During my connection with it a complaint was never made of me, either in point of morals or ability – my name was never brought before my superintendent but in terms of commendation – all useless however. Capt. Tucker, by an arbitrary exercise of the power committed to him, has dismissed the flower of his division, because they had the hardy honesty to state their grievances' (RI MS JT/2/13a/2).
2. *the thunderer*: probably a reference to the Roman god Jupiter (Jove).
3. *dismissed forever from the legion of Ordnance Surveyors*: see Appendix for Tyndall's chronology of the events leading to his dismissal.
4. *last letter*: letter missing.
5. *our Superintendent*: the Superintendent of the Ordnance Survey of Ireland, Colonel Thomas Frederick Colby.
6. *the order*: see letter 0262.
7. *Lienster*: more commonly spelled Leinster. A large province in eastern Ireland.
8. *the proceedings which led to my dismissal*: i.e., the Memorial to Sir Robert Peel and Tyndall's other protests against the treatment of the surveyors; see letter 0236.

From Jack Tidmarsh 9 November 1843 0265

Liverpool 9th Nov. 1843 | 59 Bedford Street | /Toxteth/ Park

Dear Tyndall

I can scarcely bring myself to write to you, nothing but the present circumstances[1] induces me to do so now, not as much as a note have I received from you during the last three or four months – can't tell the reason, perhaps you will be able to do so when I see you about the 21st at 59 Bedford Street. Well and so you and 'Jim'[2] are off the chalks. Oh the devil mind the pair of you, couldn't have better luck! Ginty and another were the only two persons of note that 'walked' here. You are sure to get some sort of labourers work to do here as to go home it would be madness, what the devil do you want at home – Man, you will see some life here, the moment you get your walking paper bundle down here every one of you and as many as '59' will hold will get shelter in it, oh be the powers what a spree we will have at Christmas. I was thinking of going to Dublin, the family being there, but if you chance to be here I'll no go – pon my soul Jack I'm glad you are discharged I'm anticipating the spree we'll have once more together. It is rumoured here that the Survey is to be broken up altogether on the 21st of Feb. that it may never miss say I. I wonder how do the fellows stand in Manchester I wrote the lodger[3] last night, tomorrow dare say I'll know I would not wonder if Cuddy was off – a great many things can be done Jack in this here Liverpool, for instance a walking theatre[4] might easily be got up making you a comedian, Ginty tragedian, 'Jim' something else, and I'd give up my 12 bobs to go as a doorkeeper. This and many other things too numerous to mention might be tried. However you can pick and choose when you come. Well I believe I have no more to say only tell Jim I congratulate him on his new situation. I long to see you me boys perhaps you would send me a slip of a note now.

Faithfully yours | Tidmarsh | Tell the beggar to write | Hurrah for the 21st![5]

RI MS JT/1/TYP/11/3870
LT Transcript Only

---

1. *the present circumstances*: i.e., Tyndall's dismissal from the Ordnance Survey.
2. *Jim*: probably Philip 'Jim' Evans, who was dismissed from the Ordnance Survey by the same order as Tyndall; see letter 0262 of this volume.
3. *the lodger*: unidentified Manchester surveyor.
4. *a walking theatre*: i.e., a company of street performers.
5. *Hurrah for the 21st!*: the date on which the dismissed surveyors would depart.

From William Ginty 9 November 1843 0266

33 Windsor Street | Nov. 9th 1843.

Dear Tyndall

Zounds what a crash! With one fell swoop they thinned our ranks I may say to a man – Here we are only three – scions of a nobler stock – we stand 'few and faint but fearless still'.[1] So you met 'twas in a crowd, I defy you to make the second line answer eh? Did you think they would have thinned you? I did not. Yes faith I am in a hobble. I hope to get over it though in a short time. This evening I received a letter of introduction to a gent in the Customs[2] here altho' I don't expect a berth in the Customs, he will probably put me on the scent of one. The letter is from Coote Sholley Esq.[3] High Sheriff and Depy. Lieut. of Roscommon Co. I have another from an Esq. and J.P. to a 'Joseph Lloyd Esq.'[4] This cove I can't find, even in a directory. Now I wish to heaven you could get a job in Liverpool. I of course will hunt, chase, search enquire, ask, beg, demand and insist like the very devil. Get some letters of introduction to some gents here. Send for such instanter without such it is no go. There is a Librarian wanted in the Athenaeum News room.[5] Salary £40 per year. £200 security (see 'Liverpool Courier').[6] This appears to be very low but it will only occupy 2 or 3 hours in the day – have you the security if so at it. The Committee at the News Room Church Street[7] is your mark! The advertisement appeared only yesterday. (The Athenaeum) (?spelling) we can't say a word until we see the contents of our roving commissions. Then – on the merits of those Testimonials we will raise a whirlwind, a hurricane that will shake the tottering edifice to the dust. I sent a letter to *[Colby]* (thro' H.)[8] demanding two testimonials of mine in his possession I have a copy of the letter I sent. I said 'They were my property' and that 'under existing circumstances were to me of considerable importance'. If I don't get these, if they are not forthcoming I will give him an emetic[9] of a forty bucket power. With 4 exceptions are the survivors of the wreck in Preston worth a bob a day on an average. Surely to God this will not be tolerated! It is said here that the survey is to be postponed, *[sine]*-die,[10] in February next. It may be so, but I don't believe it. Query, is it not a consolatory rumour to damp our fire – Phsa! it must be broken up, crushed – vanished into air, thin air. Those that had sinecures[11] will rue it most before this day twelvemonths. Is it to be tolerated that the best workmen on the Survey is to be dismissed in company with a few useless civilians and labourers. Are these men to be sent adrift and worthies of the Liverpool Borough School kept to repeat their outrageous botchery again. This, in plain terms, is a fraud upon the public and upon us, and in the end an increase of the expenses of the Ordnance Survey and not decrease – Tis scandalous to look at – tis madness to think they can stand by this. And come what will we will shove it with the highest. We have not yet heard from Manchester. If I have not a berth in one week after I'm dismissed I'll enter the workhouse – Philosophy – Now for the test – some say here that the thing is so preposterous it will not be carried into effect and that it is merely to tame us – This is all flam.[12] Our occupation is no more fixed to all intents and purposes. Hurrah for Spectator[13] tomorrow morning it is sure to appear.

Yours | Ginty

If I can fork out two corners we will be all right and at hand for future movements.

2nd Edition

I can't find out what month that number of the journal was for, I suppose it was September. You will be able to judge from the date of Spec's[14] first letter and Tucker's speech. Taylor has written an appeal against his dismissal to be forwarded by the Revd. Mr Carpenter[15] to the Col.[16] Carpenter is a brother-in-law of Mr and Mrs T[17] and the Cols. 'My dismissal Sir' says he 'under these circumstances is to say the least of it very unfair of Captain Tucker' – flam. Sure it is old Tom in reality that is the mover or at least the seconder. Taylor does not wish it to be spoken of – take the hint. Is *[Jim's]*[18] shipped? Send me a list of the 19 draftsmen in your next. Lyndsey and Malone[19] are in the mess too I hear. I wish we were each in a situation we could then with the stoicism of an angler who plays the salmon to the death – level the fury of our fire into their teeth and beard them in their dens at leisure. As for the Ordnance Survey 'its days are numbered'. But in faith we are not 'sleeping now'. Lash, Spec, as usual – leave this occurrence for the last letter, 'walk into' the present system in your next – let it be short, otherwise it won't appear as a large Free-trade meeting[20] is to be here tomorrow evening, and some M.P's. This will occupy at least one side of Friday week's paper. Towards the close of Taylor's epistle he requests if his dismissal is to take place that it will be deferred until the 21st of Dec. He has very fine interest here.

Yours | Othello[21]

See what it is to have a loose leg. Our hats cover all our worldly effects. Geo[22] must go preach. Jim sell matches. Tyndall paste bills and Ginty sing ballads.

I suppose Cuddy and Dingan[23] and Holmes[24] are martyrs too 'the great superseded'.[25]

W.G.

RI MS JT/1/TYP/11/3621
LT Transcript Only

---

1. *'few and faint but fearless still'*: T. Moore, 'After the Battle. Air – Thy Fair Bosom', i.3–4. On Moore, see Biographical Register.
2. *a gent in the Customs*: not identified.
3. *Coote Sholley Esq.*: not identified.
4. *Joseph Lloyd Esq.*: not identified.
5. *the Athenaeum News room*: the Liverpool Athenaeum, a gentlemen's club founded in 1797. The club provided a reading room of newspapers and magazines for its members.
6. *Liverpool Courier*: a Liverpool newspaper established in 1808.
7. *Church Street*: the nineteenth-century address of the Liverpool Athenaeum.
8. *thro' H.*: possibly William George Hamley.
9. *an emetic*: a substance that causes vomiting.
10. *[sine]-die*: without a specified day (*OED*).
11. *sinecures*: positions requiring little or no work but receiving compensation (*OED*).
12. *flam*: a fanciful notion (*OED*).
13. *Hurrah for Spectator*: Spectator was Tyndall's *Liverpool Mercury* pseudonym.
14. *Spec's*: Spectator; see n. 13.
15. *The Rev. Mr Carpenter*: not identified.
16. *the Col*: probably Thomas Frederick Colby.
17. *Mr and Mrs T*: Louisa Tyndall annotation: 'Taylor?'. Possibly Alexander Taylor and his wife.
18. *Jim*: Phillip 'Jim' Evans.
19. *Lyndsey and Malone*: James S. Lindsay and Dennis Malone, Civil Assistants dismissed by Henry Tucker's orders on 7 November 1843; see letter 0262 of this volume.
20. *Free-trade meeting*: a movement to abolish restrictions on commerce, often including protectionist tariffs.

21. *Othello*: a nickname for William Ginty.
22. *Geo*: possibly George Latimer.
23. *Dingan*: not identified.
24. *Holmes*: George Holmes.
25. *'the great superseded'*: not identified.

To John Tyndall, Snr. [10] November 1843 0267

Preston, Friday night

My dear Father

I had no time last night[1] to give you an idea of our present position. It is a fact universally admitted on the Survey that the aim of those into whose hands it has fallen is not to raise their men to a position equivalent to their merits, but to induce on their part a spirit of the most slavish submission to the mandates of their taskmasters – to crush at its first appearance the bud of honest independence – and thus enable themselves to lord it without control over their prostrate vassals. You are by no means to judge of the character of the engineer officers on the English Survey by the specimen afforded you in Mr Wynne. He was a gentleman, and could appreciate the feelings which have an existence in the breast of a man of sensibility. He would not think of stooping to acts of petty tyranny which are of every day occurrence here. With one exception, the Superintendents of the English Survey would grace the character of slave drivers. That exception is our Division Officer[2] – a son of the Bishop of Norwich.[3] But he is young, and has no power. If he had, things would change. I cannot think that all the other officers are inherently bad. It's a kind of reflected tyranny which they exercise, the source of which is the person by whom we were dismissed, namely Captain Tucker. He is the senior officer on the Survey and has the control of everything connected with it. From speaking to this man you would imagine that he was of the most lamblike disposition, but this insinuating exterior shades a heart impervious to the beams of charity or the appeals of misfortune. You have by this time read our Memorial.[4] That document was not the offspring of a momentary impulse. It contained a statement of grievances and a record of feelings which had an existence long prior to the circumstances which called them openly forth.

I told you we had written to the Master General.[5] It would be most absurd to submit silently to the stroke inflicted by Captain Tucker. Our cause is a just one, and while ever there remains a means of asserting its justice we shall do so. In the order which contained our dismissal[6] it is said that notice was given last June. This is an attempt to stop our mouths on this score. Captain T. in justification of his hasty proceeding will no doubt allege that we had been warned last June – a copy of that warning lies beside me, I may as well give it to you:–

'Preston, 7th, June, 1843
'The Division Officers will warn the C. Assistants[7] that a further reduction will be made and will forward names of 25 of the least useful or efficient men who are to be discharged'.

This warning refers to a certain class of individuals – to the inefficient men. Can it be said that we got notice – we, who have ever ranked among the best men on the Survey? Are men who have spent from 5 to 16 years on this employment, with unblemished characters, to be thus treated, merely because they had sufficient honesty to state what they felt in an open and respectful manner? In our letter to the Master General this was stated. We went into a detail of our proceedings from the first – showed how the Memorial originated, forwarded specimens to prove that we could not be ranked as inefficient men, and stated that if further proof were required we could lay before him testimonials of the strongest nature both as to ability and integrity. We begged for an investigation of our case prior to the 21st., feeling assured that it would result in a manner creditable to us and satisfactory to him. I can form no idea of the reception which the letter will receive. I fear he is prejudiced against us, and to remove whatever prejudices may exist, it is deemed necessary to send a delegate to London. I have been pitched upon, and to London I will go and carry the matter to its utmost limit – I'll leave no means untried – I'll endeavour as far as in me lies to find whether Captain Tucker can exercise such an irresponsible sway. Emma[8] knows Mr Martin, we drank tea at his place one evening. He has been connected with the Survey from its commencement. 16 years of his life have been devoted to this employment and he has proved himself trustworthy and efficient in all the duties of his situation. He is now discharged and deprived of the means of support for a wife and helpless family. Think not however that in pleading our cause before the Master General i.e. Sir R. Peel[9] that I will be actuated by personal animosity to Captain Tucker. I know such a proceeding would be a deathblow to our interests. I go there to enquire into the justice of the principle on which Capt T. has acted. I go without spleen or acrimony – fully persuaded of the justice of my cause, and supported by the hope that I will be able to make that justice apparent. I intend to apply to some of the London Engineers for employment. I will have a very good opportunity of doing so. Has not Mr John Alexander a brother bound to an Engineer in London? I wish you would get a letter of introduction for me from Mr John. It would be of great service, for though the Engineer to whom his brother is bound might not have employment for me himself, still there is a probability of his being able to direct me where I would find employment. Apply to Mr John without delay. Every moment is precious, as waiting for your reply is the only thing which keeps me here. I hope the Col[10] will give you, or has given you the testimonial.

Ginty has been discharged. Emma knows Mr Latimer,[11] he goes too – all first rate men. I shall require the printed form of recommendation along with your reply – don't neglect this.

Your affectionate son | John.

RI MS JT/1/TYP/10/3304–3305
LT Transcript Only

1. *last night*: possibly a reference to letter 0264 of this volume.
2. *our Division Officer*: Charles Edward Stanley.
3. *the Bishop of Norwich*: Bishop Edward Stanley FRS (1779–1849), Bishop of Norwich from 1837–49 and father of Charles Edward Stanley.
4. *our Memorial*: the letter of protest sent by the workers of the Ordnance Survey of England to George Murray, Master General of the Ordnance on 23 September 1843; see letter 0236 of this volume.
5. *I told you we had written to the Master General*: see letters 0245 and 0246 of this volume.
6. *order which contained our dismissal*: see letter 0262 of this volume for a copy of this order.
7. *C. Assistants*: Civil Assistants.
8. *Emma*: Emma Tyndall.
9. *the Master General i.e. Sir R. Peel*: the Master General of the Ordnance Survey was George Murray, but Tyndall hoped that the letter would eventually reach the Prime Minister, Sir Robert Peel.
10. *the Col*: probably Henry Bruen (1789–1852), the Colonel of the Carlow Regiment of Militia. A wealthy Protestant, Tory landowner and man of influence, he represented Carlow County in Parliament from 1812 until 1831 and then stood again in 1835, but the result was contested. He returned to Parliament later that year after a contested by-election but was defeated in the General Election of 1837. The letters in this volume relate to the By-Election he won in December 1840, following the death of the standing Liberal MP, and his re-election at the General Election in July 1841.
11. *Mr Latimer*: probably George Latimer.

From John C. Chadwick 14 November 1843 0268

Kinsale 92nd 1800 and boiled milk[1] | Nov. 14th 1843

Dear Tyndall

Curse of Molly Moran[2] and three score of blind bitches' bastards on my dilatory pen and more dilatory intention which have left your undeservedly kind letter so long unanswered, truth to tell you I have been paving pandemonium with the best of intentions during the silent interim of six months after the receipt of your Ossian.[3]

I require a little time to recover from the terrible effect produced on my nervous system by that magnificent effort of genius and indeed I felt that there was such a melancholy deficiency in my poor paltry style of dictation that I would be quite unable to answer such a subject in the way I should until my intellect gathered strength and power sufficient to do myself credit and justice before your critical eyes, but woe is me! I waited so long that the tide of intellect flowed and ebbed and now you must be satisfied with a plain matter of fact scrawl as I have nought else to offer – it is a lamentable fact (as the district man[4] said) that our poor earthly minds should be so dreadfully inclined to fall into the gulf of forgetfulness where they become so obfuscated that the consequences are a neglect of friend and an utter disregard of all the common decencies of correspondence. Such has been my case and I regard your last letter which came like the whisper of conscience making itself heard in the innermost parts, as a merciful interposition of providence to prevent an utter extinction of our old well cherished chumship; for my silence, which began in putting off, ended in shame and had not your kind unreproachful letter arrived you would never have I suppose heard from me until I appeared before you in person and challenged you to breakfast with me in Preston when I am sure your kind gentle heart would easily be prevailed on to extend a pardon – but enough of this. I sincerely repent of my errors towards you and as I have been unfortunate enough to err so often if you think that I have not offered a sufficient excuse look to some of my former letters if happily you have not sent them to blazes and you will find that they all begin in the same strain as this and then cull out of them a few of the best concocted excuses they contain and it will fully answer, so amen. You ask me do I care for you. I deserve the reproach but still my answer is I care for you and that with a warmth of feeling which though my actions do not evince it yet is cherished and felt in my heart with intensity and should the circumstances of life ever again throw us together I will hail the event with joy and gladness – my soul is like a flint which requires often to be struck twice before it gives a spark and though the principle of combustion is within it must be beaten out and your letters flagellate me so well that the effect is at length produced.

Your talk of the mess account brought back old recollections very forcibly, many a time has it been my care to render unto Caesar the things that were his[5] otherwise to clear up the butter and bacon account for the week[6] which generally left me but a sixpence to chase about in my pocket until I nearly wore *[the brogues]* off my fingers but those times have left but a mere shadow and are gone I often regret them and shall always think of my survey life with fond recollections – The Survey has as you say become an arena where the higher and lower powers meet in conflict. I should most particularly like to see Spectator's letters, bye the bye I know well who he is but do not wish to put aught on paper about him for fear of accident. All I say is more power to him and I hope all the helots[7] of the Survey may join with him in giving a long strong pull until they pull the whole rotten fabrick about the ears of the red dressed oppressors who have lorded it over them so long, damn them, they richly deserve it and I sincerely hope that the work may fall into the hands of the Civil Engineers when it would be much more creditably carried on. All that frets me is that I am not among you, joining heart and hand in the glorious opposition. – Do not forget to send me the letters – With respect to your observations about America, my intentions are made up to go there. My plan is if nothing interferes to prevent me that I will go out in the Spring of 1845. John Abbot and family will go next Spring and after he has been there a year, during which time he will try and find a situation, he will acquire a knowledge of the ways of the people and the details of business there so that on my following him out we could enter into business together and I can bring out a suitable stock from this country, Cincinnati is our ground. I may hope to meet you there and it would indeed be pleasant if we were again located together – it is indeed the only country in which a person has any chance of success. I may drag on a life of exertion here but all I will have for it is a mere support nothing beyond it still toiling for the food which perisheth. So I hope we may touch palms in Ohio. John Fudge[8] is getting on very well, he has now £80 a year, and he merely two years in a business of which he knew nothing before. Abbot is fully determined on going next Spring.

I should like to hear more about the Prometheus[9] that has warmed Latimer's[10] marble into life your proofs of his passion are unrefutable as Sinnock's donkey[11] still yells his guts out after the she asses.[12] Lavender Lane[13] still stinks and cods congregate to the World's End[14] with their usual pertinacity, let me hear more anent it,[15] and as I am experienced in these matters, I will give a better judgment of the affair. I hope Latimer's case is not so desperate and that he will not have to migrate to the green hills of America with a double alias as wife beware beware let not the atmosphere of love so entirely surround you that you will have no outlet of escape but like the man going into battle be always looking behind you and be sure how you stand lest you fall. Love, like a whirlpool whose circle diverge to a centre, will insensibly draw you in until you come to the apex of fornication or matrimony – My lady[16] is quite well and desires to be remembered to you – Lizy[17] is thriving like a young kid on a mountain and my last edition revised and corrected which has appeared in duodecimo.[18] I have called Sarah Kate[19] also bids fair to be stout slip from the old stock. Miss Donoclift[20] and all old friends are well. The holy stone sent up to me for the loan of an old coat as he says he cannot stand the frost as well as he did in his young days, I referred him to the poor house, poor old devil he has a bad attack of the stone also the *[whoxes]* set on him so much that it cannot serve him, but I think bad as he is he will live to see the last sun of prosperity set on Kinsale forever. You Survey chaps are likely to give me a busy time with your McCarthy[21] business. If I was to go there as often as I am desired she would not have as much time as she could spare to go to the poe.[22] I really do not think you should even in thought be poaching on Ginty's manor.[23] I spoke to the old Mrs[24] about you and she was so affected by your kind recollection that she wet a table cloth drying her eyes I was thinking of getting her made into a town pump but as she is not temperate I fear she would produce nothing but grog so gave up the notion but be satisfied with your English venuses and do not be hankering after Ginty's one little lamb – I suppose you heard of Fenwick's death[25] in Bermuda – peace to his tallow[26] he was the best among them in point of ability and I believe in other points but death's point has laid him in Bermuda's dust there let him rest – I fully expect to visit Liverpool in the Spring and of course Preston I hope I will not find the place vacant but that Butler Street will retain you among its bricks until I see you. I am longing to take a scamp[27] over to you and shall only wait for the fine weather to do so I have no news as this wretched town is not able to produce any or like the mountain in labour all it brings forth is mice. Kinsale Streets at present are a type of the path to Paradise being narrow and rough they have been covered with broken stones over the pavement. In all other respects matters continue in the old way and I suppose if you were to come here in ten years time you would find all the same except a few more houses untenanted. We have a party of strolling players here at present I saw Richard 3rd[28] for 2d which was a rich sight as you may suppose. When we meet we will be like the town mouse and the country mouse in the fable[29] telling our experience in our separate stations – have you got much better pay now than when in Ireland? Give my love to Latimer and all friends and let me hear from you as soon as you like. I suppose you are glad that they are going to ding the soul out of Dan O'Connel[30] I wish they would keelhaul[31] him well – do you ever take a trip to Liverpool they have pleasant times there. So now my dear old crathur[32] wishing you every health and happiness and a good appetite for your Christmas dinner apropos do you remember our day in Dunderrow[33] – May blessings fall on you like rain and every hair of your head be a mould candle to light you to glory is the sincere prayer of your steadfast friend and well wisher

J.C. Chadwick

It is no joke to write a big devil of a letter like this besides wearing ½d worth of pens but I am ready to make any sacrifice to atone for my negligence all I hope is that you will not split your eye balls trying to make it out. Have you any theatre in Preston do you go much into company or are they a party giving people have you worn out much pumps on your light fantastic toes? I would have answered your last sooner but was slightly ill which obliged me to delay but am now in rough health. My boat is on the shore for the winter so that there is an end of that amusement and all I have to do now is to sit by the fire and warm my shins and nurse the youngsters passing my time quite quietly having given up raking and blackguarding. Au revoir.

RI MS JT/1/TYP/11/3503
LT Transcript Only

---

1. *Kinsale 92nd 1800 and boiled milk*: not identified.
2. *Curse of Molly Moran*: this may refer to the curse of Molly Malone, a homeless widow who was blamed for the dismissal of various Irish employees. W. Carleton, *Rody the Rover; or, the Ribbonman* (Dublin: James Duffy, 1845), pp. 173–4.
3. *your Ossian*: i.e., an epic letter. An allusion to James Macpherson's anonymous forged poem, *Ossian* (1760), a cycle of epic poems. The letter from Tyndall that Chadwick refers to is missing.
4. *the district man*: not identified.
5. *render unto Caesar the things that were his*: Matthew 22:21.
6. *the butter and bacon account for the week*: i.e., paid for his necessities.
7. *helots*: serfs, workers whose class status is between that of a slave and an ordinary citizen (*OED*).
8. *John Fudge*: an acquaintance of Tyndall and John Chadwick who lived in Cincinnati. See

also letters 0306 and 0314 of this volume.
9. *Prometheus*: a figure from Greek mythology who is known for creating man from clay and for giving mankind the gift of fire (*OED*).
10. *Latimer's*: either George or William Latimer.
11. *Sinnock's donkey*: not identified.
12. *the she asses*: female donkeys.
13. *Lavender Lane*: not identified.
14. *the World's End*: not identified, possibly a local public house.
15. *anent it*: in line with, in company with (*OED*).
16. *My lady*: Mary Chadwick, Chadwick's wife.
17. *Lizy*: Elizabeth Chadwick, Chadwick's elder daughter.
18. *duodecimo*: a book in which each page is the size of one twelfth of a sheet (*OED*).
19. *Sarah Kate*: Chadwick's younger daughter.
20. *Miss Donoclift*: probably Sophy Donaclift, daughter of Samuel Donaclift, a shipwright who worked on Kinsale Harbour, and his wife Sophia (née Leggatt). In the summer of 1844 Miss Donoclift married 'Barry the tutor'; see letter 0307 of this volume.
21. *your McCarthy business*: see letter 0259 of this volume; Chadwick is referring to a Miss McCarthy with whom Ginty was enamoured.
22. *go to the poe*: not identified.
23. *poaching on Ginty's manor*: Chadwick seems to be suggesting that Tyndall has expressed interest in Miss McCarthy, Ginty's love interest; see letter 0259 of this volume.
24. *the old Mrs*: not identified.
25. *Fenwick's death*: Robert Fenwick (1801–43), a Royal Engineer who enlisted early in 1825. He progressed to 1st Lieutenant in May 1829, 2nd Lieutenant in January 1826 and 2nd Captain on 14 September 1841. Fenwick had considerable experience working on the Irish Survey, including being in charge of the 3rd and 4th Divisions, C District, before taking command of the 5th Division at Kinsale after Whittingham's departure in July 1841. Later that year he was based at Grassington, Yorkshire. After leaving the survey, he commanded the 4th Company of Royal Engineers in Bermuda, where he died of yellow fever in August 1843.
26. *tallow*: fat (*OED*).
27. *scamp*: i.e. a trip, a scamper (*OED*).
28. *Richard 3rd*: W. Shakespeare, *Richard III*.
29. *town mouse and the country mouse in the fable*: 'The Town Mouse and the Country Mouse' is one of Aesop's fables.
30. *they are going to ding the soul out of Dan O'Connel*: Daniel O'Connell. In 1843 O'Connell was arrested, charged with conspiracy and sentenced to imprisonment and a fine.
31. *keelhaul*: to haul a person under the keel of a ship, a form of severe punishment (*OED*).
32. *crathur*: Irish slang for 'creature'.
33. *Dunderrow*: a small village in Cork County, Ireland.

From William Wright 14 November 1843 0269

Cincinnati | November 14th 1843.[1]

My dear John

I once more resume my pen to address you after a delay of some weeks, which I did not intend when last I wrote to you[2] since then trade has been rather brisk and I have been constantly engaged. I trust you have received my letter which I sent you immediately after I received your welcome letter.[3] I hope you have made up your mind to start in Spring. I repeat my advice and invitation for you to come out. Every day convinces me that you would be well suited for this country and that you need have no doubts but that you would easily obtain a situation that would return you a handsome recompense, I would not have you set out with the idea that you would be able to step into a situation they day you would arrive. Weeks or probably months might elapse before you would obtain one. But that should not discourage or prevent you from coming out as I offer you a hearty welcome and an earnest invitation to remain with me and partake of my fare as long as you may be unemployed. I do not believe it would be long. But then I would not wish to have you disappointed when you would arrive. I would also wish you to recollect your stopping with me would be no expense to me or should be none to you. As I have a family to board an additional member would not increase the expense six cents a week, provisions of all kinds are so cheap. And once more and again I would advise you to come out to this land of promise and freedom. But once you get under way the favouring breezes of fortune will swell your sails, the current of enterprise will bear you onward, and with perseverance and industry at the helm you would soon arrive at the Harbour of Independence. All the difficulties lie in coming up to the starting point; that once gained I might say all the difficulties are past. I have advised my folks to come out and they are satisfied to do it. I only wrote for Mary Ann[4] as yet, but mother[5] thinks she would be able to bear up against the dangers and fatigue of a sea voyage herself and she would like to come out I have a reply to a letter of Mary Ann's advising Thomas[6] and all to come. I do not know whether it will be practicable this season as it is I that will have to raise the principal part of the funds that will defray their expenses. However if I can accomplish it they shall come here in Spring I would be much obliged to you if you would let them know that you had a letter from me and that we are all in good health and in fair prospects of doing well. I now commence to address you upon a subject which here as well as at home creates a great deal of excitement I mean a repeal of the Legislative Union between Great Britain and Ireland I have no doubt it will surprise you to learn that I am a strong advocate for a repeal of the iniquitous union. The last act of tyranny committed by the English Ministry[7] has added many to ranks of the repeal. An act as unlawful as it is unprecedented. If it was unlawful for the people to assemble and petition for a redress of their grievances why not have put them down 12 months ago it would seem that it was just as unlawful then as now, one would suppose that they expected it would have worn itself out ere this or that it only existed in the brain of some wild theorist. But No it will never subside until its aim is accomplished and old Ireland is what God and Nature intended she should be:

> Great glorious and fair
> First flower of the Earth
> And first gem of the Sea.[8]

Like the mighty river whose main fountainhead has attracted little notice, but whose course is onward aided by its numerous tributaries, swells into the majestic river throwing aside every obstacle and surmounting every difficulty until it mingles with its mother Ocean. So has been the career of the repeal movement, and now that its object is near accomplished do the English Ministry really expect that a mere Bull Rush[9] – their opposition – impede its progress. As well might they expect that solitary Snagg would turn aside the Father of Waters.[10] True it may, it may create a ripple on the surface, but it only helps (as in the present instance) to increase its velocity. It is not possible that the Ministry will be able to sustain their grounds or that the people of the 19th century are to be told that after centuries of wrong and oppression they are not to assemble and petition for their redress. Every man who feels or thinks that he has any right worth preserving should at once stamp it with his indignation and contempt. If such was attempted here every man woman and child in the land would rise up against it in arms. But I forgot that Irishmen are adopting a new and untried experiment (and I trust in God a successful one). Peaceful means and striving to convince their oppressors of their cruelties. But I fear little is to be expected from the justice of England towards my down-trodden country. Oh Ireland, all the woe and wrong that has been heaped upon you for the last 7 centuries would have annihilated any other people under heaven. But no, the Irishman's heart is ever with his country in weel or woe. See them here often a residence of 30 or 40 years and hear them talk of old Ireland in the most affectionate and endearing terms, in every instance calling it by the endearing name of Home. Now along with the news of O'Connell's arrest[11] arrived a reply of his to an address from the Cincinnati repeal association[12] on the subject of slavery. A meeting was immediately called to hear the reply of O'Connell. To understand it rightly you may recollect that some time ago Mr O'Connell received an address from a Pennsylvanian gentleman[13] who stated some facts upon what Mr O'Connell predicated some remarks of American Nation (not very courteous) and called upon his country men to come out of her. Some of the association addressed a communication to Mr O'Connell excusing the institution of slavery. The address was principally drawn up and written by a native American,[14] a man generally opposed to the abolitionism and consequently contained a great deal that would have been better left out. Mr. O'Connell's reply was more severe than his former remarks and it would give a good deal of offence to those opposed to abolition. However it does not contain one word that can be contradicted. It is one of the best arguments that can be made in favour of abolition. The American part of the Association endeavoured not to receive the reply, but it was received with approbation.[15] The finale of the meeting was 600 Dolls. was collected to aid the furtherance of repeal. I do not expect to see you a repealer[16] though I ardently wish it. But I ask of you only to look back at the course England has pursued towards her unfortunate sister it is but one long tale of suffering on the one part and oppression and tyranny on the other. Not only over their bodies and property but even the mind must be in subjection to their standard or else forfeit all right to its most important privileges. Look at it with an impartial eye, and I have no doubt I will see you a repealer yet. In conclusion I would say repealer or no repealer may heaven shower its choicest blessings on the head of you and yours, remember me to all of them and I hope they sometimes think of the wanderer.

Yours affectionately | William Wright

Direct your letters to me to 5th Street between Vine and Dace Street Cin. | and do not be long about it if you please.

RI MS JT/1/TYP/5/1813–1815
LT Transcript Only

---

1. *November 14th 1843*: Louisa Tyndall annotation: 'Forwarded from Preston to Leighlin Bridge'.
2. *when last I wrote to you*: see letter 0237 of this volume.
3. *your welcome letter*: letter missing.
4. *Mary Ann*: Mary Ann Wright, William's sister. She lived at No. 12 Nichols Street in Dublin; see letter 0237 of this volume.
5. *my mother*: William Wright's mother, name unknown.
6. *Thomas*: Thomas Wright, William's brother; see letter 0237 of this volume.
7. *The last act of tyrrany committed by the English Ministry*: probably a reference to the arrest and trial of Daniel O'Connell.
8. *'Great glorious and fair … gem of the Sea'*: T. Moore, 'Remember Thee', lines 5–6. On Moore, see Biographical Register.
9. *a mere Bull Rush*: a bull rush (or bulrush) is a reed that sways easily in the wind.
10. *that solitary Snagg would turn aside the Father of Waters*: not identified. 'Father of Waters' may be a reference to the Mississippi River.
11. *O'Connell's arrest*: Catholic emancipation leader Daniel O'Connell was arrested in October 1843 and charged with seditious conspiracy. The trial was running at this time.
12. *the Cincinnati repeal association*: more formally called the Irish Repeal Association of Cincinnati, an organization which supported the movement to repeal the 1801 Act of Union and restore the independent Kingdom of Ireland.
13. *a Pennsylvania gentleman*: not identified.
14. *a native American*: not identified.
15. *To understand it rightly … received with approbation*: for more on this exchange between O'Connell and the Irish Repeal Association of Cincinnati, see *Daniel O'Connell and the Committee of the Irish Repeal Association of Cincinnati* (Cincinnati, Ohio: Catholic Telegraph Office, 1863).
16. *repealer*: a supporter of the Repeal Association, which campaigned for the repeal of the Act of Union between Ireland and Britain.

From William Ginty 16 November 1843 0270

Thursday evening | Nov. 16th 1843.

Mon Cher Jack

Your ballad letter[1] lies 'overright' me. You wonder I have not got a situation yet – so do I, it's not my fault old boy. I am busy canvassing for one. I got a letter yesterday from an office I wrote to, offering a bonus if they got me an engagement for a situation of £2 per week £10 – 35s £7 – 30s £5 – 25s £3 – 20s 2–10 18s–2–0 15s £10. The letter was a writ[2] for my body I went and lo the sum and substance of the offer was this to look for properties for them to sell and if they, or when they, sold them I would get 2½ percent on the whole amount. On this I argued thus – It would certainly take all my time, without any certainty of even gaining a living. He said I would realize £100 a year by it – if very active. Now I don't know but the fellows have 3 dozen fellows going it for them – Looking for a needle in a bundle of straw, and so I declined his kind offer – being convinced he was codding me.[3] I wrote to another of these 'Intelligence offices' last night, a letter met my view at dinnertime today, a writ for my body and saying he had one of £52 per annum on his books. I bolted my bacon (I'm wrong it was ham) and cabbage and off I bolted to the office – it was closed, and 20 minutes since the proprietor visited this our domicile telling me to be at their office by nine in the morning 'as they had a very good situation in view'. It can't be the £52 one, else he would not say 'very good' I am topping the bonus or bribe one month after the date of my employment! I dreamt this mode of obtaining a situation, it has succeeded very well so far. I would recommend you to adopt it in London. Find out an 'Employment office' and bribe them, that is offer them a remuneration for their trouble and services. I hope I'll succeed. I don't wish to make the stratagem generally known, as I was the inventor I have a right to the preference, that is I want no opposition!

The omnibuses[4] are cheap enough! Cabs[5] are for the nobs[6] and are consequently dear. The omnibuses run regular stages, in thousands, learn the nearest place of note to where you want to be 'set down' you will hear these fellows bellow such names as 'Bank' 'Charing Cross' 'High Holborn'[7] (this is pronounced without the L) &c. &c.

I will see 8[8] as soon as possible and do your wishes and apprize him of your visit. Do you mean the 'Weekly Chronicle'?[9] Damn you and your belly, that's all troubles you. Glory to Lindsay and McGowan[10] and all the rest of you. Will you come this way?

Yours in | great haste | W. Ginty

Hurrah for O'Connell[11] – he has opened their eyes on the indictment, it is all but quashed and the Attorney General[12] is nearly crazy.

RI MS JT/1/TYP/11/3622
LT Transcript Only

---

1. *Your ballad letter*: letter missing
2. *letter was a writ*: Louisa Tyndall annotation: '?'.
3. *being convinced he was codding me*: i.e., fooling or cheating him (*OED*).
4. *The omnibuses*: road vehicles designed to carry many passengers; they ran regular routes around London picking up passengers at designated stops (*OED*).
5. *Cabs*: small Hansom cabs, designed to carry just a few passengers.

6. *the nobs*: slang for a person of wealth or high social status (*OED*).
7. *'Bank' 'Charing Cross' 'High Holborn'*: locations of omnibus stops.
8. *8*: see letter 0232, n. 13 of this volume.
9. *the 'Weekly Chronicle'*: a London newspaper in existence from 1836–67, owned by Henry George Ward M.P.
10. *Lindsay and McGowan*: James Lindsay and John McGowan, Civil Assistants dismissed by Henry Tucker's orders on 7 November 1843; see letter 0262 of this volume.
11. *Hurrah for O'Connell*: Daniel O'Connell's conviction was being challenged in Parliament; it would be revoked by the House of Lords in 1844.
12. *the Attorney General*: a nickname for Ginty.

To John Tyndall, Snr. 17 November 1843 0271

11 Butler Street, Preston | Nov. 17th, 1843.

My dear Father

This Morning's post brought me your letter together with the Dean's[1] valuable testimonial. So you have been unsuccessful with the Colonel,[2] disapproval knocks on the head my contemplated interview with Sir Robert.[3] The reception which the Memorial[4] met annoys me not a little, the Colonel is sorry that so much talent should be misapplied. How has it been misapplied? Is it in the object sought or the means used towards its attainment? Was the former unworthy or the latter illegitimate? The statements contained in the Memorial are undeniable truths, the feelings recorded there are not of recent birth, they had an existence long prior to the order which called them openly forth. So the Colonel deems me the tool of some designing villain – some priest or papist in disguise – I feel assured that the Memorial per se has no connexion with this or that creed, but he may think that the object of my prompter was to injure me, to blast my reputation on account of my religion. On this point I can say that I had no prompter and therefore could not be a tool. My way was clear before me, my conduct was the result of an unwavering conviction induced by circumstances unknown to you – a conviction common to many honest, protestant hearts, and which remains to the present moment unimpaired. You tell me that my friends regret that I had anything to do in this matter. There are gentlemen in the County Carlow whose exertions to serve me shall ever be remembered with gratitude, and I am grieved to find that my conduct has not met with their approval. I feel assured however that were they acquainted with the circumstances which led to our proceedings my behaviour would furnish nothing calculated to injure me in their estimation. You write as if we were in open rebellion against our Superintendents.[5] This is not the case. We have never relaxed our assiduity at our various duties; we have never set authority at defiance; we have always evinced the greatest respect and submission for those above us; our proceedings were carried on in our own time and in a quiet, orderly manner. Capt Tucker himself expressed his approbation of the straightforward way in which we acted. You may consider this inconsistent with his late conduct. A moment will explain it. He wishes the Ordnance Surveyor to be a mere machine, the murmur of discontent strikes harshly upon his ear, and he has struck us down lest the feeling we have evinced should become general and limit his power. The Survey however is diseased, and Capt Tucker's present act is merely an application for a local symptom. It may arrest the progress of the malady, but as it does not remove the primary cause it will eventually prove ineffectual. Discontent will dwell on the Survey until Capt Tucker changes his system, or makes mere animal instinct supersede the rational faculties of those employed on it. You advise the men to submit, submission is vain. Their doom is sealed. However, the principle which guided them is too honest to be forsaken now. If similarly circumstanced they would act the same part over again. Many of them have already obtained situations, the emoluments[6] of which will be double that of the situation they are about to leave. You tell me 'to submit and return to my work if my crime be at all pardonable'[7] – my crime! – have I then been guilty of a crime? And a crime the pardon of which is doubtful? Yes, you stung me there – do you believe that my conduct has been criminal? Nothing but the testimony of a good conscience could cheer me under this accusation. I have that testimony, and sooner than barter it to win the smile of a worthless man I would resign the drawing pen for the spade, and live on potatoes and salt. The result of what I have done may afford grounds for saying that my talent has been misapplied, but from no other premises can this conclusion be drawn. The motives which actuated me were worthy. The cause which I espoused was worthy and the course which I adopted was open, manly and respectful. We had no notion whatever to bring the matter before the House of Commons; to the Master General[8] our first and last appeal was made.

I start for London on Monday or Tuesday. Not to see the Premier[9] but to seek employment. The fare from this is two pounds. I will be enabled to go by means of a fund which we established among us six months ago. Were it not for this I'm not master of five shillings after spending five years on the Survey.

A letter of introduction from Mr Alexander to the Engineer to whom his brother is bound[10] would be of the greatest service or even his address. Try and get this. Direct it here as usual. It will be forwarded to me. Despatch is necessary.

This day I saw an advertisement in the Mercury[11] stating that a Clerk was wanted to assist in receiving money and keeping books at the Treasury office Ph of Liverpool[12] – Salary £120 per annum. Testimonials as to the character will be required and security by one or sureties in the sum of £500. I have just posted a letter making application for the Clerkship. When I was in Kinsale[13] you sent me word that the Dean would be my surety for this sum. Since that time my integrity has not been diminished. I would fain hope the same of the Dean's kindness. Will he be my surety now if my services be accepted?

I would defer going to London until the result of my application would be known, but if I remain the fund will be divided so I wish to take advantage of this. I'll receive your reply in London – direct to me here.

RI MS JT/1/TYP/10/3307–3308
LT Transcript Only

---

1. *the Dean's valuable testimonial*: 'the Dean' was Richard Boyle Bernard. The Dean's testimonial arrived later that day. From Tyndall's journal, 17 November 1843: 'Received letter containing the Dean's testimonial' (RI MS JT/2/13a/3). Testimonial missing.
2. *the Colonel*: probably Henry Bruen (1789–1852); see letter 0267, n. 10 of this volume.
3. *Sir Robert*: Robert Peel.
4. *the Memorial*: the letter of protest sent by the workers of the Ordnance Survey of England to George Murray, Master General of the Ordnance on 23 September 1843; see letter 0236 of this volume.
5. *our Superintendents*: the Superintendent of the Ordnance Survey of Ireland was Colonel Thomas Frederick Colby. Tyndall is probably also referring to Henry Tucker.
6. *the emoluments*: the salaries, benefits (*OED*).
7. *'to submit and return to my work if my crime be at all pardonable'*: letter containing this quotation is missing.
8. *the Master General*: George Murray.
9. *the Premier*: i.e., Robert Peel.
10. *Mr Alexander to the Engineer to whom his brother is bound*: see also letter 0267 of this volume.
11. *the Mercury*: the *Liverpool Mercury*.
12. *the Treasury office Ph of Liverpool*: possibly an abbreviation for the Treasury office in the parish of Liverpool.
13. *Kinsale*: a city in County Cork, Ireland.

To John Tyndall, Snr. 18 November 1843 0272

11 Butler Street, Preston | Nov. 18th, 1843.

My dear Father

Your letter[1] of this day helps to account for the reception which our Memorial[2] has met among my Carlow friends. Their regret is founded on a misapprehension of my conduct, and you write under the same misapprehension. I dare say the partial view I gave you of our proceedings has led to the mistaken view which you take of them. You tell me 'for God's sake to give up all further opposition to the Government'. Why the idea of opposing the Government was the very antipodes[3] of my intentions. Opposing the Government! Do you deem me mad? Think you that I am so presumptuous as to imagine that my puny opposition, even if I had the will to exert such, would be successful? Do you think that the whirl of agitation has reduced my intellects to a state of phrensy? I have been impelled by no passionate appeal, nor have the dreamings of young ambition had any influence on me. You deem me the pliant instrument of a designing villain, who perchance has painted in glowing colours the importance to be acquired from a contest with the powers that be, and decked out to suit my young imagination a visionary bauble which operates on my actions as the moon upon the infant that would pluck her from the sky – Believe me, I have not been so duped; I never raised my voice in opposition to the government, but I did raise it, in a respectful tone of remonstrance, against the arbitrary proceedings of an individual who acts without reference to the government. You are not aware perhaps that we are employed under the Board of Ordnance, that the Master General[4] is our highest officer. To him we appealed, knowing that he was unacquainted with our condition. We did not even appeal directly to him, our Memorial was first forwarded to Lieut Stanley our Division officer. He handed it to Captain Tucker, the latter forwarded it with his own notes and comments to Colonel Colby. Col C. put an extinguisher on it with the Master General. Had we acted in a less courteous manner had we not afforded to Capt T. an opportunity to blacken our proceedings – had we appealed the Master General without reference to his subordinates, the result would I dare say have been more favourable. We did not deviate from the strict line of propriety. Captain Tucker himself has given us credit for this. He expressed his approval of the straightforward mode of proceeding which we followed; and so he ought. He had plenty of time to digest his cogitations and to nullify all we had done. The Memorial did not leave his hands for three weeks after he received it. The right of the men to petition the Master General was asserted by Col Cunningham, the Master General's private secretary, during a late investigation in Liverpool. This however was only a recognition of the abstract principle, and Capt T. has found means to connect our proceedings with objects foreign to our intentions and which it appears render our use of the privilege worthy of punishment. I have not abused my abilities 'to please a wolf in sheep's clothing';[5] I have done nothing, which, if you were acquainted with its attendant circumstances, would not receive your approbation. I would wish you particularly to remember that our movement had nothing political in it. We have conventional laws of our own – laws manufactured by Captain Tucker without any reference to the government. It was for an alteration of these laws we petitioned; the most influential of those concerned were staunch protestants, and some of the best men on the Survey, whose dismissal, inasmuch as it seriously injures the progress of the work, would not I'm sure be sanctioned by Sir Robert Peel. Religion or politics however had nothing to do with our proceedings. You speak of submission. I am not stubborn, but what do you want me to do? Is it to unsay what I have said? I have said nothing but truth shall I then crawl like a guilty reptile to the knees of Captain Tucker and stain my conscience with a falsehood by telling him that I am sorry for what I have done? Does my father tell me to act thus? No, from you I inherit the principle which guided me, if I be wrong, you should not have taught me to be honest, as honesty and truth have been my guiding light in this transaction.

You ask me to calm the storm I have raised. A figured contained in my letter[6] to Mr Conwill may lead you to suppose that such was raised – I now speak literally – there was no storm. All our proceedings were calm and tranquil nor have I been the originator of these proceedings there were older and cooler heads than mine connected with them I am one of the youngest concerned. I may have lent my humble aid in the expression of feelings which were entertained by all, but those feelings were not of my creation.

There are but two reasons why I should be sorry for what I have done – one dwelling in the consciousness of having done wrong; the other drawn from the consequences resulting from my conduct, without respect to its propriety or impropriety. The former I do not feel, and the latter I do not fear. I have thank God, a steady hand, a clear head, and a sound constitution, and with these I will launch fearlessly upon the ocean of life, cheered by the hope that though rough weather may sometimes attend me, still I shall eventually be able:–

> 'To beat the surges under me
> And ride upon their backs'.[7]

A word in conclusion, remember I have not opposed the government. My highest wish at the present moment is to have our conduct investigated by the government – being assured that such an investigation would result in a manner creditable to us and satisfactory to every well-wisher of

Your affectionate son | John Tyndall

My next from London

Don't forget the letter from Mr Alexander. It is most important. I expect one or two from a nephew of the celebrated Mr Edgworth,[8] son of Major Edgworth,[9] Edgworthstown Co Longford. I made his acquaintance in Cork. We have exchanged several letters since

J.T.

RI MS JT/1/TYP/10/3309–3310
LT Transcript Only

---

1. *Your letter*: letter missing. From Tyndall's journal, 18 November 1843: 'Received a strong letter from my father entreating me to submit to Capt. T. wrote a tickler in reply' (RI MS JT/2/13a/3).
2. *our Memorial*: the letter of protest sent by the workers of the Ordnance Survey of England to George Murray, Master General of the Ordnance on 23 September 1843; see letter 0236 of this volume.
3. *antipodes*: the opposite of a given thing or person (*OED*).

4. *the Master General*: George Murray.
5. *'to please a wolf in sheep's clothing'*: quotation seems to be from Tyndall Snr's previous letter; letter missing.
6. *my letter*: letter missing.
7. *'To beat the surges under me | And ride upon their backs'*: W. Shakespeare, *The Tempest,* II.i.815–16.
8. *Mr Edgworth*: John Bridgeman Edgeworth (1823–61), a surveyor in Cork. Later First Lieutenant of the Royal Artillery.
9. *Major Edgworth*: Major Thomas Newcomen Edgeworth (1778–1857) of the Royal Limerick County Militia.

From George Latimer 20 November 1843 0273

Monday Night.

My dear John,

Phill's[1] may be a few yards before me, but I think I shall still overtake him, at least you will have this by the following post, and now, as all remains as when you left, only that a little more anxiety prevails, in consequence of our reprieve, which gives us another week to humbug. I am no good for Queen, Country, or myself. I presume Phill told you of the very kind manner indeed in which I have been treated by the Prestonians. I cannot tell why such should be, but such is the fact. In the first place, I spent Wednesday evening at Mr Lowe's,[2] and after our usual Bible Meeting, it was agreed, though in direct opposition to my desire, that on the following Saturday there should be a tea meeting for the purpose of showing some mark of respect and esteem to your poor 'Alfred'.[3] Of course, I was at liberty to leave the room while these things were undergoing due consideration; however, they voted me out of the room, when it was finally agreed upon, after no doubt lavishing whole bushels of praise, etc. on their friend. The meeting dispersed, I was in the parlour downstairs, when Wm Lowe came up to me, saying, 'Mr L, I have been commissioned by the ladies (!) to present you with this small token of their Christian esteem'. I received it, of course, with the necessary thanks, and why it should be so (?) from the shape I thought it was a penknife, and after leaving Mr Lowe's, coming down by the theatre, I thought these are truly English, for surely an Irish lady would not present me with a knife – no, it would cut love!! I reached home, looked at my gift, which was enclosed in a letter and found within that 'the Female Teachers of the Wednesday evening meeting prayed my acceptance of the enclosed Penknife, as a small token of their Christian esteem'. I looked at my knife – it is a first-rater, <u>mother of pearl handle, a silver plate inlaid, and my name well engraved on it in old English</u> – on the other side of the pearl is written <u>Search the Scriptures</u>, in running hand, but cut. This passed off. Same night (Wednesday) I was invited to spend the following Sunday in Mr Lowe's – this was done by Miss Lowe,[4] as a matter of course, being mother of the house on the present occasion, and mistress of the ceremonies.

Saturday night arrived, I went, and found myself surrounded 'by a galaxy of lips and eyes',[5] Tyndall! After taking three cups well got up and a whole lots of small chat, the signal was given 'bright were the joys of the table, before the cloth was removed'.[6] But, oh, John, who could stand it – joy and grief were now mingled – joy from hearing from one and another expressions of regret at my departure, which I believe to be sincere, and partial grief at being obliged to remove, though positively I cannot think that I am going. During this meeting I was presented with Mr Wesley's Notes on the New Testament,[7] a valuable work, and also an adieu written by our countryman Barnes,[8] a brother poet. If ever you see it, you may pronounce it middling enough. Well, that night passed over, I was told by one lady, 'Well, Mr L., I felt for you'! 'Yes', said I, 'I was going to cry out: Spare my blushes'! Sunday arrived, I dined and spent the evening in Mr Lowe's, and thus my time is consumed. You had better not be fighting me with my own weapons again for this, that would be savage.

But now to come to more sterling stuff. I had a long interview with Charley[9] after office time this evening, and the following is the amount of our cogitations. That you should if possible have an interview with the M.G.,[10] but how to accomplish this was the question. In the first place he would advise you to write, but if you put your name to it it might be productive of some bad fruits, as your name is already before him, and on this account you might be refused, as he would hear nothing from you or any other against the Superintendents of the Survey.[11] Then in that case, he would advise you to write anonymously stating that you had sufficient reasons for withholding your name. In case an interview is granted, to use some of the following implements as your weapons of attack, as being in themselves the most formidable and fatal to the 'concentrated sagacity of the R. Engineers'. You are aware, I presume, that Captain T.[12] is getting specimens drawn by the Sappers[13] in York – for what purpose? Remember there are only about 12 Sappers at work on the Survey, able to draw anything of a plan, and only 4 or 5 out of this number able to produce specimens equal to those of ours, in the M.G's possession. It is presumed by some cool heads, these specimens are for the purpose of superseding ours, and shewing to the M.G. that the Sappers are by no means inferior in point of ability to the C.A's[14] and that the work could be carried on without them. Grant the former part of the argument for a moment, the question arises, how many such were on the Survey as drew these specimens? I answer from experience – produce more than 4 or 5. Now, except this mode of argument be adopted on these premises, the M.G. will be left under the impression that enough can be found to execute the work similar to those specimens. Again, take the character of the work into account. It is a fact notorious on the Survey, and especially on that part or division of it executed under Captain T's superintendence, that the expenses for re-surveying amounts to as much as would pay good men who would be well adapted for the work; for the towns that have been surveyed in the County of Lancaster have mostly to be re-surveyed. Then take into account the length of time and money, spent by the most experienced hands, endeavouring to put this work to right, but after all their experience, finding it utterly impracticable, it has to be surveyed again, and even during its examination errors are still found to exist, and of such a character – for instance, 6 links in the width of this Street, and 10 in that, 15 links in that range of houses, and 7 in that – the error is thus equated, but the work on the whole is inaccurate, and such must be the case, no alternative, while things remain in their present condition – for how can a man be expected to do good work, who is bound to return so much per day, which is impossible while he is limited in the means – for instance, one chainman, this we have had proof of. Saw George Scott,[15] and many others who in order to come up to the scale have surveyed their work in the houses, measured their lines from paper, etc., etc., which was afterwards found wrong after the labour of experienced hands been tried to correct it in the office. Thus these unfortunate men are checked for bad work, although they were compelled to do it, and under such circumstances it should be bad. It comes to this, if they do not return enough, they are checked; if they do bad work, they are checked. And in what does this check or stoppage consist? Why, in this, some men have perhaps from 2/– to 3/6 per day, and these may only be allowed 4 or 5 days in the week, but having the name of 2/– to 2/6 a day. How frequently does it occur that during the stoppage, those unfortunates are compelled to run in debt for the purpose of keeping themselves in existence, and are not able – it may be for 12 months or more – to pay the debt thus incurred, after having made good the sum levied on them by their hard task-masters. Again, some of those men who after being from 5 to 18 years employed on the Survey are discharged and owing to their miserable pittance they are unable to lay by anything for the evil day which often approaches them in a direful form, when they are obliged to solicit at the hands of the office men a few pence, to transport them to their native soil or support them on their way or buy a few little things to put in a basket that they might be by this means saved from starving altogether. And now is Col. C.[16] acquainted with those particulars? Surely the M.G. is not. The fact is, Col. C. knows nothing about the expense of the re-surveying, for he is supplied with vouchers containing certain sums of money laid out for certain duties, but these are under the head of Surveying, no re-surveying. Again the drawing department of the Surveying is in the greatest arrear – when will this be brought forward, and by whom? Seeing the number of efficient hands among the Sappers are very few. Again, owing to the manner in which Capt. T. has treated the men all through, they can place no confidence in him, either Sapper or Civilian, not one in one hundred could do this conscientiously, and the men cannot feel an interest under such a government of things. And now these, with many others, have been suggested as data for you in your anticipated interview, by one with whom you have had frequent cogitations before. You are acquainted with all these before, but it is more for the purpose of reviving them in your mind, and some of them – in fact, all of them are choice, and I hope in case of an explosion will prove very unwelcome missiles. We had a letter from Mr Manaway,[17] no possibility of getting there, as he lives nearly under the same roof with a Sapper, and they don't run well in tandem, so this is ed.[18] Phill had a letter from Cuddy who says that all the men in Manchester have procured situations, mostly connected with the Sheffield Railway. Digan[19] and some others are only making 4 guineas a week, while O'Connell[20] and Conran[21] are from 16 to 17 shillings a day. All very fine, this, bright hours will come. So Johnny you have heard from Nancy![22] I saw her brother[23] this evening, he was axing[24] after your welfare.

Remember me to McDermott,[25] and believe me your once sanguine friend but now cool, determined | George. (Latimer)

[To Mr. J. Tyndall | Post Office | High Holborn | Postmark – Preston. Nov 22/43][26]

RI MS JT/1/TYP/11/3739–3741
LT Transcript Only

1. *Phill's*: possibly Phillip Evans.
2. *Mr Lowe's*: William Lowe, biographical details unknown.
3. *your poor 'Alfred'*: a nickname for George Latimer.
4. *Miss Lowe*: not identified; possibly William Lowe's daughter or sister.
5. *'by a galaxy of lips and eyes'*: T. Moore, *Lalla Rookh, An Oriental Romance* (London: Longman, Hurst, Rees, Orme, and Brown, 1817), p. 38: 'The Eastern world in every region blest | With woman's smile sought out its loveliest | To grace that galaxy of lips and eyes | Which the Veiled Prophet destined for the skies'. On Moore, see Biographical Register.
6. *'bright were the joys of the table, before the cloth was removed'*: J. O'Keefe, 'How Bright Are the Joys of the Table', ll. 1–2: 'How bright are the joys of the table; / I mean when the cloth is removed'.
7. *Mr Wesley's Notes on the New Testament*: J. Wesley, *Notes on the New Testament* (London: William Boyer, 1755). Wesley was an Anglican cleric and the co-founder of the Methodist movement.
8. *our countryman Barnes*: an unidentified Irish poet.
9. *Charley*: possibly Charles Edward Stanley, although the nickname 'Charley' suggests a level of informality not indicated in previous mentions of Stanley.
10. *the M.G.*: the Master General of the Ordnance Survey, George Murray.
11. *the Superintendents of the Survey*: includes Thomas Frederick Colby.
12. *Captain T.*: Henry Tucker.
13. *the Sappers*: see letter 0232, n. 10 of this volume.
14. *the C.A's*: Civil Assistants.
15. *George Scott*: not identified.

16. *Col. C.*: Thomas Frederick Colby.
17. *Mr Manaway*: not identified.
18. *so this is ed*: not identified.
19. *Digan*: not identified.
20. *O'Connell*: not identified. Not to be confused with Daniel O'Connell, the Irish politician.
21. *Conran*: not identified.
22. *Nancy*: not identified.
23. *her brother*: not identified.
24. *he was axing*: regional pronunciation of 'asking'.
25. *McDermott*: not identified.
26. *'To Mr. J. Tyndall ... Nov 22/43'*: address and postmark given by Louisa Tyndall.

From William Ginty 28 November 1843 0274

Friday night.

As I live 'the clock strikes twelve' – 'tis now drear midnight' Jack! 'Sleep death's counterfeit'[1] 'swings his leaden scepter o'er a slumbering world' – Liverpool – tumultuous bustling Liverpool is still and silent as the dismal tomb and oh God as if to add a sterner gravity to thought – an old sea captain that sleeps in the room above me is groaning horribly with the nightmare. Powers eternal what a solemn hour is this, what a time for calm contemplation – the din of this mighty mart – 'the hum, the noise, the shock of men'[2] has subsided and the 'spirit of silence reigns unbroke' save by the solitary footsteps of the constable that paces in measured tread beneath my window. Aye! this is a solemn hour

'This is the hour when fancy wakes
  'Visions of joy that did not last
'This is the hour when memory takes
  'A survey of the past

Oh that I could recall that past
  In fancy's vision sweet and clear
Oh that those happy days could last
  Till chased by others yet more dear

But no! to me they're ever lost
  The hopes that lit them – all have flown
A heart by sorrow's tempest tossed
  Remains for me – and I'm alone (God keep me so)[3]

Now I'm just far enough. My thoughts overwhelm me! Well you have seen Spec's[4] handwriting I s'pose. It was placarded on the walls of Liverpool this day on the same sheet with the 'Repeal Question'[5] 'China tariff'[6] 'Welsh riots'[7] &c &c.!!!

Sinnett swears the final is a – that is he says 'the officers never exd., the plans before the revision' d'ye hear! Don't libel the wretches!

Do you intend to criticise on the policy of sending the old experienced officers away and replacing them by 'Gawkys'[8] that is after you have tossed up the 'Detail Survey'[9] 'which is' as Veritas[10] says 'now being carried on in great perfection'

I did not see S[11] yet – tis not my fault faith – but I have plenty of time. I will post No 3[12] for him along with this and then when I call on him for the memorial[13] I will hear what his literary highness has to say about the matter and report accordingly to the prototype of *[*Junius*]*[14] (!) No 3 by the way is far superior to No 2.[15] One fellow here is about writing to Dublin to Griffiths[16] men to find out the author – No less – !

Good night | Yours | The 'Attorney General'[17]

Answer this

Do you get those letters posted by me in the morning or the evening? Look at the cover!

I find I have not a copy of this affair, so must excite my mental powers to give you summat in the shape of the original.

Few brook abuse and don't resent it
Then by St. Mark you shall repent it
Had I a prouder – sterner heart
Silent contempt would be your part
But since the love of right is stronger
Be this my plea 'truth stoops to conquer'
Be mine the task our rights to shield
My pen my sword – this sheet the field
I'm not the first of fighting lords
So here goes 'parrys' 'points' and 'guards'
Resolve on giving your wrath full fling
You plucked a pearl from Byron's string
Ill-suited every line to me
But what of that – it answered thee
Rage fancied fitness in the scroll
For dark vindictiveness of soul!
Oh noble relic oh nobler muse
Can Time thy beauties thus abuse
Bright spirit Band you little dreamed
Before your mortal visions faded
When from your souls deep essence teemed
Your laurels thus by Time degraded
'When first I met thee warm and young'[18]
Twas thus the sullied gem begun
Not over young – nor yet too warm
But that must pass it means not harm
'There shone such truth about me' too

In truth I can't say that of you
Not <u>now</u> at least and even then
I never sought your truth to ken
I tried not – cared not – proved not how
'Twas then[19] no more than now
'Tis mine to hope that lucent beam
Of Truth may ever round me gleam
<u>Thus</u> may it shine forever there
To guard my path from hidden snare!
Next in your passion's bitter torrent
I'm styled Deceiver – term abhorrent
Now by my faith – the charge's untrue
The term is false – aye false as you
÷ ÷ ÷ ÷ ÷ ÷ ÷ ÷ (forgotten)
'Go – go Deceiver' – powers of song
What means the 'go' when I am gone
St. George's Channel broad and deep
Its vasty billows 'twixt us sweep
May every billow swell a sea
To roll its waves twixt you and me
May tempests lash those waves on high
To wall the clouds, the heavens – the sky
As barriers fierce twixt you and I
I've done – farewell! but no! not yet
I cannot pass your green vignette
Eolian harp[20] – with strings unstrung
Emblem of ruins on rose tree hung
And then to damn it in effect,
The motto 'ruined by neglect'
In blazing letters round it burn
Like torch-light pointing out the sun
Alas sad task! O enough Good Bye
I wish thee all a wish can buy
That's simply nothing and to you
There's something more than nothing due
But I forgive thee at the shrine
Of loves own patron – Valentine.[21]

A dreadful bull. Subsequent to the transmission of this to the injured fair one (?) I bethought me that Moore and not Byron was the sufferer. A very ignorant mistake! I intended it to be at least four times as long. I however expunged nearly one half of it and of that[22] half this is, 12 or 14 lines short my memory having so much failed me

Yours | Ginty.

[Enclosure: a letter and poem from Martin Cuddy]

My dear Ginty, [23]

It was with infinite pleasure I received yours long, long ago. I hope you will forgive me for not answering ere this. Neither am I at present prepared to argue with you the cause of our slavery. You cannot deny but that Ireland is in worse bondage than ever the Israelites of old were under Pharaoh. For instance what is the cause of so many miserable beings leaving the land of their birth to seek the little food that's denied them at home, curse upon the cause, I say, and to hell with bad men, and bad measures. If Repeal was granted tomorrow could we be worse off than we are – whatever better, we cannot be worse. It is that infernal religious bigotry is the cause of Ireland not getting her just *[1 word illeg].* Government knows it well, and why not when men are paid well to add fuel to the fire.

I send you a few verses of mine, in favour of liberty but as you do not choose to be free you'll not regard them.

Oh! Erin[24] my country awake from thy slumbers
The trumpet of liberty, re-echoes, the land
Shake off the vile shackles that does thee encumber
And crush'd thee, for years, with a death dealing hand

Arouse then brave sons of freedom benighted
And hail the loved morning of liberty's dawn
Remember the curse by which you are blighted
Arise then, and vanquish the wolf and its fawn

Men of Erin unite, round the common cause rally!
Let your voices go forth on the wings of the wind.
Let repeal, be the word, through mountain and valley,
And starvation to the Saxon, that would you dare bind.

Too long have you suffered, by Saxon oppressors,
And your country laid desolate by fire and sword
But vengeance ere long, will o'ertake the aggressors
For heaven, and justice their aid will afford.

The sun its brilliant beam on Tara[25] is shedding
The spirits of our forefathers look smilingly on
And the music that once enlivened that dwelling
Will burst with fresh ardour when our freedom is won.[26]

Then Erin arise from thy heart rending sorrow
Behold that bright orb, proclaims victory's *[1 word illeg]*
May each coming day give him strength for the morrow
To fight with fresh vigour for the rights of our land.

If you like this, perhaps in due time I may send you more but you must not expect that anything in the strain of poetry from me will be sublime, but 'I give thee all I can no more'.[27]

[Ginty:	What direful thoughts of sickening gloom,	
[30–40	Around me gathers, this evening Noon! [Question]	[Ginty: *[3–4 words*
words illeg]]	Enough to quench, the little ray of joy	*illeg]*]
	That grew with me, ere I was a boy	
	Like waves, that roll, with thundering tempest tost	[Ginty: a miserable
	And dash their fury on the sea beat coast.	attempt stealing
	With madning force, and mountains of white foam	another's *[1 word*
	Engulfing men into one vengeful doom.	*illeg]*]
	Like tyranny – base that demon power	
	A.F.R. Griff[28] – dark as midnight hour	
	With nostrils gaping!!! See behold they come	[Ginty: G flat at
	With fiendish strides and hellish brimstone fume	fortissimo]
	To suffocate devour and dethrone	
	All poor Surveyors and hunt them home	
	For why the millions, T.R.[29] days is growing slack	
	Those beggars I'll hunt – Griff your knapsack	
	To hold the surplus of the Government Stock	[Ginty: Hooray!
	'Tis little enough the thing will shortly dock	more music!]
	Of all my vestures maps and title deeds	
	For robbing the Surveyors of their fees	
	But what for that I to myself must look	[Ginty: Oh Lord!]
	I'm almost blind Youth hath me forsook	
	My ears grow cold, a naceous juice doth flow	
	From out the pores oh! misery, sorrow, woe!	
	To be deprived of life. Oh! God what shall I do	
	For satan claims his own. I cannot look to you	
	See the fiend comes while blasted spirits attend	
	To waft my soul to H—[30] most wretched is my end	
	So ends the life of one a blasted race did run	
	A Robbing thieving knave. A father's 'cursed Son'	
	Philanthropist.	

These last 8 defy all common sense![31] The 'Scotch reviewers'[32] would throw down their scourging quills in despair and set about writing an ode to the memory of 'damned nonsense'. It excels all I ever seen or imagined in the way of literary hotch-potch! Certainly it is a philanthropical effusion with a heavy vengeance! in the letter speaking of it he says: I 'must not publish it' as the subject is rather a 'domestic one'. This is the first letter I had from poor 'Blackthorn'[33] since I came to England, so it is quite evident that it is his own high opinion of 'the poem' has induced him to write. I'll keep him to it! until I get a few more from him. Byron could not afford as much amusement![34]

RI MS JT/1/TYP/11/3623–3625
LT Transcript Only

---

1. *'Sleep death's counterfeit'*: W. Shakespeare, *Macbeth*, II:3.75: 'Shake off this downy sleep, death's counterfeit'.
2. *'the hum, the noise, the shock of men'*: this quote appears in several publications from the 1800s but the original source is not identified. See, for example: J. G., 'Remarks on the Natural Productions of Lexden and its Neighborhood', *The Magazine of Natural History*, 7 (1834), pp. 17–9, on p. 19.
3. *'This is the hour ... God keep me so)*: Mrs C. B. Wilson, 'The Evening Hour' (1821).
4. *Spec*: Spectator, Tyndall's *Liverpool Mercury* pseudonym.
5. *the 'Repeal Question'*: the debate over whether to repeal the 1801 Acts of Union and restore the independent Kingdom of Ireland.
6. *'China tariff'*: a reference to the newly signed Treaty of Nanking, which brought an end to the First Opium War and set fixed tariffs for trading at Chinese ports.
7. *'Welsh riots'*: a reference to the Rebecca Riots, which took place from 1839–43 in Wales. The riots were protests by local famers in response to perceived unfair taxation.
8. *Gawkys*: Gawkeys is derogatory slang for an incompetent person (*OED*).
9. *the 'Detail Survey'*: the branch of the Survey tasked with surveying and mapping topographical details, rather than boundaries.
10. *Veritas*: a pseudonym under which one of Tyndall's critics wrote to the *Liverpool Mercury*.
11. *I did not see S yet*: unidentified initial – possibly Charles Edward Stanley.
12. *No 3*: this probably refers to Spectator's third *Liverpool Mercury* article.
13. *the memorial*: the letter of protest sent by the workers of the Ordnance Survey of England to George Murray, Master General of the Ordnance on 23 September 1843; see letter 0236 of this volume.
14. *the prototype of Junius*: the Letters of Junius, published between 1769 and 1772, were a collection of open letters from an anonymous polemicist who used the pseudonym 'Junius'. The letters stridently criticized various British politicians, mainly King George III and his ministers.
15. *No 3 by the way is far superior to No 2.*: possibly a reference to the second and third Spectator letters.
16. *writing to Dublin to Griffiths*: possibly Richard Griffith.
17. *The 'Attorney General'*: a nickname for Ginty.
18. *'When I first met thee warm and young'*: T. Moore, *Irish Melodies Volume IV* (Paris: A. and W. Galignani, 1823), 'When First I met thee', ll. 156–7. On Moore, see Biographical Register.
19. *'Twas* then: this blank is faithful to the typescript.
20. *Eolian harp*: a harp that is played by the winds, named for the Greek God of the winds, Aeolus (*OED*).
21. *'Few brook abuse ... loves own patron – Valentine'*: this poem appears to be an original one by Ginty.
22. *and* *of that*: this blank is faithful to the typescript.
23. *My dear Ginty*: Louisa Tyndall annotation: 'the notes are by Ginty'. Louisa Tyndall copied Ginty's annotations onto her typescript; however, owing to the tight binding of the typescripts, many of these annotations are now illegible.

24. *Erin*: a modern derivative of Éirinn, the Irish-language name for Ireland.
25. *Tara*: the former coronation place of the Kings of Ireland.
26. *Oh! Erin ... freedom is won*: this portion of the poem also appears in letter 0255 of this volume from Martin Cuddy to Tyndall.
27. *'I give thee all I can no more'*: T. Moore, 'My Heart and Lute', l. 1. On Moore, see Biographical Register.
28. *A.F.R. Griff*: not identified.
29. *T.R.*: not identified.
30. *H—*: likely 'Hell'.
31. *These last 8 defy all common sense!*: this final paragraph appears to have been written by Ginty, not Cuddy.
32. *The 'Scotch reviewers'*: possibly a reference to Lord Byron, 'English Bards and Scotch Reviewers' (1809).
33. *Blackthorn*: apparently a nickname for Martin Cuddy.
34. *as much amusement*: Louisa Tyndall annotation: '[LT: 29. Ginty ('No 3' posted a clue)]'.

To William Ginty [Late November 1843][1] 0275

Dear Ginty,

I am much obliged to you for your kind offer regarding the Athenaeum[2] but I regret to say that I shall not be able to avail myself of it. I shall have so much to do between June and October[3] that the course you propose is quite beyond my ability – Circumstanced as I am here you will readily see that it is out of my power to get any other than a written opinion from Capt. Wynne and I feel as strong a reluctance as you do to asking him for it. I shall be in town at the end of May and shall see him and ask his verbal opinion but this perhaps will be too late for your purpose. Has the rail not yet been tested?[4] You know he is not a man to do a thing under the rose and the statement that no public use would be made of his opinion would I am afraid have little weight with him – indeed I hardly think he would do anything privately in such a matter that he would shrink from doing publicly. Just think of the character of the man and I am sure you will be of my opinion.

Affectionate regards to Mrs Ginty.[5] | most truly yours | J. Tyndall. | Thursday.

RI MS JT/1/TYP/11/3723
LT Transcript Only

1. *[Late November 1843]*: this date is uncertain; it was inferred from letter 0266 of this volume.
2. *your kind offer regarding the Athenaeum*: see letter 0266 of this volume, in which Ginty suggests that Tyndall apply for a position at the Liverpool Athenaeum news room.
3. *much to do between June and October*: possibly a reference to Tyndall's plans to emigrate to America; see letters 0234, 0251, 0254, 0259, and 0281 of this volume.
4. *Has the rail not yet been tested?*: not identified.
5. *Mrs Ginty*: probably William Ginty's mother; Ginty did not marry until 1846.

To George Murray [24 November] 1843[1] 0276

Sir,[2]

I have been employed on the Ordnance Survey of Great Britain and Ireland for the last five years – during which time a complaint of any kind was never lodged against me. My constant aim has been to gain the approbation of my Superintendent[3] by unremitting assiduity, and I have every reason to believe that my efforts have been successful. I was led to hope that by a steady adherence to this principle, I would ultimately arrive at a creditable position in my situation. This hope however has proved fallacious. My services on the Ordnance Survey have been lately dispensed with,[4] and I am now a stranger in the City of London, without friends or employment.

With respect to my character both as a man and a civil assistant I beg to refer you to the accompanying copies of testimonials from the persons whose names they bear: should the originals be required I can produce them.

Lt. Wynne had means of judging of my character not only by personal observation but he also had an opportunity of finding his own opinion of me corroborated by the testimony of gentlemen of the highest respectability who have known me from my childhood. I send with this copies of testimonials[5] from two of the latter. One is from the brother[6] of the Earl of Bandon[7] and uncle to Lord Bernard,[8] Member for the town of Bandon, Ireland. The other is from a gentleman of property[9] – the friend of Colonel Bruen,[10] M.P. His not having any personal knowledge of me is the only thing that prevents the latter from adding his testimony – he is well acquainted with the character which I have ever bourne in the neighbourhood of my birth.

Capt. Tucker's testimonial contains nothing adverse to this, on the contrary, he has corroborated the report made of me on my departure from Ireland, which was that my character was very good. He assigns as a reason for my dismissal that a reduction was required to lessen the expenses: but this cause alone would hardly have induced him to choose for this purpose a number of good and efficient men while many of a contrary character are retained. It may, however, be motives of good will prevented him from stating the true cause – a dislike to injure them may have prevented him from going into a detail of the circumstances which led to their dismissal. There are about a dozen persons similarly circumstanced, the majority of whom, I would venture to affirm, have been as efficient in their situations as any to be found on the Survey. The necessity of reducing the expense would in itself hardly induce Capt. Tucker to fix on such persons, while numbers of a contrary character are retained. Motives of good will may, however, have actuated him in assigning this cause. I have come to London for the express purpose of obtaining an interview with you – this favour I earnestly beg. I am a young man unacquainted with the world and altogether unaccustomed to a contact with persons of distinction – still, Sir, the character you have encourages me to appeal to you. I do so supported by the hope that when my case is laid before you I shall be rescued from many disagreeable circumstances which are naturally consequent on being suddenly dismissed from such a situation as I have held on the Ordnance Survey.

Private circumstances compel me to request an answer as early as convenient.

I remain, Sir, with feelings of the deepest respect,

Your most obedient humble servant, | John Tyndall.

RI MS JT/1/TYP/10/3904
LT Transcript Only

---

1. *[24 November] 1843*: dated by relationship to the following letter, 0277 of this volume.
2. *Sir*: Louisa Tyndall annotation: 'Rough sketch of letter sent on *[2–3 words illeg]* to the M. G.'. 'M. G.' refers to George Murray, Master General of the Ordnance Survey.
3. *my Superintendent*: the Superintendent of the Ordnance Survey of Ireland, Colonel Thomas Frederick Colby.
4. *dispensed with*: see letter 0262 of this volume for a copy of Henry Tucker's order dismissing Tyndall.
5. *copies of testimonials*: testimonials missing.
6. *the brother of the Earl of Bandon*: Richard Boyle Barnard (Bernard).
7. *the Earl of Bandon*: James Bernard (1785–1856), 2nd Earl of Bandon in Ireland.
8. *Lord Bernard*: Francis Bernard (1810–77), 3rd Earl of Bandon in Ireland.
9. *a gentleman of property*: not identified.
10. *Colonel Bruen*: Henry Bruen (1792–1852); see letter 0267, n. 10 of this volume.

To George Murray 30 November 1843 0277

Sir,[1]

The liberty I now take would be an unwarranted one were the causes which prompt me to take it of an unimportant nature. I wrote to you on the 24th,[2] requesting the favour of an interview and laying before you at the same time copies of testimonials[3] which I trust have divested your mind of doubts which might naturally be attached to the statement of a stranger unsupported by such testimony. I perchance have presumed too far in asking this favour. I should not have done so were not the interests of friends who are dear to me involved in a fate similar to my own, and did I not imagine that my exertions would be conducive towards rescuing them as well as myself from the embarrassment consequent on our dismissal from such a situation as we held on the Ordnance Survey. I, as one of those who signed the letter to you on the 9th,[4] have been informed that if we wait till December we shall be made acquainted with your opinion of the document. The favour for which I prayed may be denied me: still the vindicating of my conduct is no less my bounden duty and compels me, probably for the last time, to crave your kind attention to a few facts which, owing to misadventure or other causes, may not have been submitted for your consideration. You are already acquainted with my character from my first connexion with the Survey to the latest hour of that connexion, you have this confirmed by gentlemen of the highest respectability who have known me from my infancy. I am not the only one of those lately dismissed[5] who can produce such creditable testimony: there are other men similarly circumstanced whose moral characters would not suffer by a comparison with that of any man in existence. I speak strongly, Sir, but not more strongly than is merited, and with respect to our qualifications I can only say that I feel assured we would be found upon trial equal if not superior to the same number selected from any Division of the Survey. Before my departure from Preston, I, with others, forwarded you three small specimens.[6] I beg of you to recollect that these were executed under very unfavourable auspices, they were done in a great hurry, with unsettled minds and by candle light. Necessity compelled us to lay these very imperfect scraps before you. They would, however, be a very inaccurate test of the ability of those who drew them – I would repeat, Sir, that upon a fair trial we will not be found inferior to the same number selected from any Division on the Survey.

RI MS JT/1/TYP/10/3905
LT Transcript Only

---

1. *Sir*: Louisa Tyndall annotation: 'rough sketch of *[*letter to the M Gen*]* on Nov. 30.' 'M Gen' refers to George Murray, Master General of the Ordnance Survey. It is unclear whether this letter was sent.
2. *I wrote to you on the 24th*: probably the final copy of the letter drafted in letter 0276 of this volume.
3. *copies of testimonials*: testimonials missing.
4. *the letter to you on the 9th*: possibly letter 0236 or 0245 of this volume, although neither letter is dated the 9th.
5. *those lately dismissed*: see letter 0262 of this volume for a copy of the order dismissing Tyndall and other workers on the Ordnance Survey.
6. *three small specimens*: i.e., specimens of Tyndall's surveying work.

To George Murray 6 December 1843 0278

Sir,[1]

I have been just informed that the fate of the individuals who signed the letter forwarded to you on the 9th ult.[2] has been settled – that they have received the final decision of Colonel Colby, which amounts to a confirmation of his original order for their discharge.

I am one of those affected by this decision. I am dismissed on the charge of 'disobedience of orders in not signing Captain Tucker's order[3] and being concerned in a combination to force more pay to be given'. To the former part of this charge I plead 'guilty'. I declined signing Captain Tucker's order because I con-

sidered that by signing it I would injure my future prospects in the Survey. I am quite aware that disobedience of orders, taken in an abstract point of view, is a serious crime: but were the circumstances attending the appearance of the order in question taken into account much would be found to ameliorate even this. I laid before you my reasons for acting as I did, hoping that those reasons would be my justification. With respect to the latter part of the charge, namely being concerned in a combination to force more pay to be given, I, Sir, in the name of my fellow sufferers most solemnly declare that our combination extended no further than a common consent to lay before you, as our highest officer, a plain and truthful statement of our condition; and that whatever may be the grounds for supposing the existence of such a combination, the conclusions drawn from them are completely at variance with the motives by which we were actuated.
I leave London on Friday morning. Should you, Sir, in the mean time think proper to honour me with any communication, the address stated outside will find me. My address afterwards shall be: –

Leighlin Bridge, | Ireland.

I am, Sir, | Your most obedient humble Servant, | John Tyndall

RI MS JT/1/TYP/10/3906
LT Transcript Only

1. *Sir*: Louisa Tyndall annotation: 'Sketch of letter to M. G. refd to in 6th entry on last page Dec 6th 1843'. 'M. G.' refers to George Murray, Master General of the Ordnance Survey.
2. *the letter forwarded to you on the 9th ult.*: possibly letter 0236 or 0245 of this volume, although neither letter is dated the 9th.
3. *Captain Tucker's order*: see letter 0236 of this volume.

From John Hebdon 15 December 1843 0279

Old Bank | Preston Dec 15th 1843

My dear Tyndall

I am afraid you will think I have quite forgotten you but I will assure you it is not the case, we have been so very busy at the Bank preparing for our Christmas balance, that I have not had time to write to anyone. I received your letter of Saturday night[1] Sunday morning and was glad to hear of your safe arrival in London. I am afraid from your being now in Ireland that you have been unfortunate with the Master General of the Ordnance;[2] – had you stayed in London I had flattered myself that I should have had the pleasure of seeing you shortly shaking you by the hand, but I hope it will all be for the best. Mr Evans left here this day week to Liverpool on his way to Ireland, probably you may have seen him ere this. Mr Latimer[3] also left here for Birmingham last Monday. Mrs Mulhall[4] is quite desolate since you all left, for Mr Parker is leaving here and going down into Bow Lane to Mr Marquis. We have had Mr Carter the Lionking[5] here for the last fortnight, he seems to have given universal satisfaction. I was there last Friday and Saturday night + was very much pleased, I saw your friends Mr and Miss Banister there, I think Mr B. is a constant attendant. Mr Bonwell[6] preached his farewell sermon on Sunday last to a very crowded audience I think I never heard a more affecting sermon in my life. Miss Braithwaite[7] is quite well and begs to be remembered to you. I think I have no more news at present but must conclude with my kind love to you and best wishes for your welfare.

I remain dear Sir | your most faithful and ever | attached friend | John H. Hebdon[8]

P.S. Shall be glad to hear from you as early as convenient.

RI MS JT/1/TYP/11/3727
LT Transcript Only

1. *your letter of Saturday night*: letter missing.
2. *the Master General of the Ordnance*: George Murray.
3. *Mr Latimer*: probably George Latimer.
4. *Mrs Mulhall*: possibly Margaret Mulhall.
5. *Mr Carter the Lionking*: Carter was a circus performer who worked with lions and other large cats. 'Mr. Carter and His Animals', *Illustrated London News*, 17 September 1842, p. 293.
6. *Mr Bonwell*: probably the Reverend James Bonwell, a curate in Preston who was the subject of a critical 1843 pamphlet. See A. Clissold, *A Letter to the Rev. James Bonwell, Priest of the Church of England, and Curate of the Church of the Holy Trinity, Preston, Lancashire: Upon the Subject of His Sermon on the Perishing in the Gainsaying of Core* (London: Hayward and Company, 1843).
7. *Miss Braithwaite*: not identified.
8. *John Hebdon*: not identified. The contents of this letter show that Hebdon was a friend of Tyndall's who lived in Preston and worked at the Bank there. Possibly John Henry Hebdon (1822–?) who appeared on the 1851 census of Preston along with his wife Alice Hebdon (1824–83); his occupation was listed as 'Banker's clerk'.

From Robert Allen 18 December 1843 0280

Preston 18th. Dec. 1843

My dear John

Having some leisure time I think I cannot pass it in a better way than acknowledging in a few lines the receipt of your kind but short note[1] and likewise letting you know all the news that I can furnish you with. I have just returned from having a long walk with Sayers during which I communicated your kind regard but did not read your sentiment as you desired; for this reason – at the time he mentioned the matter to me he charged me particularly not to make it known, however I am very happy to inform you that our opinions with reference to his applying to be re-employed are of little importance now as a better prospect lies stretched before him. I think it was about last Thursday he received a letter from Captain Stotherd[2] informing him that a situation of no mean character had been promised for him in some honourable department at Woolwich where he should receive the pay of a second Lieut until a trial of his abilities merit a higher rate of pay – this letter to him appears only to have conveyed the intelligence without the appointment as Stotherd mentioned that another would shortly be forwarded from the officer superintending the department – I know you will be glad to hear this, but not one in the office is aware of it; at least so Sayers tells me. Will that settle Garryonne?[3] Accept his kindest and best remembrances to you, the next in course is Old Bob, or in more respectful language Mr Martin.[4] A letter came here yesterday from the Capt.[5] stating that Martin and Carroll[6] were to be re-installed at a reduction of 3d per diem each, and that if their conduct in future were satisfactory or something to that effect, they would at no distant period be restored to their former pay or it advanced to them which you please and also a little promise by the way of things better awaiting them not of course directly stated in such plain terms as if entering into a 'Virtual Contract'. I have not seen the letter but from what accounts I received as far as I can discern you have the substance of it in this. Well, but you are anxiously waiting the effect it produced and that you shall have. Mr Martin has returned to his former occupation accepting the proposed terms and was at work this day, now he might have waited for some time longer as Halliday[7] the Surveyor was giving him a job or about doing so, but he says the wife thinks that even less wages is preferable in a settled situation, of course the man ought to know better than I how to manage his own affairs. Yet it is the general opinion that by his acceding to Tucker's proposals in the case, he has done more to strengthen that obnoxious principle which would change a justified mutual request into an unlawful demand, than all we did to uproot and destroy it. Carroll yet sensible of the wrongs enforced and conscious of the justice of his remonstrances, hesitates not to inform the Capt. that on no condition will he return except he has granted to him his former rate of pay as that barely afforded him the means of subsistence. I was speaking to him this morning and these are his sentiments – now he made no application whatever according to what he states but merely requested a regeneration of his character. See then how Tucker like a serpent coils – had Martin not accepted the offer, I am convinced and so is Walkem that before another week a circular would have been issued to affect a more honourable compromise. I had a private discourse with Walkem today on the matter and he seems to think that if a proper document were drawn up by the parties dismissed stating that as Col. Colly[8] kindly proposed to employ them at reduced wages in cases of not meeting with situations in other quarters that they now offered their services for their usual wages after being out of employment for — weeks considering that time unemployed a sufficient loss sustained without a further reduction. This he thinks in some degree would as it were allow them the satisfaction of carrying out their principles and afford the men a chance of employment on honourable terms, without an actual submission by either party – at the same time he would take advantage of the great falling off in the progress of the work for the last month and see Lieut Stanley himself on the matter when he returned from leave of absence showing him the awful consequence of such a discharge of useful men and then by using his influence with Tucker, Walkem is of opinion that all might again be settled satisfactorily. This he wishes me to write to you as his views on the matter, as I told him I would be writing this evening – of course you will have your own opinions and I hope you will not consider me as dictating but my desire and prayer is for your welfare and for all those who suffered in the same cause – the independent struggle for a glorious emancipation for glorious it would have been had we succeeded. I had a long letter from Latimer[9] on Friday he was then (when he wrote the letter) in Manchester expecting to proceed to Birmingham no doubt you have heard from him since therefore I need say nothing of what he communicated. I hope he may have succeeded in procuring a good gentlemanly methodistical, situation a honest-hearted methodical reformer! – I wonder what he is doing tonight the idea is a queer one but I was about exclaiming in what circle does his 'nose' form a 'segment'! He like yourself stated that he felt no regret for what he had done. I was indeed glad to hear that 'Tyndall is still' himself a spirit bounding son of Erin;[10] well but I hope you have before this received your Yankee epistle.[11] I am rather uneasy concerning it, I am sure I re-directed it in the post office correctly but most likely they forgot to send it by the same post which conveyed my letter and the paper. The latter came to the office, the letter went to Mulhall's[12] but they did not keep it so I had to see Johnston the letter carrier[13] at night when I got the letter from him and returned it assured that it should be forwarded. I will enquire about it as I put in this but I hope you are not until now without reading its contents – dear John art *[thee]* bound for America? I hope you may have encouraging intelligence from the land of liberty. I must conclude, good night, the Lord bless you, bright prospects yet await you perhaps now bursting the suspended gloom.

I hope so with best wishes I am | yours ever faithfully | Bob.

RI MS JT/1/TYP/11/3482–3483
LT Transcript Only

---

1. *your kind but short note*: letter missing.
2. *Captain Stotherd*: Richard John Stotherd, a Captain on the Ordnance Survey of Ireland who supervised the B District.
3. *Garryonne*: not identified.
4. *Mr Martin*: Robert Martin.
5. *the Capt.*: possibly Henry Tucker.

6. *Carroll*: Peter Carroll; see letter 0262, n. 7 of this volume.
7. *Halliday*: not identified.
8. *Col. Colly*: probably Thomas Frederick Colby.
9. *Latimer*: William or George Latimer.
10. *Erin*: a modern derivative of Éirinn, the Irish-language name for Ireland.
11. *your Yankee epistle*: probably a letter in which Tyndall told Allen of his plan to emigrate to America. Letter missing.
12. *Mulhall's*: not identified, possibly a reference to Margaret Mulhall.
13. *Johnston the letter carrier*: not identified.

From George Latimer 27 December 1843 0281

Newtown Gore[1] | December 27th 1843[2]

My dear Tyndall

I am glad that a letter from either of us is so short a time en route, yours of Christmas day[3] arrived here the day following. I am happy to hear that you have enjoyed good health since you went to Leighlin. I admire the bold determined stand you have taken, relative to your future destination,[4] without reference to America, but mind you by this, I by no means despise America, more of this hereafter. Your circular is truly first rate, and not exaggerated, no, I believe you could not exalt yourself above measure in those attainments I think the smattering of Radicalism you got from Charley[5] has served you no small share, as appears from the general favourite you are become with both parties, to whom you refer. I have no doubt but you will ere long, get some calls, which shall raise you to such an eminence, that you will become not only a conspicuous character but one whose services shall be sought for, with greediness. More particularly as I hope there shall be a jealousy in other minds on account of the specimens that you will have executed – I was so tossed about in mind, when I last communicated with you, that I entirely forgot to say that previous to my leaving Preston I got another character from Capt. T.[6] much milder than the former, leaving out the obnoxious paragraph with which he concluded before, but still retaining as much of the virus as poisons all. I also forgot to tell you to acquaint Phill,[7] that Sergeant Rodgers[8] got his also, and will keep it till he sends for it. I gave Rodgers his address, I know Phill will be raging with me about this neglect, but I leave him in your charge to manage and curb him. I must write him a short note this time, as I have nothing to communicate but what you can acquaint him of. Your American correspondence is quite consonant with the general character which I always heard of your friend Wright.[9] I cannot say about going to the States in Spring, I am not prepared in the exchequer line[10] for such a journey now. Remember I only speak for myself in this matter – I fully agree with you, that there is a continual drain on the purse when out of a *[situation!]* I don't know whether I told you that two of those to whom I showed my specimens in Manchester requested me to leave my address, which I did. I am making some preparations to draw a plan from the Ordnance map as a specimen, to a scale of either 12 or 16 perches,[11] I think the latter may be the best, it will not be so large and yet large enough. I hear some talk of one of my neighbouring gents getting his estate surveyed, a part of which he is getting done now. I have sustained a loss by not being at home a month ago, as I would most assuredly have fallen in for this, but I don't give up hopes of the estate and I hope to have my specimen well on next week. I find it difficult to work on account of this person, and that person coming to see me – In what style should I execute it?, in crow pen, or colour? Let me have your advice about this, I would do two, but I am afraid I would fall short in time. However if time permits I will have one of each, but let not this prevent you of complying with my demands. On the present, I go tomorrow please God with my Father to look at a new line of road which is to be laid out in some of Colonel Clement's[12] property. The county surveyor wants to run it one way, the tenantry another, and as it is intended for an accommodation, of course the latter must be pleased, and for this purpose my Father wishes me to go, and see it. I could manage it with a circumferentor[13] rightly, as I have got a level[14] in Liverpool which will answer all the purpose of a level attached to the instrument. Besides the intended line will be nearly a direct one, who knows yet I spent today or at least a young men spent with me in my Father's who has been seven years back and forward in America he gives a very favourable account of it, and invited me to his place, where he will give me all the information he can on the subject, I think you misunderstood me about writing to Bob.[15] Now I did not write to Bob since I came here until yesterday. I wrote to him from Manchester to which I referred in my last[16] and you have understood it to be from here, not so. I heard something in the way of rumour about Bob's reemploy but only a rumour, nothing official whatsoever. I wrote yesterday to <u>Bob, Charley, and Otty,</u>[17] and expect to hear a great deal of news the end of the week. Tyndall, I have no desire to go back on my old pay.[18] I was there long enough, perhaps too long, <u>I got a hearty welcome home and will have the same as long as I wish to remain</u>, besides there is a great desire to keep me, hoping that something may arise which will be an America at home but it must be good that will compensate. I never regretted doing what I did, I never had an uneasy 5 minutes about the matter since its first proposition, and Jack, I did not go to Kingstown! I am sorry in one way, that I did not go, to behold the idol of your soul though Jack requested me to go, but I really had not time. I bought a chain that cost me 9/6, 6lbs and at 1/6 per pound, no less could be taken so I had to down with the cash – And now for the whole truth – you ask me about my leaving Preston – why you puzzle me indeed for 'tis more than I can tell, it would require a poet's pen to describe that scene, you are possessed with all the truth, and description of a Cowper.[19] Suffice it to say that I got the Testament, and worthy of being looked upon, as proceeding from the house it did, and as for the penknife if only you could see it! and then about the parting of friends. I saw them all, and you don't know the half of it, what do you think I was to have fallen in for after Xmas had I remained? No less than being best man!! at a wedding, of Miss Fothergill[20] and a dissenting Minister the name of Fielding.[21] Oh Jack how unfortunate, but what? it is all for the better, and then the morning I started I bade farewell to Mr Lowe,[22] and family though last of my friends not least! First Mr Lowe, then Miss Lowe,[23] I pause, 'but oh! I never can forget, the parting look of sorrow',[24] then with William, and here the scene drops which may, and I doubt not but has drawn a veil, between us, which may never be rent – I never raised hopes to crush them! Not like don't tax me with inconsistency! else the burden devolves on your own devoted head! Now have I not spun a very long yarn, you know my address put me in mind of anything. I forget, but I did not forget the tin case, nor the screw nail boots. You will likely see Phill show him this, really I don't know what to say to him let this be for you both for the present, I said in the former part of this I would write to him, but I have nothing to say to him, but what I have said to you, let him not be jealous for this time, ask him did he hear anything about the farewell to Miss Simpson[25] I forgot to ask Bob about it.

believe me dear | yours ever | Geo. Latimer | 20 minutes of 12!! and all in bed

RI MS JT/1/TYP/11/3742–3743
LT Transcript Only

---

1. *Newtown Gore*: a village in County Leitrim, Ireland.
2. *December 27th 1843*: from Tyndall's journal, 29 December 1843: 'Received letters from Payne, Sayers, and Latimer, the latter is still subdued and has not the least desire to return to his old rate of pay' (RI MS JT/2/13a/10).
3. *yours of Christmas day*: letter missing.
4. *the bold determined stand you have taken, relative to your future destination*: see letters 0234, 0251, 0254 and 0259 of this volume for further information regarding Tyndall's plans to emigrate to America.
5. *Charley*: possibly Charles Edward Stanley; see letter 0273, n. 9 of this volume.
6. *Capt. T.*: Henry Tucker.
7. *Phill*: probably Phillip Evans.
8. *Sergeant Rodgers*: not identified.
9. *your friend Wright*: William Wright.
10. *I am not prepared in the exchequer line*: i.e., Latimer does not have the savings.
11. *12 or 16 perches*: a perch was an archaic unit of length, largely discontinued in the 15th century but still used in some professions. In Ireland, one perch was equal to approximately 6.5 metres (approximately 21 feet), although conversions could differ by local custom.
12. *Colonel Clement*: not identified.
13. *a circumferentor*: a surveyor's compass, used to measure horizontal angles (*OED*).
14. *a level*: a surveyor's instrument for measuring relative heights of land (*OED*).
15. *Bob*: Robert Martin.
16. *my last*: possibly letter 0273 of this volume, though Latimer did not refer to writing to

Robert Martin in this letter.

17. *Charley, and Otty*: not identified, though 'Charley' could be a nickname for Charles Edward Stanley; see letter 0273, n. 9 of this volume.
18. *back on my old pay*: Latimer had been dismissed from the Ordnance Survey in November 1843; see letter 0262 of this volume.
19. *a Cowper*: possibly a reference to William Cowper (1731–1800), English poet and author of *Olney Hymns* (1779), *The Task* (1785), and English translations of Homer's *Iliad* and *Odyssey*.
20. *Miss Fothergill*: not identified.
21. *Fielding*: not identified.
22. *Mr Lowe*: William Lowe, no biographical details known; see also letter 0273, n. 2 of this volume.
23. *Miss Lowe*: not identified; see also letter 0273, n. 4 of this volume.
24. *'but oh! ... parting look of sorrow'*: possibly a reference to G. H. Rodwell, 'Thy Parting Look' (London: Goulding & D'Amaine, 1829), ll. 18–19.
25. *Miss Simpson*: not identified.

From W. Marquis — 31 December 1843 — 0282

Preston | 31st Dec 1843.

My dear Tyndall,

I cannot say the degree of pleasure I felt in receiving a line from you.[1]

It is Sunday and the last day of the old year. Parker is gone to visit the Banisters[2] and I sit alone by the fire enjoying the social companionship of my old friend the pipe – Allen pops in 'I am just writing a line to Jack'. 'Are you? will you remember me to him and tell him I will write to him in a few days'? 'I am going to a wedding tomorrow a clergyman is to be married, he is 40 the bride 30 I am to be groomsman!' 'and I suppose – Smith[3] is to be Bride's maid? No, no that would be too good a thing – but she will be at the party in the evening'.

This was some of our conversation. Allen and your French tutor Sindal[4] took their Christmas with me – Sayers and Parker were at Banisters. It is surprising to me the effect of this moment's interruption – I am quite driven out of my course – When Allen came in I was closely perusing your note – believe me my dear Jack I appreciate – I feel the goodnatured – kind feelings that pervade your – I mean that dictated your letter.

Damn this annoyance! I say – here comes Banister[5] – He has a temporary engagement at Blackburn came home today to see his family and returns by the coach at 7 o'clock. His mother[6] went to London with Sayers yesterday morning 20 to 10 o'clock he desires to be kindly remembered to you. Miss Banister often asks me for you I assure you that you are not lost in the remembrance of any of your acquaintances, they all speak in the highest possible terms of you and regret exceedingly your leaving Preston.

Well, this is too bad – that I cannot communicate with my old and well-beloved friend – even in silence, without being annoyed by repeated interruptions – Here comes Fitz Patrick[7] with his God save – and how the devil are you &c.

– 'I say Marquis will you come to the Ball tomorrow evening there will be some fine girls there – Do you know Marquis that the finest girls in Preston are Roman Catholics – ay my boy and some of the richest too – there's Miss – ,[8] her father is one of the richest men in Preston'.

I am sorry I can't go – I have a very sore foot – I could not go to the office yesterday at all – for which you know Stanley[9] gave me a blowing up – James Martin[10] was checked by order – made public 5 days for absenting himself 4 days last week.

In come Allen – Is this you Fitz – how are you Allen &c. &c. He has a letter in his hand – the seal speaks its contents – 'Without you I die'!!! he sits down and opens the letter very carefully and reads – but inwardly – his countenance brightens – his eye sparkles with ecstasy. 'Good news I hope Allen'? 'A-y' – but will you come to meeting tonight and usher in the New Year. I am sorry I can't call for you as I said I would – but I will see you there I hope – My services are required in another quarter or else I should be most happy to go with you. I went to Lune Street Chapel at 10 o'clock – heard a very God-like discourse – a humiliating exposition of our past follies – a solemn supplication to heaven for forgiveness – Three minutes before 10 o'clock all was in solemn silence and serious prair – Even in the awfulness of this moment I perpetuated a – no I won't say a crime – I only looked at a pretty girl – but she looked at me and so had the sin to answer for. As the clock struck 12 upwards she darted her eyeballs and thus for some time seemed to be in communion with the Spirit – I left her so!!

My dear Jack you must recollect that I am finishing my letter to you – Fitz is busy preparing for the Ball – he was here just now and he desired to be most kindly remembered to you. I haven't seen Allen today, so I cannot tell you how the wedding came off. Holland was employed to day Mr Gowan[11] left for Ireland on last Friday. Fitzpatrick and Bob Small[12] are living at 11 Butler Street. Barker[13] and I at No 5 Bow Lane. Parker had a letter from Phill today will you please to remember me to him, and tell him I felt disappointed at his not calling on Mr Macklion[14] – he does not speak of having done anything in Dublin. Did he call at Griffith's[15] /himself/? As to yourself Jack I must say that all your travels and adventures are a perfect mystery – at least to me – N'importo[16] – as you say will talk of love and Goosnaugh[17] matrimony. I haven't been there since but when I do go you shall not be forgotten to them – to Miss Lucy[18] – I beg your pardon – it's 'poor Lizzy' I mean, the 'star of your soul'!! Eh! Jack you're a rum 'un – Come now Jack you must not think that I am disposed to trifle with your better feelings, no far from that – I've a sore foot or else I would go out to Goosnaugh next Sunday and represent you to her in the most loving manner – Yes and if opportunity would permit I would give her half a dozen kisses for you.

Now about the writer – I was taken in beautifully the last evening then. You have heard of the Mistletoe Bow I happened most unwittingly to get under it – and maybe I hadn't kissing you ever saw. The girls kiss first, the boys kiss after of course – they told me it was Lancashire fashion when they saw me blushing and embarrassed – the dear creatures what a nack they have for soothing difficulties – Note the following Receipt – Take a bough of Mistletoe and hang it up in a warm place. Put beneath it a young lady with whom you are rather sweet. When she begins to melt, which you may know by her change of colour, you should run your rigs and mould your tips for kisses. We had a splendid party the last night I might say the best we ever had. Need I say that I shall always be happy to hear from you – I have an affection for that part of the country near which you reside I have lived in Bonnie[19] a fortnight the girls there wanted to know I would Goke.[20]

very sincerely yours | W. Marquis

To John Tyndall Esqre | Leighlin Bridge | Co Carlow[21]

RI MS JT/1/TYP/11/3750–3751
LT Transcript Only

---

1. *a line from you*: letter missing.
2. *the Banisters*: though there were many Banisters in Preston, this probably refers to the family of Mary (*b. c.* 1801) and Samuel Banister (*b. c.* 1791). Samuel Banister was the High Constable of Preston. His children Emily (*b.* 1820) and Frederick (*b.* 1823) are mentioned in the letters; see letter 0284 of this volume.
3. *Smith*: not identified; possibly Robert Allen's future wife, who had many relatives with the surname 'Smith'.
4. *Sindal*: not identified.
5. *Banister*: Frederick Dale Banister; see n. 2.
6. *His mother*: Mary Banister; see n. 2.
7. *Fitz Patrick*: not identified.
8. *Miss –*: this name does not appear in the transcript, though it may refer to Miss Emily Banister, see n. 2.
9. *Stanley*: possibly Charles Edward Stanley.
10. *James Martin*: not identified.
11. *Mr Gowan*: not identified, possibly John McGowan, a Civil Assistant on the Ordnance Survey who had been dismissed alongside Tyndall; see letter 0262 of this volume.
12. *Bob Small*: Robert Small, a civil assistant on the Irish Ordnance Survey. He worked in the A District 1st Division under the command of Captain Tucker. Small joined the Survey on 30 August 1838. His station was at Waterford. He primarily worked as a surveyor, but was also examining groundwork at certain junctures. His daily rate of pay rose to 2s 8d by the time he left the Survey. Small left the Survey along with the majority of the A District workers during the autumn of 1841; his reappearance in Preston suggests that Small had joined the English Survey.
13. *Barker*: not identified.
14. *Mr Macklion*: not identified.
15. *Griffith*: probably Richard John Griffith.
16. *N'importo*: not important (Italian).
17. *Goosnaugh*: more commonly spelled Goosnargh; a small village near Preston.
18. *Miss Lucy*: not identified.
19. *lived in Bonnie*: not identified.
20. *I would Goke*: the meaning of this is unclear. Goke means at the heart or core (*OED*).
21. *To John Tyndall ... Co Carlow*: address given by Louisa Tyndall.

1844

From Archibald McLachlan 1 January 1844 0283

New Years Day

Dr Tyndall[1]

Procrastination – just as usual, and what other way should it be – were I to be punctual you might suppose something was wrong. But its all right, just what might be expected – And now allow me to wish you a good New Year – with us here of the O.S.[2] it is dull enough god knows. I heard one fellow express himself 'If I was at home I would think it a Disgrāse to spind[3] the day this way'. If I don't tell you now I will forget it – I had two letters from Bob Bloomfield[4] he is in Scotland on some railway between Paisley, Dumfries, & Glasgow[5] – he tells me that a C.E.[6] an old acquaintance of mine has been asking him concerning me. Bob says, 'I put in a good word for you' – so peradventure I may fall in for a good job – Two extra causes has delayed me in writing this letter – One is, I expected to get a peep at all the Tidal observations. I am disappointed – But I have got what is much better, the opinion of a learned man[7] who calculated and arranged the whole thing. and secondly I expected to steal a copy of the Tide Wave,[8] I have been able to pilfer only one Trace of one Station, another Trace I am sure I can get during this week, and, this, or the next week I am told I can manage to trace of the 'Shannon Wave'.[9] I look on this wave as the most important & interesting –

– A great wave follows the course of the Moon, Sir Isaac Newton and others calculate the moons mean wave at 9 feet[10] – The Suns wave at 2 feet – This wave forms what we call the tide – there are 2 waves, one due to the moon, and one to the Sun – the Moons wave follows after the moon, no matter where the Sun is – the Suns wave follows the Sun no matter where the moon is – so the 2 waves seldom coincide – Suppose the Sun & moon to be at right angles, or, suppose the moon to be over head, and the sun at that moment to be just setting – we will at this moment (or nearly so) have high water of neap Tide[11] – the moon loses ¾ of an hour every day – the Sun gains upon her ¾ of an hour every day – till at last he comes up with her – If he is gaining ¾ of an hour on the moon, of course his tide of 2 feet is catching or following her tide of 9 feet – & when the Sun has come up with the moon (as is the case in the eclipse of the Sun) the tide which followed the Sun will be on the top of the moons tide, the 2 waves are then run into one wave and form what we call Spring Tides[12] – The height of those waves will vary according to the season – or the distance of the earth from the Sun – The true form of the Tidewave cannot be determined, except by measurements taken in the open, wide Ocean, free from the obstruction of headlands Continents &c – The object of our O.S. measurements is of a local nature, to determine the mean rise or fall of the Tide at any given place, and thus fix a Datum line for our O.S. levels – round the coast of Ireland there are, or were 22 Tide gauges (Poles or Boxes, with Feet and inches marked on them and sunk or placed the most convenient place) at each gauge a man was placed with a proper book, and chronometer – the observations were made at all the Stations on the same days – they all started at the same moment and ended at the same moment – each man entered in the book the height of the tide on the gauge & that was all he had to do – he entered the height every 6 minutes –

Thus!	Hour-M	feet	in
Augt 25. – – A.M.	6.0	14 –	4
	6.6	14 –	0
	6.12	13 –	9
	6.18	13 –	5
	6.24	13 –	0
	&c	&c	

Now you will see with ½ an eye that unless the observations are regularly entered, the whole is useless – suppose the observer gets tired of watching – or – falls asleep – and wakens after 2 hours sleep! He could by having the first half hours observations carefully made he could construct or fill up the book by allowing a fall or rise of Tide of 4 inches for every 6 minutes!! this may, and has been done to an extent unknown – however after all the 22 stations have been compared, Proffessor Airey of Greenwich[13] is fairly floored & will give it up in despair!

– One grand object in view was this. It was expected that the <u>mean</u> tide <u>mark</u> between Spring & neap tide would be in the same point or would coincide – whether from Bad observations or otherwise nobody knows, but the points wont agree.

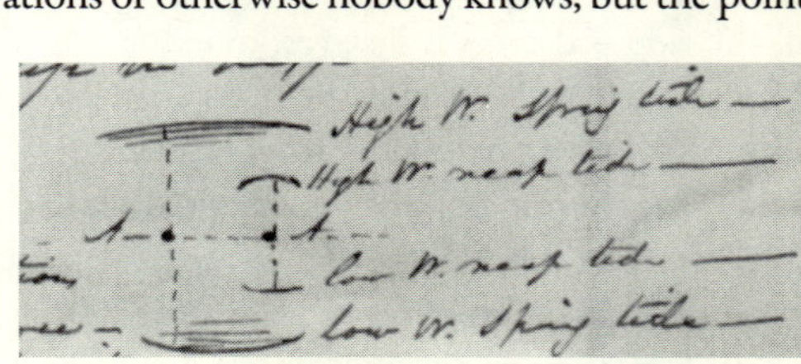

Divide the distance between high & low water of Spring tide into 2 parts – call the point A – divide the distance between high & low water of Neap tide into

2 parts, call this point also A – Query, will these 2 points be on the same level – will they coinside – or will they be equidistant from the earths centre? Our observations say No – Sometimes one point would be below the other nearly 2 feet – at others nearly 2 feet above! – So far as we have gone & so far as I can discover – our experiments &c have ended in a bottle of smoke, so far as the grand point is concerned – & must remain undecided til we can find Trusty men – men who feel a personal interest in the results – and who will spare no pains to note and mark regularly the different rises & falls – it wont do to fall asleep – get tired – go on the Spree and then Fudge a Book!!! –

'Howsomever' from what has been done we can see the form of the wave – nearly – – We can't discover the form or quantity of the suns wave – we only see the moons – keep in mind Sir Isaac's calculation of the suns wave of 2 feet – Now suppose the the sun & moon to be at right angles – it is evident the suns wave of 2 feet must occupy the hollow between the moons waves and will raise that hollow (or low water) 2 feet higher than it would be – or, prevent the tide from falling so low at that place as it would otherwise do – this is theory, and our observations don't throw any light on this point – You will see on the Tracing curious irregularities in the curve – but these are not due to the Sun or gusts of wind – or anything else except the carelessness of the observer – we might expect some irregularities from the obstruction the wave meets with – but not such as are on the tracing – The R.S.M.[14] observers knew, & expected irregularities – and not to disappoint 'Filosofers', they made irregularities accordingly!! –

You perceive that from low water on one side of the Earth, to low W.[15] on the other side is 12.000 miles – the wave is, say 20 feet high – we have then a Base of 12.000 miles and a perpendicular of 20 feet – how can we plot that?!! – only one way, make the Base on one scale, the perpendicular on another – which will be done –

Our observations include from 6 AM to 6 PM – taken on one gauge only – at one place – Draw parallel perpendiculars equidistant – See Tracing[16] – Call each space between the lines = one hour – divide the hour into 10 parts each = 6 minutes – Look to the right of the Tracing there you have your Tide Gauge – Zero is down in the mud below low water – the perpendicular scale is any size you please – but on the trace the perpendicular scale of feet, is made to correspond with the scale of minutes – this is owing to accident – (as a piece of paper was found by the plotter, ruled, and nicely divided, he adapted the lines merely to save trouble, because if he did not use that paper, he would have to rule & divide a sheet for himself) – You commence at 6 oclock – it is evident that the space between 6 & 7 oclock represents so many miles – that is, so many degrees or miles, according to the latitude of the place, – The wave passes over you!! or suppose it so – you are supposed to stand under low W. mark, and as the Tide rolls over you – you push up your staff and take offsets to it – Just the same as if instead of walking along a crooked fence & taking offsets, you stood in one place and made the crooked fence pass by you and as it passed by, you took offsets to the crooks; if the fence was made to pass by at the rate of so many feet per second – you could take an offset at every second or every 6 seconds, and afterwards turn your seconds into feet – It is all the same in the end, whether you fly like a Thunderbolt over the Ocean at the rate of 1000 miles a second, and take offsets at every mile, down to the Sea – or whether you sit quietly down and let the Roaring ocean pass over you &c

– Each line on the Tracing is taken at the same place but each line on a different day – each colour or form of dot, represents a different tide – Now let us take one Tide, it will explain the rest, take the Black indian ink line – At 6 AM – Tide gauge read – 13 f –0 in – and there the Black line begins (on the left); we will throw away the minutes and state the hours only – –

				f	in	
At	7	P.M.	Tide gauge read	15.	0 –	
"–	8	"–	"– "– "–	17–	0 –	
At	9	"–	"– "– "–	18–	0 –	And this 18 feet is high water – the tide begins to ebb – and
At	10-Oclock	AM–	Tide gauge =	16 –	9	
"	11 –	–"	–" =	14 –	0	and ebbs till we come to 3 oclock P.M.

when it is low water – It then begins to rise and at 6 P.M. we have high water = 12 – 2 – Thus the form and quantity of the wave at that time and at that place is plotted – Observe high waters on the left hand, how they shift their places – The high red line is farther to the left than the top of the Black line, this is caused by the wave following the moon – The Sun loses no time – 6 oclock is 6 oclock, or nearly so – the moon however loses ¾ of an hour every day – & so must her tide shift – were the series of observations carried on, you would see the tide (top of the wave) moving to the right hand ¾ of an hour every day – You will see a number of small little bends or irregularities in the curves, I can't account for them otherwise than by supposing bad observations – were they caused by Islands, or obstructions of any kind, each wave would be similarly affected, but you see they are not all affected at the same place – You see the words – 'Standard level of Meads Quay' – This was only an assumed point – it does not interfere or explain anything – that standard was to be used but is to be abandoned as impracticable and (it is said) something else is to be adopted!! –

General

It was the intention of Proffessor Airy to have all the high & low waters of the 22 stations laid down on a map of Ireland, just in the same manner we lay down our heights on a 6 inch plan – by that he could have a general view at a glance, of all the different rises – their time – and distances from each other – he could balance the quantities, or, nicely adjust them – and then strike a 'Mean' – he would reduce their differences to one common Datum line, and this was to have been the datum for the O.S. levels – At present this is undecided & probably will remain so for some time longer, for our observations are not very satisfactory. –

On some parts of the Coast the rise or fall of the Tide is hardly perceptible – in the C°. Antrim[17] near Island Magee[18] the fall is not more than 2 feet –

At Courtown, C°. Wexford[19] – low water is higher than high water!! i.e. – when by the regular course of the tides round the Coast we would expect to have high water at Courtown, we have Low Water; & when we should have low W^{r}. we have high water, & the difference between high & low is only 4 inches!! Low water is 3 inches higher than high water –

The difference between low water at the S end of Sligo[20] & Dublin Bay[21] is only 1/100 part of a foot – so we make it –

The tide rises at Limerick[22] one foot higher than it does at the mouth of the Shannon[23] –

Generally speaking, the tide falls or Rises 3 or 3½ inches in 10 minutes – At Limerick the tide rises 20 feet – and falls at some parts – (or during a certain portion of the fall) the tide falls one foot in 10 minutes –

Wishing you grace, good luck & patience to read my scientific scribble – & hoping you may require some explanations! I wait in peace for <u>your</u> Cogitations

Yours sincly | <u>AMcL</u>[24]

RI MS JT/1/TYP/11/3820–3824
RI MS JT/8/1/4a

---

1. *Dr Tyndall*: Dear Tyndall.
2. *O.S.*: Ordnance Survey.
3. *Disgrāse to spind*: McLachlan was probably using the underlines and overbars to indicate a particular regional pronunciation of these words.

4. *Bob Bloomfield*: a civil assistant who joined the Irish Ordnance Survey on 12 September 1840, working in the Examination Office at the C District headquarters at Cork. He was transferred to the 3rd Division, C District on 14 February 1842, with his principal duty being drawing. By August 1842 he was back at the Examination Office in Cork, where he worked alongside John Bridgeman Edgeworth, although he was transferred to the English Survey in the following month, where he was involved in the surveying of York.
5. *Paisley, Dumfries, & Glasgow*: the Glasgow and Paisley Joint Railway Company was formed in 1837.
6. *a C.E.*: a Civil Engineer, not identified.
7. *a learned man*: not identified.
8. *a copy of the Tide Wave*: possibly a reference to George Biddell Airy's 'Tides and Wave', *Encyclopedia Metropolitana* (1841), pp. 241–396.
9. *the 'Shannon Wave'*: the rise and fall of the Shannon river in Ireland.
10. *Sir Isaac Newton and others calculated the Moon's* mean *wave at 9 feet*: for Newton's calculation, see I. Newton, *Principia Mathematica* (1687), Proposition 66, Corollaries 19 and 20.
11. *neap Tide*: neap tide occurs when the sun and moon are at right angles to one another as observed from earth; this is the lowest level of high tide each month (*OED*).
12. *Spring Tides*: spring tide occurs when the sun and moon are in alignment, as seen from earth; this is the highest level of high tide each month (*OED*).
13. *Proffessor Airey of Greenwich*: George Biddell Airy (1801–92). Airy was a renowned astronomer who served as Astronomer Royal from 1835–81. He moved to the position from Cambridge University, where he had held the Lucasian professorship in mathematics and the Plumian professorship in astronomy. Airy's best-known accomplishments include the establishment of the Greenwich meridian, supervision of several expeditions to observe eclipses and other planetary phenomena, and the creation of highly detailed tables and maps of planetary motion. He also oversaw the collection of a tremendous amount of data about British tides, especially on the coasts of Ireland, as this letter illustrates.
14. *R.S.M.*: Royal Sappers and Miners; see letter 0231, n. 32 of this volume.
15. *low W.*: low water.
16. *See Tracing*: tracing is missing.
17. *C°. Antrim*: a county in northern Ireland.
18. *Island Magee*: Islandmagee is a peninsula on the east coast of county Antrim.
19. *Courtown, County Wexford*: a town and county in Northern Ireland.
20. *Sligo*: a city on the western coast of Ireland.
21. *Dublin Bay*: an inlet on the eastern coast of Ireland.
22. *Limerick*: a town in Ireland on the River Shannon.
23. *the Shannon*: a major Irish river.
24. *AMcL*: Archibald McLachlan.

From Robert Allen 3 January 1844 0284

Preston 3rd Jan. 1844.

My very dear Jack

Depending upon the forgiving character of your disposition I venture to send a few lines after a week's delay in reply to your very kind letter of the 25th Xmas day,[1] on which I dined with Marquis and your late French master.[2] Tyndall's prayer availed on behalf of young Allen's appetite. Had the disfigured remnants of the roast joint and pudding, been visible at L.Bridge[3] I am convinced your opinion would coincide with my statement of the devouring fact. During the festivity I was highly amused with the master of French language, despoiler of its manners but an adept at Ballroom etiquette. An ensample of polite literature indeed. In my mind I compared the Tutor with his absent Pupil, and what a contrast realized itself. Without flattering, Tyndall appeared enthroned on the most exalted views of my imagination. 'Oh! land of my forefathers',[4] seat of intelligence, bright adornment of the universe – ever remain 'first flower of the earth – first gem of the sea'[5] – still continuing to shine in the light of thy gifted sons. But I must subdue the rising spirit of patriotic pride that would fain expand my soul at the remembrance of past days spent with the brave and generous men of Erin;[6] and try to give you the particulars at least an account of passing events. Since I last wrote Sayers is gone, I saw him off on last Saturday morning with Mrs Banister, patroness of —[7] for London. She desired her kind remembrance to be sent by me to Mr Tyndall. Sayers had the honour and privilege of witnessing the tender smiles playing upon the lip of the amiable Miss B—r[8] as he left, either to his pleasure or pain, perhaps both – poor fellow (soon by all accounts to become rich) I felt very much at his leaving, as I might almost term him 'the last of the Mohicans'.[9] He promised to write in a week or ten days; had a letter from bro. Lat.[10] on Friday, is well, and safe at home; his love I am sure is not waxed cold, neither are his affections gone from his old help-mate of Butler St.,[11] so that I conclude you have ere this heard a long detail from him of his tour by land and sea. I had also a note on last Monday morning from that little rebel of Carlow[12] reproving me for a neglect of his letter of eight or ten days antiquity which his impatience had summed up to a fortnight. I had just the night previous to the date of his letter, sent six pages of cross written scribble[13] which I hope for the present has set his mind easy. You state that an appeal for the last time was made by him concerning his notorious character as an outlaw; a very great inconsistency appears to be connected with those characters given at the same time. You and Sayers have got Protestant ones, previous to the general delivery. Now Griffin being a pious Roman came with the rest under the condemnation of the unrighteous law of Tucker and suffered the penalty thereof with his unholy accomplices. Yet a facsimile of your dismissal was handed to him at the same time with that of the Out Cast Evans, as a quick passport thro' out of work purgatory. This information I send considering it to be a great injustice done to the remainder. Caroll I believe has got a situation as schoolmaster. Holland has come back on the reduced wages of 2/9 per diem. Since the 1st good for nothing sneak, never made any exertions except exercising his limbs six nights out of the half dozen on the boards of a factory ball room (alias) the floor of a public tap room. McGowen[14] left on last Friday for Manchester to meet his Papa. Todd[15] received a letter today conveying no better news from his family than that he refused to go home – intended to try Bolton,[16] and determined to dance on next Tuesday night at St. Austin's Ball in Preston, sweet man. Again, nought else of importance has since transpired save that your humble servant on yesterday took the place of George[17] at Miss Forthergill's[18] wedding. The boys in the office all were certain that it was myself who bore the name of groom, but if you add man to it a correct idea of my situation will be had by you. Needless to mention that all went off well. Felt myself an Irishman throughout the whole, consequently satisfactorily performed my duties to the approbation of the idle spectators and to the delight of my charming Miss Smith,[19] and to my own gratification, spent a pleasant day and a happy evening enjoying the rapturous emotions that ever swell the blissful bosom when seated close to the darling of the heart, you may guess and rightly too that my arms more than once encircled her graceful form and pressed it gently yet —[20] to my beating breast. I have little time to say more on the subject at present, but if spared the next will be on the heavenly theme of devoted love. Goodnight my dear Tyndall write soon and

believe me yours most sincerely | Bob Allen

I will be most happy to hear of your proceedings at London[21] if prudent to make them known I should never ask had you not intimated *[for]* having received more than one or two insults from some of the fellows here, for not acceding to a proposal which I conscientiously refused at the time of your leaving, not being made acquainted with it in a satisfactory manner, nor at all in a way pleasing or agreeable to sensitive feelings. I have been called to my face a coward in the matter before the men of the room when discussing upon other topics not at all connected with it, *[but]* I feel myself ever the same, ready to act consistently at all times upon the principles of propriety and justice. I may not have sacrificed as much as many on the occasion yet I have a certain sense more than persons are aware of. I care not for this, what I have done is done without regret and was performed without fear. I should not let you know anything of this only I did feel a little hurt by the insinuations which have in the manner referred to been thrown out without any cause. Laterly there has been no excitement of any kind.

R.C.A.

RI MS JT/1/TYP/11/3484
LT Transcript Only

---

1. *your very kind letter of the 25th Xmas day*: letter missing.
2. *your late French master*: not identified.
3. *L.Bridge*: Leighlin Bridge.
4. *'Oh! land of my forefathers'*: G. N. Reynolds (1770–1802), 'The Emigrant', l. 9.
5. *'first flower of the earth – first gem of the sea'*: T. Moore, 'Remember Thee!', l. 6. On Moore, see Biographical Register.
6. *Erin*: a modern derivative of Éirinn, the Irish-language name for Ireland.
7. *patroness of —*: this word was omitted deliberately from the typescript.
8. *Miss B—r*: probably 'Miss Banister'.

9. *'the last of the Mohicans'*: a reference to J. F. Cooper, *The Last of the Mohicans: A Narrative of 1757* (New York: H. Carey & I. Lea, 1826).
10. *bro. Lat.*: brother Latimer, probably George Latimer.
11. *his old help-mate of Butler St.*: probably a reference to Tyndall.
12. *that little rebel of Carlow*: not identified.
13. *cross written scribble*: a reference to the practice of cross-hatching a letter, in which the author rotated the letter 90 degrees and wrote over the top of the existing text.
14. *McGowen*: John McGowan.
15. *Todd*: John Todd.
16. *Bolton*: a town in Greater Manchester, England.
17. *George*: George Latimer.
18. *Miss Forthergill*: not identified; see also letter 0281, n. 20 of this volume.
19. *Miss Smith*: possibly Allen's future wife, who has not been identified but had many family members with the surname Smith.
20. —: Louisa Tyndall has intentionally excised this word from her transcript.
21. *your proceedings at London*: i.e., the attempts to meet with George Murray about the Ordnance Survey.

From John Sinnett 3 January 1844 0285

Liverpool 3rd January 1844.

My dear Tyndall

I received yours of 29 ult[1] in due course. I need scarcely say that I was glad to hear that you arrived safely at home. Your disasters in the deep, so graphically described, came in the end more fortunate than otherwise, although of course poor Phill[2] felt annoyed at parting with the 'garbage'.[3] And now I would notice a matter that I had thought of and intended to do before, and that is that in your future communications (which I hope will not be few) that you will dispense with the frigid mono-syllable with which you usually commence your letters to me, I mean 'Sir', for chilly it is, albeit the corners are somewhat rounded off by the prefixed 'My dear'. I hate it, it involves a degree of respect which between you and me is absurd, and always was, so 'no more of it, Hal, an thou lovest me'.[4] I send you the specimen[5] and as I love modesty have extended the size from 'one half crown' to somewhere about 2, I am as busy as a nailor[6] during the day therefore was compelled to do it by candlelight. I will not be so affected as to make any apology about it, although it was mere guesswork by candlelight, however it has turned out well and is I believe the best bit of wood I have done yet, you cannot but see its great superiority over the Cork work (I mean in style). You perceive it is quite a landscape view and shows the trees in groups as they ought to be, and not in single trees as in Cork. The colours are precisely the same but there is no blending of one into the other as we did in Cork: first there is a light wash of green (sap of course), then a working up of the groups sufficiently strong only to tell upon the ground, leaving some spaces for the burnt sienna which you must soften off with another half dry brush. Then a deeper tint of green to touch up and a finishing touch of lake mixed with sepia which produces the rich brown colour. The best 'hint' I can give you is to study the style, observe how the rotundity of the groups is brought out. The piece appears to me to be quite successful in this point. I have done it in colours although our practice here is sepia and if you can spare time I recommend you to try a small specimen or two all in sepia, it is a beautiful colour, and there is none works up so well. Larry Eivers was so fond of this style that he took away a piece of it done by me. I would not have had this so very green had I let daylight to it, but as far as the touching goes I could not do it better. I believe I have said my say on this subject; if there is any point upon which I can further assist you let me know and I shall answer without fail. You perceive I was tolerably prompt about this business, as I wanted to assist you in astonishing the natives. Ginty went I think on Sunday week, R. Hunter[7] had a letter from him, he has arrived safe in Carrick.[8] There has been a letter from Taylor in Dublin he has it seems got into Kildahle,[9] the Land and House Agents office at £50 per annum. There is another wanted, and he has been enquiring about Ginty, who did not see him as he staid only 20 minutes in Dublin. I believe R.H.[10] is writing to him to night about it. I wish you had passed as many years in the Dublin *[County]* House and school as I did, if you had, I verily believe you would not now need a specimen from your poor friend nor yet trouble yourself about gleaning after those trusty reapers Hodges[11] and Smith:[12] Yes I have seen Adam and Eve[13] and was entranced gazing at it. Was there ever a woman so lovely as the Eve in the temptation? If our good(?) mother was so lovely I fear much many, very many, of her descendants could scarcely help following the example that brought 'death into world and all our woe'.[14] So fare thee well friend John. I need scarcely say that you have my warmest wishes for your success.

Yours faithfully | John Sinnett

I recommend you not to follow Captain T's[15] system of putting in single trees thus without *[stems]*. Captain F.[16] and Mr Hamley, both men of refined taste, utterly abhorred it, and so does every one but its inventor who of course admires his own bantling. Don't forget to have a plan in sepia you will find it much easier for yourself and indeed the colours one is obliged to use in plan drawing are too rich and would certainly offend the eye of a person of taste. The sepia is chaste and would be preferred by all except the vulgar. At all events let people see both and if they prefer the sepia so much the better for you. | Adieu.

You must not forget that this system of wood is very much more difficult than the Cork plan and if badly done will look hideous therefore take two days hard work at it before you venture to put it on your specimen.[17]

RI MS JT/1/TYP/11/3864
LT Transcript Only

---

1. *yours of 29 ult*: letter missing.
2. *Phill*: Phillip Evans.
3. *the 'garbage'*: unidentified reference to something in Tyndall's missing 29 December

1843 letter.

4. *'no more of it, Hal, an thou lovest me'*: W. Shakespeare, *Henry IV Part I,* II.iv.202.
5. *the specimen*: specimen missing.
6. *as busy as a nailor*: 'as busy as a nailer' is an expression indicating hard work and tenacity (*OED*).
7. *R. Hunter*: Richard Hunter.
8. *Carrick*: there are multiple Irish locations that have 'Carrick' in their name, though this probably refers Carrick-on-Shannon, a town in County Leitrim.
9. *Kildahle*: possibly a variant on Kildale, a village in North Yorkshire.
10. *R.H.*: probably Richard Hunter.
11. *Hodges*: not identified.
12. *Smith*: not identified.
13. *Adam and Eve*: the precise image of Adam and Eve that Sinnett is describing has not been identified.
14. *'death into world and all our woe'*: J. Milton, *Paradise Lost*, book 1, l. 3.
15. *Captain T*: Henry Tucker.
16. *Captain F.*: not identified; possibly Robert Fenwick; see letter 0268, n. 25 of this volume.
17. *take two days hard work at it before you venture to put it on your specimen*: from Tyndall's journal, 6 January 1844: 'Received letter from Sinnett containing the bit of wood, practised on it all day' (RI MS JT/2/13a/11).

From Edmond B. Sayers 6 January 1844 0286

E.B. Sayers | Mr. Goodalls House | Samuel Street | Woolwich.

Saturday[1]

Here I am my dear Jack at as nice duty as you could wish superintending docks &c., and drawing plans – Not six inch plans,[2] but ground plans and elevations at 5s a day, 7 days a week, and a promise of 7/6 a day when I am more advanced. Do you call that nothing? Hours from 9 to 12. 1½ to 10 minutes to 4, and they are not particular to ½ an hour or 1 hour. 4,500 men under our superintendence, we have two docks making now that will cost 540,000 and more buildings than we can do almost. We are going to add 34 acres more to the yard, to pull down 43 old houses to build two barracks, a marine and foot one, and to survey the Thames from London to the mouth – 60 miles. That will be a job. Only three of us and Captain Danivon.[3] We are treated like gentlemen not like the survey. Can go in and out of the office when we like, no one to ask where we are going. Two of the Lords of the Admiralty[4] were here today and I had a long chat with them. I could not give much information to them being so lately joined. We have a splendid office, plenty of every kind of stores, and the best, we have H.H. pencils,[5] harder than the 4. H. ones of the Survey. I showed them some of them and they said they never heard of such things before. Are you doing anything? I came from P.[6] last Saturday with Mrs Banister.

How is Philigeen *[Evans]*?[7] remember me to them and George,[8] also, don't forget yourself. All quite well at Preston. When will I see it again?

Yours as ever | E.B. Sayers. | Samuel Street

RI MS JT/1/TYP/11/3858
LT Transcript Only

1. *Saturday*: Louisa Tyndall annotation: 'Jan. 8th 1844 | Envelope'. 8 January was a Monday, meaning this letter was probably written on 6 January 1844.
2. *six inch plans*: maps drawn at a scale of six inches to one mile.
3. *Captain Danivon*: not identified.
4. *Two of the Lords of the Admiralty*: not identified.
5. *we have H.H. pencils*: an 'H.H.' grade pencil is a very hard pencil, as opposed to 'B.B.', the softest grade pencil.
6. *I came from P.*: Preston.
7. *How is Philigeen Evans?*: probably a nickname for Phillip Evans.
8. *George*: George Latimer.

To Sir Robert Peel via the *Liverpool Mercury* 16 February 1844 0287

LIVERPOOL MERCURY, February 16th., 1844

Letter V.

To the Right Hon. Sir Robert Peel, Bart. &c., &c.

Sir,

Duties of a pressing nature requiring for the last few weeks my undivided attention, I was unable to continue my promised series on the Ordnance Survey. I now propose to address you, as briefly as possible, on the treatment which the persons employed on this great public work receive from those to whose superintendence it is entrusted.[1]

It might be reasonably expected that an undertaking so expensive, and requiring such an amount of talent for the effectual discharge of its various duties, would afford a decent means of subsistence to those employed on it; yet such is not the case; for though the resources placed at the disposal of the superintendents of the survey are quite sufficient, if properly applied, to render a situation on it comfortable, – nay, respectable, as it should be, – still, owing to mismanagement, the poverty of those connected with it has formed 'a proverb and a by-word'[2] in every locality where they have been stationed.

Why do not the superintendents of the survey stand by their 'Form of Recommendation,'[3] where six shillings a day, and upwards, are said to be awarded to efficient draftsmen? Is it a lure – a decoy – first to woo the young aspirant, and afterwards to furnish him with the 'flattering unction'[4] that he shall one day receive the pay of an efficient draftsman? I could point this moment to some of the best draftsmen on the survey who are compelled to barter health and talents for a pittance which was refused, not long ago, by the common labourers in the docks at Liverpool – namely three shillings a day. This 'Form of Recommendation' then is no standard. What then is? I reply, that the information obtained from a savings bank established by the superintendents of the survey in 1826–7 at Castleknock,[5] near Dublin, together with reports collected from herring vendors, potato merchants, &c., form the reputable premises whence the minimum expense of the Ordnance Surveyor's subsistence is inferred, and by which his pay is regulated. Is there another situation under the British Government where such means are resorted to? No, Sir. In this respect the ordnance survey stands alone in its glory; on it alone exists the anomaly of a first-rate draftsman working for three shillings a day, while his superintending captain of engineers[6] – the brightest evidence of whose ability or usefulness probably exists in his statistics of a huckster's[7] shop – makes the public purse available to himself for fifteen times the sum. If the unfortunate assistant fall sick, his pay is altogether withdrawn, and he may die like a dog, if his poverty stricken comrade assist him not. I am sure I am under the mark when I say that the average number of appeals from destitute ordnance surveyors to their brethren amount to seventy yearly. I may add that the ordnance surveyor has died while actually engaged in the duties of his situation, and his body, altogether unheeded by his employer, was left to the tender mercies of the parish authorities.

The wages of the great majority of those employed on the survey vary from twelve to fourteen shillings a week, a sum altogether inadequate to afford even a single man the means of repairing the bodily waste incurred by the muscular exertion which some of the ordnance surveyors undergo. I have seen men in the county of Lancaster[8] walk ten miles to their work, return after a day's incessant toil, and sit down to a meal which an engineer officer would think unfit for his dog – a mess of bruised potatoes and melted hog's lard formed the oft-repeated repast. I have seen men, after having been discharged, offer their services to the recruiting sergeant, and rejected on account of their functional debility induced through over exertion on the survey. I have seen some apply to the poorhouse for relief, and repulsed as impostors. But why continue these details? The whole system is made subservient to the ambition of a few engineer-officers, who, to secure a quiet continuance of their own lucrative success, have reduced those beneath then to a state of obsequious and abject vassalage.

Surely, Sir, it will be admitted that there is enough to excite the disgust of a respectable assistant, and to induce him to look beyond the survey for employment. Every possible means, however, is used to deter him from doing so. Witness the following hue-and-cry form, which is circulated through all its divisions and outposts, whenever such an event occurs: –

Name	Age	Height	Complexion	Eyes	Nose
A.B.[9]	23 years	5ft.9in.	Dark	Dark	Hooked

Marks	Address	Native of	Remarks
pock-pitted.	good	Limerick, Ireland	Insubordinate.

Thus is a respectable man reduced to the position of a common felon! Is there credit to be gained by the survey? – the engineer-officers have it. Is there blame incurred? – it presses on the men. Is a bundle of erroneous documents committed to the flames? – the officers may smile at the conflagration, their emoluments[10] are undiminished. The efficient assistants and sappers[11] are the only sufferers. In the eulogiums[12] heaped upon the survey before Parliament and learned societies, those are quite forgotten to whom the credit is justly due. Look at an engraved map, 'drawn' forsooth 'by Captain Tucker' who never placed a pen on it, engraved by someone else, while the name of him who first gave the thing an existence is consigned to utter oblivion.

The effects of this system are such as every reflecting mind must expect. In a moral and religious point of view they are most pernicious. Nor can the plea of economy be urged in its vindication; the fact of the officers being so well provided for, ought, in itself, to render their use of this plea suspicious, especially when we recall the doings of 1832. But my sixth letter,[13] in which I shall have occasion to remark on the history of the English survey from 1841 to the present, will prove that if the funds now wasted in the revision of botchery, resulting from the combined causes of apathy and inability, were applied to the encouragement of talent, the ordnance survey would be what it assuredly is not at present – a credit to the nation.

Spectator.[14] | Dublin, Jan. 13, 1844.

RI MS JT/1/TYP/11/3916–3918
LT Transcript Only

1. *those to whose superintendence it is entrusted*: the Superintendent of the Ordnance Survey of Ireland was Colonel Thomas Frederick Colby.
2. *'a proverb and a by-word'*: Deuteronomy 28:37.
3. *'Form of Recommendation'*: the form suggesting payment for members of the Ordnance Survey; see letter 0231 of this volume.
4. *'flattering unction'*: W. Shakespeare, *Hamlet,* III.iv.147.
5. *Castleknock*: a village west of Dublin.
6. *superintending captain of engineers*: likely Henry Tucker.
7. *huckster's*: a person willing to make money in petty ways (*OED*).
8. *men in the county of Lancaster*: Lancashire.
9. *A.B.:* initials not identified; they are presumably pseudonymous.
10. *emoluments*: i.e. salaries (*OED*).
11. *sappers*: see letter 0232, n. 10 of this volume.
12. *eulogiums*: praiseful talk (*OED*).
13. *my sixth letter*: Tyndall planned a sixth Spectator letter, but never wrote it.
14. *Spectator*: Tyndall's *Liverpool Mercury* pseudonym.

From William Ginty 23 January 1844 0288

Carrick on Shannon | January 23rd 1844.

Mon Cher Tyndall

'Man' says the Rationalist's creed 'is the creature of circumstances'[1] and with this for my text I propose to give such an exposition of it as shall shew that my long silence is not a fault of mine. The question is not a religious one – it is not whether those circumstances were ordained by the Immaculate One or not, but it is this – did they not prevent me from answering your letter[2] in proper time and was not my conduct the inevitable result of which those circumstances were the cause – emphatically Yes! I'm far enough in this strain I ween, suffice it to say that together with being out of Carrick for four days many other things combined to frustrate my intentions which I now so gladly execute – perform and so forth. I received No 5[3] in due time and without losing a post transmitted it to his literary highness of Lord Street[4] who was *[conning]* its contents on the 15th inst. – of state trial notoriety. I conjured him to all possible speed. The 'statistics of a hucksters shop'[5] takes my admiration to all intents and purposes – probably it has appeared on last Friday, but in truth this state-trial is monopolizing sundry columns. I received the 'Despatch' 21 days after date – which said Despatch was my prayer book on Xmas day last.

McGowan has joined the Survey again. 'The dog shall return unto his vomit again and the sow that was washed to her wallowing in the mire'.[6] T'is too great a triumph for the despotic vampire-blood-sucking herd! Oh degenerate sons of 'Erin's Isle'[7] how I blush for your spiritless conduct! a dive into the depths of the Shannon were preferable – The workhouse would be a heaven, the blue canopy of our planet an elysium contrasted with the scoffs and the sneers of the sapper-mob[8] and their degenerated satellites.

> 'A worthless crew to honour lost
> Who know them best despise them most'.[9]

Do you know a person named 'Garrett'[10] in Carlow he is a quill driver or accountant of some kind? I'll tell you why I ask. Some time ago an advertisement appeared in the 'Boyle Gazette' to the effect that the books of an Agent who had suddenly died were to be adjusted. 3 months were allotted for the execution and tenders would be received by the executors. This Mr Garrett proposed for them and offered to come here and do them, what his proposal was I can't say, but faith if he imagines that he can do them in 3 months and has tendered accordingly, he will be most damnably bit and his securities too. There are 252 accounts some not settled for 9 years – some for six &c &c. There were 18 large account books and no mortal men could do them in 12 months. 4 of us here have joined and put in a tender I think we will get them. £50 boy o! among four of us for 3 months would pay well enough, particularly as 3 of us are chaps I'm the only man among them. Mr Garrett would not get them if he offered for £10. The Executor[11] is a friend of ours.

Your correction of my French is no doubt just, but faith I seen it in 'Prent' that-a-way! —?[12] is not Adios[13] Spanish? How is Jim?[14] I was talking to a Captain Duckworth[15] the other day (the executor) he was after returning from Dublin and he told me that 'some fellow in a Dublin paper says that the Ordnance Survey of Ireland is "all wrong"'. 'He says right' says I '—' and so forth. Pray your Reverence is this fellow's name John Tyndall? If so, I have an ordnance map of a part of Roscommon Co.[16] joining this Co. I have to enlarge a farm for a friend the gratification of scolding them this way would be some compensation, for I'm only to get thanks for the job. I will of course examine it and if you like send you an affidavit of the errors!!! I know they will be numerous for I have glanced over a comparison with the engraved one and the ground I'll tell you a good */bull/* – the Shannon is the Co. B.Y.[17] consequently the Shannon is on it and the heights or 'altitudes above low water mark' were given at a certain point it is shown to be 138 feet and about a mile and a quarter nearer to the source another height is given viz 136 so that it seems it is running up the hill!!! Good the Shannon running up an inclined plane. What would Sir Isaac Newton say to this? – How would he reconcile it with his laws of gravitation?

I want you to write me an acrostic[18] – don't stare! tis not for myself! tis for a Mr Harrison[19] – the case stands thus – he loves her deeply, but oh he's afraid it is not reciprocal. Now then love but love without return a heart given but not exchanged! this is enough for your poetic imagination! The name is Dorothea Evans.[20] She spells her name he says Dorathea, this however I think is wrong enquire! Let me have it if possible in your next. I showed him some of your compositions and he is all anxiety to get you to do this. If you think tis for me I'm damned if I won't send you his certificate assuring the contrary!

Omega[21] Aubrey![22] eh! Oh you Ronayne,[23] Chadwick, Maryanne,[24] Tidmarsh, Deeble,[25] McCarthy,[26] Donoughcliffe,[27] Hebdon,[28] Wilden[29] humbug!!! Dare you thus rake up the memory of the past – look at that catalogue of martyrs thou most incorrigible of all incorrigible miscreant and call from it the many pure and unsuspecting hearts your wiles and artifices have consigned to the sear and yellow leaf – gaze again heartless monster, and cease thy idle prattle for ever! 'Oh there are some &c.'

I have not had a letter from Liverpool this fortnight. I wrote to George Latimer 3 weeks since and the sanctified rascal has not yet responded. May he be crushed neath the ruins of a tract repository or may his big bible trip him and break his neck for this treatment. Valentine's day is coming are you preparing a summat for Preston!!!

I will expect to hear from you in a day or two. My kindest remembrances to Emma[30] and Evans (and to Ellen[31] if you are winging a messenger to that quarter)

Yours ever faithfully and sincerely | Wm. Ginty.

RI MS JT/1/TYP/11/3627–3628
LT Transcript Only

---

1. *'Man ... is the creature of circumstances'*: an aphorism attributed to Robert Owen (1771–1858), the socialist and educational reformer. The phrase occurs often in discussion of Owen's work but does not appear to be a direct quote from his writings.
2. *your letter*: possibly letter 0275 of this volume.
3. *No 5*: Ginty is referring to Tyndall's fifth pseudonymous letter (written under the name 'Spectator') in the *Liverpool Mercury*; see letter 0287 of this volume.

4. *his literary highness of Lord Street*: not identified. Louisa Tyndall annotation: 'Does he mean the Mercury office'.
5. *'statistics of a huckster's shop'*: this is a quote from letter 0287 of this volume.
6. *'the dog shall return ... wallowing in the mire'*: 2 Peter 2:22.
7. *'Erin's Isle'*: 'Erin' is a modern derivative of Éirinn, the Irish-language name for Ireland.
8. *sapper-mob*: see letter 0232, n. 10 of this volume.
9. *'A worthless crew ... despise them most'*: R. Burns, 'Lines on Stirling', lines 7–8: 'An idiot race, to honour lost; who know them best despise them most'.
10. *Garrett*: not identified.
11. *The Executor*: likely Captain J. Duckworth; see n. 16 below.
12. —?: this term has been intentionally left out of the transcript.
13. *Adios*: goodbye (Spanish).
14. *Jim*: likely Phillip Evans.
15. *Captain Duckworth*: Captain J. Duckworth, full name unknown.
16. *Roscommon Co.*: a landlocked county in Western Ireland partially bordered by the Shannon River.
17. *Co. B.Y.*: County Boundary.
18. *acrostic*: a form of poetry in which the first portions of each line combine to form a word or phrase.
19. *Mr Harrison*: not identified.
20. *Dorothea Evans*: not identified; possibly a relative of Phillip Evans.
21. *Omega*: an occasional alias for William Ginty.
22. *Aubrey*: possibly Aubrey O'Brien.
23. *Ronayne*: not identified.
24. *Maryanne*: possibly Maryann Commins, a friend from Ireland.
25. *Deeble*: not identified.
26. *McCarthy*: possibly Michael McCarthy, one of the civil assistants fired from Tyndall's division; see letter 0262.
27. *Donoughcliffe*: not identified.
28. *Hebdon*: John Hebdon, a friend from Preston; see letter 0280, n. 8.
29. *Wilden*: not identified.
30. *Emma*: Emma Tyndall.
31. *Ellen*: a love interest and romantic ideal of both Tyndall and William Ginty, whom they presumably met while stationed in Kinsale in County Cork in the summer of 1841. Although she remained a regular subject of Tyndall and Ginty's letters and verse, Tyndall's ardour seems to have waned by July 1842.

From E.B. Sayers — 26 January 1844 — 0289

E.B. Sayers | No 11 Charlton Terrace | Woolwich

My dear Jack

I am very much obliged for the papers – Letters Nos. 1, 3 & 4[1] – I did not get No. 2. Capt. Stothend[2] desired me thank you in his name for them – but he says that it would be very easy to disprove every statement in the first one (he has not read the others yet) and that it is some person not on the Survey he is convinced who writes them! Is there a No. 5 yet? I hope they won't stop, with the discharging match. Did you get the paper safe? I can't go to London on a weekday to buy nice paper, so I hope that will do, it was the smoothest I could get here.

I hear Parker is going to Liverpool, under Hamley, he is foolish to leave Preston, I think. His mother says he is very lonesome without me.

Banister has got a situation in Blackburn £2.2.0 a week – no great things after all. How is the country? any fear of a rising? 'Put your trust' in Providence and keep your powder dry'.[3]

Anything doing about the petition to the Master General?[4] or is it dropped? after all the trouble and expense you have been at, I wish you had all your money in your pocket, and that the devil had the M.G.[5] by the nose.

So you are going over to the repudiating gentlemen?[6] I am sorry you should be thrown away on them. I hope your sister[7] has quite recovered her health. Poor Dan![8] he is in a fix I think, will he get out of it think you?

Does George[9] ever hear from his lass,[10] remember me to G-e-o-r-g-e – Tell Fill[11] Miss Simpson[12] is as large as life – I hope he is *[too]*.

My paper says stop. If you should come in this direction before you are transported, I will give you ½ a bed and 10/11 of a feed for a month with a 100.000 welcome (that should be in Irish).

Yours as ever | E.B. Sayers | 26th Jan.

RI MS JT/1/TYP/11/3859
LT Transcript Only

---

1. *Letters Nos. 1, 3 & 4*: Tyndall's letters to the *Liverpool Mercury*, which were written under the pseudonym 'Spectator'; see letter 0228, *Correspondence of John Tyndall, Volume 1*; and letters 0248, 0252, 0261 and 0287 of this volume.
2. *Capt. Stothend*: possibly Captain Richard John Stotherd; see letter 0280, n. 2 of this volume.
3. *'Put your trust in Providence and keep your powder dry'*: a saying attributed to Oliver Cromwell.
4. *the Master General*: George Murray, Master General of the Ordnance Survey.
5. *M.G.*: Master General; see n. 4.
6. *the repudiating gentlemen*: unknown group; possibly a reference to Tyndall's interest in moving to America.
7. *your sister*: Emma Tyndall.
8. *Poor Dan!*: probably a reference to Daniel O'Connell.
9. *George*: probably George Latimer.
10. *his lass*: not identified.
11. *Fill*: probably Phillip Evans.
12. *Miss Simpson*: not identified.

From Archibald McLachlan 31 January 1844 0290

31st Jan:

Dr T.[1]

I suppose you are in the Country – else I should have had a letter from you before this – I will send you a long letter tomorrow[2] with a tracing in it – I got a letter the other day from a Scotch C.E.[3] (a Scotchman) an old acquaintance he makes me an offer – or if I don't except it myself he asks can I recommend him a man – To Survey Estates in a Beautiful part of Scotland – the Job to last at least 6 months – he must be a good L. Surveyor[4] & Draftsman – he will give 20 to 25 shillings a week!!! – that is the Total amount he will give – I won't except it myself & I can't recollect anybody who would accept it – Nobody would close with his offer except one who is quite idle – out of work – Do you know of any Tolerable good Field & House man,[5] who would be glad of the Job? – If you know of one, send me his name & address – Of course one would not like to recommend a very inferior man, but I suppose a tolerably good man may serve his turn – whoever goes may expect to be well worked – he will take the worth of his 20 Shillings out of him – As for myself, I do not intend to accept it – Nobody could save any money, on it – and none should take up the job, except one who is Shack-floored[6] – – Can you think of any person – and let me know soon – I have got the other parts of the Tides[7] –

Yours sinly | AM^c^L

Mr Focus! | Mr Focus[8]

RI MS JT/1/TYP/11/3825
RI MS JT/8/1/4a

---

1. *Dr T.*: Dear Tyndall.
2. *I will send you a long letter tomorrow*: letter missing.
3. *a Scotch C.E.*: unidentified Civil Engineer on the Ordnance Survey.
4. *L. Surveyor*: land surveyor.
5. *Field & House man*: i.e., a surveyor who specialized in assessing and mapping property boundaries.
6. *Shack-floored*: not identified; possibly slang for a deeply impoverished person.
7. *I have got the other parts of the Tides*: possibly a reference to George Biddell Airy's 'Tides and Waves', *Encyclopedia Metropolitana* (1841), pp. 241–396, or a reference to a manuscript copy of George Biddell Airy, *On the Laws of the Tides on the Coasts of Ireland* (London: R. and J.E. Taylor, 1845).
8. *Mr Focus! | Mr Focus*: written on the outside of the letter.

From William Wright [31 January] 1844[1] 0291

Cincinnati[2]

My dear John

On yesterday I received both your letters and in compliance[3] with each I hasten a reply. I regret very much that this will not be in time to go by the mail steamer still I hope it will reach you before the 1st of April. One of your letters I perceive was written on the 17th of January the other on the 31st, the former must have been retained until the sailing of the mail steamer, as both came to hand the same day. The newspaper[4] I did not receive, however I am as much obliged to you. It gave me great pleasure to learn that you have made up your mind to start for America and particularly Cincinnati, as it is universally acknowledged that a man who cannot get along in it cannot do it elsewhere. You I hope will never regret the choice which you have made, surrounded as you will be by a comparatively new country whose resources are scare half developed, the fertility of its soil unsurpassed, (and above all) populated by a people whose activity and enterprise and industry are unequalled – in Cincinnati with its 50,000 inhabitants and 40 years growth there is more machinery and varied manufacturies at work than there is in Dublin with its 240,000. This is a place of all others for a man to stick himself down and you particularly who are possessed of more useful knowledge and scientific information than a 10th of the young men who usually immigrate here. I showed one of your letters to my uncle[5] and when he looked over your list of acquirements he asked why you hesitated, I replied that you did not for that I expected to see you before June. He then said that there would be no difficulty at all in getting employment and in some of our numerous public schools you would be a desirable person, at all events you may rely with confidence on a mercantile situation until something better would turn up. I am glad to find that your namesake's[6] letter corresponded with mine. Do not let him stop you in Philadelphia. I am overjoyed to find men like you Jack in the list of repealers[7] for every man who is a radical must be a repealer of every iniquitous statute which would bind one nation to the will of another. Oh that all Irishmen would but feel that they were Irish and how soon would England respect their rights and privileges. Notwithstanding my quotations to the contrary that England never &c. &c. Jack bring out that young lady[8] along with you, she would be a great solace to you in this far off land. The friends and companions of youth occupy a high position in our estimation. I suppose by this time you have found out that there is one step higher to be occupied but I suppose at this time it is engaged with you. I am much obliged for the intelligence respecting our old neighbours[9] what a change has come over them in a few years. Now to answer some of your questions; I sailed from Liverpool in one of the N.Y. Packets for which I paid £5 but you can get for less by making a hard bargain for I learned while on board ship from some of the passengers who occupied the berth alongside of mine that they only paid £3 10s. The course which I came to Cincinnati was out of the way and more expensive than the one I would advise you pursue from N. York to Philadelphia is 12 hours ride by railroad, from thence you will proceed to Pittsburgh by way of the Pennsylvania and Ohio canal and railroad, the usual charge from Philadelphia is 12 Dolls. From Pittsburgh to Cin. by steam boat will be about 5 Dolls. the whole expense from N. York to Cincinnati will be under 20 Dolls. or £4.17.6. A sovereign is worth 4 Dolls and 83 cents. Gold or silver is the only money you can pay your way with or you can buy in Liverpool Mexican Dolls or Spanish 5 for one pound but you should be careful that they would be all good silver. If you could get them, I believe that they would be the most convenient. We were six weeks at sea but that is considered a long passage, it's more than probable you might reach here before that time. I forgot to state that you will be provided with board en route from Philadelphia to Cincinnati without any additional charge. I forget what the exact amount of sea store is necessary, but be sure and provide yourself with six weeks provisions at least bring a ½ Brl.[10] of good potatoes pack them well in a tight cask that will keep out water. 5lbs of good butter in a small tub, a few doz. eggs and a large share of flour and oatmeal, bring ½ lb of soda and a small portion of tartaric acid they suit in place of barm[11] – you mix one teaspoonful of soda in your flour dissolve the acid in the quantity of water which you think would be sufficient to make your dough you will find it very pleasant if you do not put too much acid I believe ½ teaspoonful would be about right you know what quantity of tea, coffee, and sugar would be sufficient. Some ham and dried beef are very requisite, a large loaf of bread sliced about ¾ of an inch thick and placed in a cool oven until they would be well brown they keep good a long time at sea. You will have to bring some sort of bed and bed-clothes which you can dispose of in Philadelphia. See that you have a stout water-tight sea chest large and with partitions to keep your provisions from getting loose and many a lock and key are necessary upon it also upon your bed, as you will find persons near the end of your voyage who would wish to get in partnership with your stores, you will be obliged to put up with a great deal at sea as you will have to mix with a rough set. Get a berth as near the hatchway[12] as possible and endeavour to find out that there are no leaks above it, avoid coming in a ship that there are many passengers in, do not be in a hurry choosing one nor mind the blarney of the ship brokers, if they get the money that is all they care for, you are fortunate in having a companion[13] coming out with you or else you would have to share your berth with a stranger as they are fitted up to contain two persons. <u>Your companion shall meet with hearty and warm welcome we have got lots of room and markets plenty so that even your voracious appetite will be appeased, come along then, your friend shall be my friend</u>. The oldest suit of clothes you have will answer on ship board the best. I would not advise you to lay in a large stock of clothing as the style which is worn here is so different. You ought to have a large lot of shirts made as the making of them here would cost as much as the whole at home, light *[tressia]* duck pants[14] and dark linen frock from Holland makes a very good coat you will not require much clothing in summer, they will all have to be linen or cotton goods. If you would bring out cloth and cassimere[15] for a winter suit you could have them made up here in a more suitable style. It would not be of any use to you to be a Mason,[16] as there are a great portion of the community opposed to them on account of the disappearance of a delinquent member some years ago in the State of New York. A number of them reside here or at least a large number of Odd Fellows[17] which is something similar. You would find a heavy cloth over coat, that would come down about to your knees, very comfortable in winter also a heavy pair of boots. Boots are almost exclusively worn. In Liverpool you will provide yourself with water, can, cups, plates &c. for use on board ship, they must all be made of tin, have a small lantern made and bring a stock of candles. I am sorry I cannot give you a more minute account of the expense but I have no doubt it will be under £12 when two are travelling together it will make it come lighter on each. You will have to keep on the lookout that no person will have an opportunity of taking you in, there is so much competition amongst the proprietors of the different conveyances that it would be well to make enquiry of all of them and find out the cheapest and most comfortable conveyance. I would also advise you to examine the transient ships of Philadelphia and N. York they are generally the longest on the passage – but I have learned that they make up in comfort and convenience on account of not having so many passengers, it would not make so much difference to be a week longer at sea. There are some of them that make the trip in as short a time. I would not mind about the size of the vessel. I hope my folks are preparing to start about this time, I wish you were coming along with them. I mailed a letter to them on the first of February which I am anxious they should receive, as it contained enclosures for £30. A draft on Charles Thompson &co. of Liverpool[18] for £20, and a £10 draft on the Provincial Bank at Waterford drawn in favour of Peter Mc'Queen.[19] I am thus particular lest they might not go punctually to hand and payment stopped they are both made payable to my brother Thomas[20] the draft on Thompson's was drawn by Geo Milne &co.[21] and the other made payable to him and by him endorsed, to my bro. Thomas. I will make a request of you which I am sure you will readily grant that is to see all our old acquaintances and make my best respects to them as you are one of my best correspondents it is the only opportunity I may have of doing. Remember me affectionately to all the Tyndall family, see young Mary Hanlon[22] and give her my best regards for her welfare and happiness strange as it may seem she it was who first created an interest in my childish affections. I will now close by commending you to kind and watchful Providence, may his almighty hand sustain you in all the perils which you will meet by sea and land may he enable you to reach here unharmed and in afterlife may he enable us to praise and bless his Holy Name for all his mercies towards us and after this weary life is past to meet with all we have left behind. I hope your father[23] will not think bad of me for urging you to come out, it is what I would have done by a brother. He nor you I trust will have reason to regret it, you need have no anxiety about a situation you will be here but a short time until you will have one – I will bid farewell and I hope I shall soon have the long wished for pleasure of grasping you by the hand. A quick and safe passage to you. May God guard you upon the Ocean and upon the land is the prayer of your

affectionate | friend and school mate | William Wright

If you can find a nice oak stick just bring it along with you.

As soon as you arrive in N.Y.[24] forward me a letter and if you wish anything just mention it. Mail me a letter before you start from Liverpool.

Adieu W. Wright.

RI MS JT/1/TYP/5/1816–1819
LT Transcript Only

---

1. *Cincinnati*: from Tyndall's journal, 4 March 1844: '... reading a portion of a letter from Wm. dated Feb. 1st. This relieves my mind from the sickening state of doubt in which it has been plunged since I saw the account of that fatal shipwreck. Wm. is safe and I am happy' (RI MS JT/2/13a/22).
2. *[31 January] 1844*: Louisa Tyndall annotation: 'letters written on 31st Jan only rec'd March 3'. However, Tyndall's journal indicates that this letter is dated 1 February 1844 (RI MS JT/2/13a, f. 22).
3. *both your letters*: letters missing; see letters 0269 and 0281 of this volume.
4. *The newspaper*: possibly a clipping about the Ordnance Survey.
5. *my uncle*: not identified.
6. *your namesake's letter*: probably the letter from William Tyndall of Philadelphia which expressed pessimism about John's chances of finding employment in America. See William Tyndall to John Tyndall Snr, 24 September 1843, RI MS JT/1/TYP/10/3312–3313.
7. *repealers*: supporters of the Repeal Association, which campaigned for the repeal of the Act of Union between Ireland and Britain.
8. *that young lady*: not identified, although apparently a romantic interest of Tyndall's. Possibly the woman Tyndall and Ginty called 'the Northern Briton'; see letter 0235, n. 31 of this volume.
9. *our old neighbours*: not identified.
10. *a ½ Brl.*: a half barrel.
11. *barm*: the froth that forms when malt liquor is fermented; used as a leavening agent in breads (*OED*).
12. *the hatchway*: an opening or hatch in a ship's deck (*OED*).
13. *you are fortunate in having a companion*: a reference to George Latimer, who also intended to emigrate.
14. *duck pants*: pants made from a heavy cotton or linen fabric (*OED*).
15. *cassimere*: a tightly woven fabric, usually made from wool (*OED*).
16. *a Mason*: a member of the Freemasons, an American fraternal organisation.
17. *Odd Fellows*: The Independent Order of Odd Fellows, a North American fraternal organization.
18. *Charles Thompson &co. of Liverpool*: not identified.
19. *Peter Mc'Queen*: not identified.
20. *my brother Thomas*: Thomas Wright, William's brother, who still lived in Ireland (probably Dublin).
21. *Geo Milne &co.*: not identified.
22. *Mary Hanlon*: not identified.
23. *your father*: John Tyndall, Snr.
24. *N.Y.*: New York, the port where Tyndall would have first docked in America.

From Jean Lancaster 10 February 1844 0292

Bolton's Court | Preston | le 10 février 1844

Mon Cher Ami

Tout le monde trouve une sorte de plaisir dans la société de leurs amis, et moi, j'ai toujours trouvé le plus grand plaisir du monde dans votre société, mais comme vous avez dit, Nous ne pouvons pas parler l'un à l'autre parceque la mer roule entre nous mais la mer ne peut pas nous empêcher d'exprimer nos pensées l'un à l'autre. J'ai reçu votre lettre du 7 ult. et je suis bien satisfait de votre arrangement des lettres. Quand je vous écrivais ma dernière lettre, je n'avais pas eu la votre, et voilà pourquoi j'ai écrit vous êtes en ma dette, voilà 'En ma dette', si j'ai écrit En je me suis trompé, je devais dire Dans ma dette.

J'ai vu la lettre que vous avez écrit à M. Parkinson, et je vous remercie beaucoup de ma part, pour votre bonté en nous donnant une histoire de vos voyages, depuis que vous avez quitté Preston, jusqu'à ce que vous êtes arrivé en Ireland. J'étais bien *[amusé]* de vos aventures dans la voiture. Sans doute vous seriez bien charmé de vous trouver dans la compagnie de deux Messrs qui parlent le Français. Je ne suis pas surpris qu'ils pensaient que vous étiez un Français, vous le parliez très bien quand vous étiez à Preston, et pour moi, j'ai grande honte en vous écrivant dans cette langue là, parce que je suis sur que vous l'écrivez bien plus bien que moi. Quand j'arrivais a cette partie de votre lettre ou une dame entrait la voiture, je sais bien que vous seriez bien poli, et vraiment elle doit payer pour soit siège, et elle ne peuvent moins payer, que de vous embrasser, mais je présume, que quand vous demandiez une telle récompense, vous tous deux seriez placé dans une situation, que n'était pas tout à fait agréable, quand elle dit que c'était son mari qui s'asseyait près de vous, il faut prendre garde mon cher Ami, ou vous serez placé dans une telle situation que le jeune Noble Français dont vous parlez dans votre dernière lettre, et je serais très fâché d'entendre que vous aviez été engagé dans un duel à cause des femmes, je pense qu'elles ne valent pas la risque. Vous dites que vous avez l'intention d'aller à l'Amérique l'avril prochain; quand vous y allez, j'espère sincèrement que vous aurez le dieu Neptune pour votre ami pendant que vous êtes sur son empire. Vous dites que vous êtes *[malheureux]* d'être forcé de quitter votre chère amie, et former une alliance avec une Janne Yankee, je vous conseillerais d'aller à l'Amérique pour obtenir une fortune, et alors revenir pour obtenir une femme, je suis sur je préférerais de rester en Angleterre, et obtenir une femme comme il faut, même si je n'étais que pauvre, plus que d'aller à l'Amérique et marier une Yankee Jaune si elle avait toutes les richesses de Péru. Vous ditez qu'on ne peut pas aimer quand on veut, qu'il faut aimer en dépit de vous même, mais moi je pense qu'on peut aimer quand on veut, c'est à dire s'il y a des objets amoureux près de vous, moi, je peux aimer tous les *[celles]* femmes, je ne suis pas particulier, mais je prends garde que ni l'une ni l'autre ne vole mon cœur. Vous et moi, nous sommes trop jeune pour marier à présent, il faut les jouir, les embrasser et tout comme cela, quand vous aves trente ou trente cinq ans il serait bien bonne heure de penser à se marier. Donnez moi votre opinion sur ma philosophie, peut être vous le penserez très étrange. Si vous visitez Preston avant votre départ à l'Amérique, j'espère d'avoir le plaisir de vous voir. Maintenant il faut dire adieu pour le présent. À propos de 'Adieu' comprenez vous ce mot là? C'est composé de deux mots français – voilà, je vous confie À Dieu et

Croyez moi toujours | Votre Ami Sincère | Jean Lancaster

Bolton's Court | Preston | February 10th, 1844

My dear friend

Everyone finds a kind of pleasure in the company of their friends, and for me, I have always found the greatest enjoyment in your company; but as you said, we cannot speak to one another because the sea rolls between us, but the sea cannot prevent the expression of our thoughts to one another. I have received your letter of 7th inst.[1] and I am well satisfied with your arrangement of the letters. When I wrote my previous letter, I did not have yours and this is why I wrote that you are *en ma dette*,[2] thus 'En ma dette'; if I wrote En I was mistaken and I should have said *Dans ma dette.*[3]

I saw the letter that you wrote to Mr. Parkinson,[4] and I am very grateful for your kindness in giving us the history of your travels since you left Preston and up until your arrival in Ireland. I was very amused by your adventures in the coach. No doubt you would be delighted to find yourself in the company of two gentlemen who speak French. I am not surprised that they thought you were French; you spoke it very well when you were at Preston. And as for me, I am very ashamed of writing to you in this language since I am sure that you can write much better than I in it. When I came to the part of your letter where a lady entered the coach, I know well that you would have been very polite, and really she must pay for being seated, and she could not pay less than by kissing you, but I suppose that when you asked for such a reward, both of you would be placed in an unpleasant situation when she says that it was her husband who was sitting next to you. You have to be careful my dear friend, or else you will be placed in a situation similar to the young French noble,[5] of whom you spoke in your previous letter, and I will be very upset to hear that you would have been engaged in a duel over women, I think that they are not worth the risk. You say that you want to go to America next April; when you go there, I sincerely hope that you will have the god Neptune as your friend when you are in his domain. You say that you are *[unhappy]* that you are forced to leave your dear friend and to make an alliance with a *[young]* Yankee,[6] I advise you to go to America to obtain wealth, and then to return in order to find a wife, I surely prefer to stay in England and get a proper wife, even if I was poor, rather than go to America and marry a young Yankee, even if she had all the riches of Peru. You say that one cannot love when one wants to, that you should love in spite of yourself, but I think that we can love when we want. In other words, if there are objects of love next to you, I can love all those women, I am not *[picky]*, but I am careful so that neither one nor the other would steal my heart. You and me, we are too young to marry right now, we must enjoy them, kiss them and all that, and when you are thirty or thirty-five years old, then it will be a good time to think about marriage. Give me your opinion on my philosophy; maybe you think it is very strange. If you visit Preston before your departure to America, I hope to have the pleasure of seeing you. Now I must say goodbye for the moment. About the word 'Adieu', do you understand that word? It is composed of two French words[7] – there, I entrust you to God and

Believe me always | Your sincere friend | Jean Lancaster[8]

RI MS JT/1/TYP/11/3737
LT Transcript Only

---

1. *your letter of 7th inst.*: letter missing.
2. *en ma dette*: in my debt (French).
3. *en ma dette ... Dans ma dette*: Lancaster appears to be correcting the grammar of an earlier letter he wrote to Tyndall.
4. *you wrote to Mr. Parkinson*: John Parkinson, a friend of Tyndall's on the Ordnance Survey; letter missing.
5. *the young French noble*: not identified.
6. *make an alliance with a young Yankee*: probably a reference to William Wright.
7. *composed of two French words*: 'adieu' is a composite of the French phrase 'je te recommande à Dieu' (I recommend you to God).
8. *Jean Lancaster*: a friend of Tyndall's from the Ordnance Survey.

From Joseph Payne 12 February 1844 0293

Monday morning | Feb 12. 1844[1]

My dear Tyndall

I received your never expected,[2] and I assure you it has given me great pleasure for more reasons than one. I thought from the first of Spec's[3] epistles that you were him, but was not sure until 'the last dearly purchased smoke wreath mingled with the atmosphere',[4] and then any little doubts I had were completely recovered. Yes Father John, nothing but an avowal from yourself to the contrary will ever persuade me that you and Spec, added together would not turn up John Tyndall (oh it's mortal cold, frost and snow to the eyes here). His letters were received here as clever truisms written in the true spirit of an independent gentlemanly style and as gaul and wormwood[5] to the gizzard of every tyrannical

Engineer Officer. There's not a true heart amongst us that does not warm at the word Spec and would shake his fingers off for the love we bear his noble spirit. May his pen never flourish in a worse cause and may better success attend his next undertaking! It is hard singlehanded to make way against such a current; but though the main object may not have been attained, yet the villainy of the Engineer officers is exposed and that's a comfort. And has Tyndall shook off the trammels of Toryism, has he conquered the demon prejudice and, if not in act, is he at heart the son of liberty the friend of the oppressed and the scourge of the oppressor? Will he confess – what he must have felt – his unfortunate country governed by a selfish despotism under the name of a British Constitution? More power such as you made America as she is and such will keep her so.

You ask me if I am a Repealer,[6] I did fear and do still that popish ascendancy would be eventually not the least important result of Repeal. I have tried long to persuade myself that we would have an impartial protection of civil and religious liberty, but I can't. I fear something Jesuitical has crept into the repeal movement; but as one evil is existing and the other only anticipated I chose the latter and I believe were both existing I would do the same. It's an old standard and a good one that the majority should not suffer for the benefit of the minority, so that as far as I morally can be I am a Repealer – The corn law league[7] has a deal for and against it and though the majority of the latter are influenced by selfish motives yet I believe there are a good many influenced by having the good of their country (as a country) at heart. There can be little doubt but it would injure the farmers at first, but the Right Hon. <*1 word excised*> this and his grace the Duke of Marquis of that[8] should knock some thousands of his Rentroll[9] and cover the loss by fewer visits to the gambling house, shortening the season in town, keeping a procuress less on the Continent or at home, and in short whirling less in the vortex of dissipation. Something is wanted to counterbalance the injustice of the grinding groping manufacturer in this country, to reduce the almost incredible number of forty thousand! prostitutes in London, and take no less number (according to their extent) of the other large towns of England. It's an old proverb that idleness is the mother of vice, poverty might be called its father, I pretend to know little of how it might affect the commerce of the country, but I know this that an English government was never yet at a loss to levy a tax, and (damn time) let them (if necessary) levy it on something that won't wring the last farthing out of the poor operatives' pocket. Beatty[10] has promised me the league Newspaper tomorrow, I will send it to you, and for goodness sake send me a few Carlow Sentinels, I am longing for the local news – Things progress as usual here 2 or 3 Journals monthly out of the office, Tithe work[11] showering in upon us. Scally[12] has got charge (absolute) of the 6 feet department; I am his right hand man, but Capt. four eyes[13] has me marked for the limb of a rebel that strove to sow the seed of dissension in his milksop[14] Division, so that my prospects on the Survey are blasted (who the devil ever had good prospects on it? What is the best of them in the end?) Scally's brother is after returning from America – could not do well there, got a few jobs on Railways and canals but nothing permanent – says none but an American indentured surveyor can work on his own account. I don't believe that Jemmy Cain[15] for instance surveyed the City of Toronto. He is trying to get employed here, and I think he will succeed as we are here over ears in office work – the thankless Tithe plans to be thanked for that – I have little acquaintance with him yet but if you require any information concerning America I will get it from Allick.[16] What with advertising and registering this last three months it has cost me nearly 30/– and I could have walked into a situation in Bedfordshire[17] of £80 per annum if my uncle[18] only sent me a few pounds. But the old knave would not even answer my letter and now I want you to find out for me (hiding nothing its very cold) his feelings towards me, and how he is determined to act, for something must be done. While I remain on the Survey I can do nothing for myself or mother.[19] I am a big fellow, remarkably handsome, 11st. 4lbs. wears my hair à la Rubini[20] and turns out a hell of a swell on Sundays. Kindest remembrance to your father,[21] mother[22] and Emma[23] I'll cherish the sentiment.

write me soon and believe | Sincerely Yours | J. Payne[24]

P.S. How is J. Murray?[25] Is it as a Surveyor you go to America? or are you going to your uncle's or do you go it mercantile? | J.P.

RI MS JT/1/TYP/11/3845–3846
LT Transcript Only

1. *Feb 12. 1844*: Louisa Tyndall annotation: 'postmark Skipton'.
2. *your never expected*: letter missing.
3. *Spec's epistles*: the letters Tyndall published under the pseudonym 'Spectator' in the *Liverpool Mercury*; see letter 0228, *Correspondence of John Tyndall, Volume 1*, and letters 0248, 0252, 0261 and 0287 of this volume.
4. *'the last dearly purchased smoke wreath mingled with the atmosphere'*: this is a quotation from Tyndall's 24 October 1843 'Spectator' letter in the *Liverpool Mercury*; see letter 0252 of this volume.
5. *gaul and wormwood*: more commonly spelled 'gall and wormwood'; an idiom for deep resentment or bitterness.
6. *a Repealer*: a supporter of the Repeal Association, which campaigned for the repeal of the Act of Union between Ireland and Britain.
7. *The corn law league*: the Corn Laws levied taxes on imported grains, making food more expensive but protecting landowners' interests. The Laws were abolished in 1846 after extensive campaigning by the Anti-Corn Law League, which was formed in 1838 in Manchester.
8. *the Right Hon. <1 word excised> this and his grace the Duke of Marquis of that*: imaginary names meant to stand in for various elites.
9. *his Rentroll*: the list of people who owe a nobleman rents.
10. *Beatty*: not identified.

11. *Tithe work*: see letter 0231, n. 19 of this volume.
12. *Scally*: not identified.
13. *Capt. four eyes*: possibly Henry Tucker.
14. *milksop*: a feeble or ineffective person (*OED*).
15. *Jemmy Cain*: not identified.
16. *Allick*: not identified.
17. *Bedfordshire*: a county in England west of Cambridge.
18. *my uncle*: not identified.
19. *mother*: not identified.
20. *wears my hair à la Rubini*: Payne may have styled his hair in the manner of the famous tenor Giovanni Battista Rubini (1794–1854).
21. *your father*: John Tyndall, Snr.
22. *mother*: Sarah McAssey Tyndall.
23. *Emma*: Emma Tyndall.
24. *J. Payne*: Joseph Payne, a civil assistant in the 5th Division, C District of the Irish Ordinance Survey. He joined the Survey on 1 April 1839, the same day as Tyndall, and worked closely with him. He was initially employed in plotting but subsequently in drawing and correcting. On 8 April 1842 he was transferred to the English Survey. He is possibly the Joseph Payne, son of William and Jane Payne, who was born on 5 November 1821 in Dunleckny, County Carlow (Irish Church Records).
25. *J. Murray*: John Murray. He joined the Ordnance Survey of Ireland in December 1839 and served as a civil assistant in the 5th Division, C District. Primarily involved in registering and calculating, in 1841 he worked in the office party with Tyndall, Phillip Evans and Joseph Payne. In April and May 1842 he was correcting and completing registers, before being transferred to the English Survey on 20 May 1842.

From John Todd 5 March 1844 0294

Preston, | 5th March 1844.

Dear Tyndall,

After a silence of a considerable length I again address you,[1] it may be interesting to know how things are going on with us at Preston, and whether the strong feelings aroused previous to your departure has yet subsided. Everything as regards the progress of the business we are employed on is going on in pretty near the usual way and the only change of importance that has taken place since you left is the removal of Corp. J *[Walkem]* from the Division, who I need hardly mention is universally regretted. That he has always carried himself smoothly through all parties and pursued a straight-forward independent line of conduct in the discharge of his duties and that he has ever acted so as to give the civilians with whom he was connected the least possible annoyance is the opinion expressed by all who know him. He left here for Woolwich on the 29th Feb. He and the party to which he is now attached will embark on the 19th I believe of this month for America, to be employed under Capt. Broughton[2] on the Survey of the Southwestern Boundary.[3] (They are to be in Quebec on the 5th of May.) Corpl. Quadling[4] succeeded him in charge of the Drawing Department here. We have got a new Free-trade paper published here called the 'Guardian' published weekly by Livesey and son. Spectator's last letter[5] was given with some remarks in its last week's number. (I send it to you). Dear Tyndall we see little hopes of any change being made for the better in our circumstances by our Superintendents,[6] willingly, and we are therefore resolved to try the effects of further agitation. We have exposed ourselves as much as we can to our officers and to the public, so much that the inhabitants of Preston can point at us in the street and say there go the poor sappers and miners[7] that have only 12 shillings a week. Whilst in some of the most influential quarters our claims are either altogether unknown or, if known, t'is to those only perhaps who would be better disposed to lay their foot on them than investigate their merits. I am directed to submit the following to you for your advice and assistance. We think that if Spectator's letters together with the other matters relating to late occurrences which you are in possession of and may be deemed expedient to add were published in the form of a neat pamphlet and copies given to the leading journals in Ireland as well as in this country, and also to those members of Parliament who would be thought might take an interest in it, it might get our case into notice and be instrumental in causing some enquiry which would end in making some change for the better in our situations. About 200 copies of it might be contracted for with some publisher in Ireland whom you can arrange matters with and then there are as many of us here anxious to have it tried as will render the individual expense trifling. Although you are now out from amongst us we thus calculate on your assistance for we believe that it would be gratifying to you to be able to render a service to your old fellow sufferers. Let us have your answer as soon as convenient for 'If when 'twas done, 'twere well done, then 'twere well it were done quickly'.[8] I hope the congenial air of home is exerting its influence to some purpose in invigorating your constitution and bodily strength and that your mind is more engaged in contemplating and looking forward with hope for better prospects and brighter days than on regretting your departure from the Ordnance Survey. How is Phil[9] he promised to write to me but did not, remind him of it in your next and

Believe me to be | My dear John | Yours faithfully | John Todd.

RI MS JT/1/TYP/11/3882
LT Transcript Only

1. *I again address you*: previous letter missing.
2. *Capt. Broughton*: William Delves Broughton (1802–80), a Captain in the Royal Engineers. Broughton joined the army aged 23 and moved rapidly through the ranks, becoming a 2nd Captain on 1 April 1841, 1st Captain on 1 April 1846, and eventually General on 1 October 1877. He served on the Irish Survey, and also served in North America, Bermuda, Hong Kong and Tasmania. He died in Surrey on 5 April 1880.
3. *Survey of the Southwestern Boundary*: a survey of the United States-Canadian border was

conducted between 1844 and 1846. 'Emory, William Hemsley', *Texas State Historical Association*, at http://www.tshaonline.org/handbook/online/articles/fem03 [accessed 11 August 2014].

4. *Corpl. Quadling*: Corporal Henry Quadling of the 14th Company of the Royal Sappers and Miners. Quadling was born *c.* 1810, and enlisted to the army in July 1830. He was a miner by trade. His division was stationed at Newtownbarry in County Wexford before moving to Clonmel in County Tipperary. His duties were protracting, registering, and drawing. His daily rate of pay stood at 1s 7d by the time he left the Survey. Quadling left the Survey along with the majority of the A District workers during the autumn of 1841 and later joined the English survey.
5. *Spectator's last letter*: see letter 0287 of this volume; Spectator was Tyndall's *Liverpool Mercury* pseudonym.
6. *our Superintendents*: the Superintendent of the Ordnance Survey of Ireland was Thomas Frederick Colby.
7. *sappers and miners*: see letter 0231, n. 32 of this volume.
8. *'If when 'twas done, 'twere well done, then 'twere well it were done quickly'*: W. Shakespeare, *Macbeth,* I.vii.1–2.
9. *Phil*: probably Phillip Evans.

From John Sinnett 15 March 1844 0295

Liverpool | 15th March 1844.

My dear Tyndall

I suppose that by this time you have given me up for a bad job but I hope before you proceed to judgment you will hear what the prisoner has to say in his defence. You must know then that I have a very heavy job in the office so much so that very often I have scarce time to look round me. Then at home I have got hold about a month since of a very valuable work on perspective and as I must give it up very soon I have been working at it hammer and tongs. I have a night sometimes I am on it still and only I fear you might slip off to Cincinnati[1] before I could write probably I would have deferred some time longer. I have unfortunately mislaid your last,[2] but I recollect that you wished me to do a specimen of wood in sepia – this I will do – this evening if spared – so that you must not attribute my silence to neglect and this you will perceive is not the usual excuse under such circumstances. Poor Larry Eivers sailed yesterday morning I believe, or at least will when the wind proves favourable, as his passage is paid and all his traps on board. He goes to New York in the Sheridan, a fine Packet ship. He called on me several times and old relations which indeed I had thought were quite broken off were again renewed. Poor fellow I never knew a worthier soul, but Cap. F's[3] removal done him up for the Survey. Old dogs can hardly learn new tricks. We have just started 15 sappers[4] from this to Cork en route to the disputed B Y[5] in Canada. I don't think you knew any of them, Lonegan[6] you did and perhaps Walkem who was at Preston. We have a young officer here under instructions. We are in hopes that if Mr H.[7] goes he will replace him, he appears a very nice man. The grant[8] I hear is considerable this year whether the commons of the 'hogs lard' *[1 word illeg]* will be improved thereby I can't say nor have I much expectation of it. I had a letter from Ginty – the fellow is doing nothing but going to orange meetings![9] He goes there however he says to blow them up. I believe he had a narrow escape of a kicking at some of their orgies. He says very coolly that he is not particularly anxious about employment as long as he gets plenty to eat! I made Larry laugh at your quotation from one of his lucubrations[10] ('twould be absurd to call them letters) i.e. about Dame Reason[11] reading the riot act to his disorganized thoughts. You ought surely to take him with you to Cin. only among our brethern Jonathan[12] will be in his element. I hope to have a letter from you before you start, if it were only to show you can return good for evil. I trust also that if you and George[13] go to the far west that that our correspondence will not cease for, independent of the wish I will always anxiously feel to hear of you both, I would like much to hear what sort of country it is, as I must now in a very few years be looking out for a new course, for surely I do not intend to remain here till I am unable to make some sort of a dash at settling myself comfortably; and why not there anywhere where I would get fairplay. I am sure you will not fail by writing now and then to assist me in this matter. I hear from William Latimer that you are likely to spend a few days at Newtongore[14] before you start. If so tell George that I do not forget him, but am hard up for time. I must write to the scapegrace in Carrick[15] tomorrow. All the lads[16] are well, vegetating on the two bobs as well as they can. So now to conclude and make an end

Believe me ever yours | Most sincerely[17]

Larry does not stop at New York but goes into the interior to some warm climate where his uncle and brother[18] are, his health is his principal object.

You know Baker,[19] he is gone to Woolwich sick and recommended for a foreign station. A happy riddance we have had of him and his rib. You also knew Newton the sapper at Leighlin Bridge he has just come out of some penitentiary where he has been for 6 months for 9 months desertion.

RI MS JT/1/TYP/11/3865–3866
LT Transcript Only

---

1. *Cincinnati*: Sinnett was referring to Tyndall's plans to move to Cincinnati; see letters 0237, 0269, 0291, 0296, 0299, 0304, 0306, 0307, 0313, and 0324 of this volume for further information regarding these plans and their outcome.
2. *your last*: letter missing. From Tyndall's journal, 24 January 1844: 'Posted letter to Sinnett in the evening' (RI MS JT/2/13a/3).
3. *Cap. F*: possibly Robert Fenwick; see letter 0268, n. 25 of this volume.
4. *sappers*: see letter 0232, n. 10 of this volume.
5. *the disputed B Y*: the disputed boundary.
6. *Lonegan*: not identified.
7. *Mr H.*: possibly William George Hamley.
8. *The grant*: a Parliamentary grant allotting funds to the Ordnance Survey; see letters

0236, 0238, 0241, 0245 and 0246 of this volume for insight into the Survey's Parliamentary grant of 1843.
9. *orange meetings*: not identified, possibly a reference to Protestant gatherings.
10. *lucubrations*: nocturnal studies, meditations (*OED*).
11. *Dame Reason*: a female personification of Reason.
12. *Jonathan*: not identified.
13. *George*: George Latimer.
14. *Newtongore*: possibly an alternate spelling of Newtowngore, a small town in County Leitrim, Ireland.
15. *the scapegrace in Carrick*: not identified.
16. *the lads*: probably the other employees on the Ordnance Survey.
17. *Most sincerely*: the signature is missing. It is not clear whether Sinnett did not sign the letter, or whether Louisa Tyndall's transcript omitted the signature.
18. *his uncle and brother*: not identified.
19. *Baker*: not identified.

From Archibald McLachlan 15 March 1844 0296

45 Swann Street, York 15th Mar: 1844.

Dr Mr T.[1]

I am glad to find that you are still in existence and have determined to quit this land of starvation[2] – I am certain you will succeed on the other side of the 'salt sea'. 4 of my old O.S.[3] acquaintances went in 1829 – at first – and for a few years they had all to work hard, as Book-Keepers &c – and each has now sufficient of worldly comforts – and a provision made for old age. But lest I should forget, here follows their address:–
(Joseph Workman)[4] Workman, Brothers & Co. Toronto – (William Workman)[5] 'Frothingham and Workman' Montreal – Besides 2 brothers Henry[6] and John Mulhalland[7] – the Workmans will tell you where they are – Should you pass near them on your journey, they would be very glad to give an old O.S. man every information you may require, on the best method likely to secure success –

– On the other matters, you must recollect that I am at present in 'pretty considerable fix' all letters coming to any of us, are by mistake, or design, taken to the R.E.[8] office, even those sent to our private address are taken by the postman to the office – even your last letter[9] passed through the inspection of the District office!!!! I am determined should this be persevered in, to write to the Post Mast Gen:[10] on the subject –

I am of opinion that a pamphlet[11] is the best method, it would be too late to be of any use in debating on the ordnance estimates, but it would be in good time to check the proposed S.[12] of London, when that is brought forward; I think that S. Crawford[13] would have a good opportunity of handling the subject when London is proposed, and I think that the C.[14] proffession have an interest in opposing that S. being done by the R.E. – and not at all unlikely, they have by this time given the subject some consideration, and perhaps have instructed some member to bring the claims of the civil proffession before the Ho:[15] I think it very unlikely that they should have read Spectators letters,[16] and yet remain indifferent or unprepared to oppose the Ordnance claims to S. London –

At all events it too late now to write or print anything with a view to the estimates – the principal Item has been passed by the Ho: 140 or 150.000 has been granted[17] 'for the Ordnance Surveys' – and before you could print the pamphlet, all the estimates will be passed. It would be necessary to print all Specs: letters; that any member who should speak on the subject should first be well informed, should have a good view of the past as well as of the present – S. Crawford is an Irish land owner, he knows his estates were surveyed twice by the O.S – he knows also of many other Estates being twice surveyed, he knows these facts and the dates also – and these are public facts – But a very great deal more is still wanting to enable Mr S. C. to make a good stand in the House – a long cool letter stating a number of the practical details – such as the pay of a man, the quantity of work he does – then showing the acreage price, the expenses of Ch: Survey,[18] plotting, Drawing, Calculating &c – The Recommendation sheet should be introduced – A proposition should be made, that the men should receive an acreage price, that the men should be paid for work done, not for time spent – officers should have an acreage price also – he should be instructed to call for a return of all the individuals Employed, their date of joining the S – their rates of pay, and the nature of the duty employed on – In fact to be prepared for opposition Mr S.[19] should be well informed, for he would be stiffly opposed, and unless he were master of a great many of O.S details, he would come to the wall in a debate – Now all this requires time, caution, cool reflection – it would be impossible to do justice to the subject in less than a week or two, at least – but a pamphlet could be ready before London comes on, for I don't think it will come on before the Easter weeks – –

Could you convince Mr T.[20] of the necessity of giving a great deal of the O.S details? I can furnish him with a good deal; all the essentials – Could you convince him of the necessity of doing the work himself – to have communication with no parties – not more than 2 persons at the utmost – & his communication should relate to the collecting of money, a trusty person could collect a few shillings under the pretence of assisting a poor brother O. surveyor, or the like – I could with ease collect £2 – perhaps more – There is a spirit of dissatisfaction – a desire to have things put to right &c but we are deficient in coolness, in steadiness, which will make sad work should our communications not be restricted to one or two individuals – If the P.O.[21] at Preston is not watched so narrowly as at York I can safely send letters to M^r^ T – but his to me would be very unsafe – and should his be peeped into and the handwriting known, the consequence would perhaps, be his discharge, he would naturally blame me for a want of caution and silence &c – I shall send him all the essentials, he can work them up – the letters which he sends to me must be well secured under an envelope & written in a Female hand!! – (i.e. the address of the letter) – – I don't make the least doubt of it, before London comes on the pamphlet will be circulated – –

I wish much for the pamphlet, it will circulate and preserve Spectators letters – they will be read & studied by parties who never look at Newspapers – –

There has appeared a letter in the London times Newspaper in favour of good maps &c – and blaming the Government for the niggardly grant this year – it is the first of a series – the author is unknown – I have not seen the paper, but must be on the watch in future – should he continue the series I shall conclude his intention is to do away with the effects of Specs: letter – if I find this to be his object, I must meet him!! and in a London paper too – Meet him by simply stating the plain facts of the case, in a plain way –

On Monday or tuesday you shall have another letter[22] – I did not observe that my paper was so nearly scribbled up – I hope you are in good spirits, in my next I will say something of sea sickness

Yours very sinly | A M[c]Lachlan

M[r] J Tyndall | Leighlin Bridge | – Ireland – | P.P

RI MS JT/1/TYP/11/3830
RI MS JT/8/1/4a

---

1. *D[r] M[r] T.*: Dear Mr Tyndall.
2. *quit this land of starvation*: Tyndall was planning to emigrate to America with George Latimer in the spring of 1844; see letters 0237, 0269, 0291, 0295, 0299, 0304, 0306, 0307, 0313 and 0324 of this volume for further information regarding these plans and their outcome.
3. *O.S.*: Ordnance Survey.
4. *Joseph Workman*: not identified.
5. *William Workman*: not identified.
6. *Henry Malhalland*: not identified.
7. *John Malhalland*: not identified.
8. *R.E.*: Royal Engineers.
9. *your last letter*: letter missing.
10. *Post Mast Gen*: Postmaster General. A member of the Cabinet tasked with maintaining the postal system. In 1844, the Postmaster General was Viscount William Lowther (1787–1872).
11. *a pamphlet*: see letter 0294 of this volume for John Todd's opinion on the possibility of creating a pamphlet as a means of disseminating information about the conditions for workers on the Ordnance Survey.
12. *S.*: Survey.
13. *S. Crawford*: not identified; later information in the letter indicates that he owned significant property in Ireland.
14. *the C.*: possibly an abbreviation for the Civil Engineers.

15. *the Ho:*: probably an abbreviation for the House of Commons.
16. *Spectators letters*: Spectator was Tyndall's *Liverpool Mercury* pseudonym; see letter 0228, *Correspondence of John Tyndall, Volume 1*, and letters 0248, 0252, 0261 and 0287 of this volume.
17. *140 or 150.000 has been granted*: this refers to a Parliamentary grant allotting funds to the Ordnance Survey; see letters 0236, 0238, 0241, 0245, 0246 and 0295 of this volume.
18. *Ch: survey*: not identified.
19. *M^r S*: possibly the same person as Mr. S. Crawford.
20. *Mr T*: possibly John Todd or Alexander Taylor.
21. *the P.O.*: Post Office.
22. *another letter*: letter missing.

From E.B. Sayers 17 March 1844 0297

Patrick's Day | March 17 1844.

Dear Jack,

You need not be at all uneasy about your letter to Walkem,[1] as I received it here. He started from this for Liverpool on Tuesday last.

I have just refused a good situation of 12/6 a day (Superintendent of a field party and making of 20–000 ecres[2] of Tythe work).[3] I think I will get on better where I am, and I was afraid I should not be quite able to fill the situation.

I have a promise of a good berth in Bermuda or America when one uccurs,[4] but I'll wait here and learn as much as I can first. How is George?[5] Remember me to him and tell him to drop me a line – tell him I'm the devil's methodist by this time.

There is hardly a decent fellow left on the survey now, all gone. The best thing that ever happened for us was that petition.[6] I dined with Colonel Colby a few days ago – He did not know me but I knew him the scoundrel. Parker rejoined and was re-employed and is to be sent to Liverpool. They are playing the devil about the 3 books[7] that are gone. 'Don't they wish they may get them'? I wish they may. The work in Preston is getting on as usual, slow and no money. I have no news, so hoping to hear from you as soon as you have time (if not from the land of the sod at least from America).

I remain dear Jack | wishing you all good luck a good | wife &c. | Your friend | E.B. Sayers.

RI MS JT/1/TYP/11/3860
LT Transcript Only

1. *your letter to Walkem*: letter missing.
2. *ecres*: i.e., acres. Sayers often employed spellings that reflect regional pronunciations.
3. *Tythe work*: i.e., work on Tithe maps; see letter 0231, n. 19 of this volume.
4. *uccurs*: i.e., occurs; see n. 2.
5. *George*: George Latimer.
6. *that petition*: see letter 0236 of this volume.
7. *books*: Louisa Tyndall annotation: 'bricks?'

From William Ginty [19] March 1844 0298

Carrick on Shannon | Wednesday 19th March 1844.[1]

My dear Tyndall,

On yesterday noon I returned from Sligo where I had been *[8]* days. That will be some excuse for my silence. I did really expect to have one from you sent after me[2] – I got it yesterday and it has struck me as a thunderbolt. I did not think your departure[3] was *[so]* near at hand. I never sat before a letter with bitterer feelings – with more real sorrow, not only for your removal to such a distant place but also my total inability to do what I would to heaven I could do. I am poor, Tyndall and being idle – in fact I cannot send it nor can I write about it. I'll send it after you when I can scrape it together or I'll send it to your father[4] as of course you're indebted to him more than that. I have not much to say. I had hoped up to yesterday that you would not go,[5] but that it seems as foolish as it was useless. I suppose George[6] goes with you, a worthy companion without a doubt. The non-appearance of No. 6[7] will look odd, but that can't be helped – trifles must submit to greater events! Yet you cannot be aware that some who knew you had a suspicion that you were guilty of the authorship.[8] The knowledge of your departure brought in superposition with this will fasten the suspicion on you to all intents and purposes. What of suspicions though – you thought in your last 'that it would be your last scroll'. Do you mean the last forever? Surely not – And I think you may find time to scrape a few lines to correct and acknowledge your error before you go. Write if you can before you go and oh do not fail to write when you cast your anchor in the strangers' clime! Goodbye my dear Tyndall. You will find enclosed an effort of my muse – tho' the short time I had prevented me from making longer or more perfect it contains the essence of my soul the breathings of my spirit. Goodbye – oh don't for pity's sake omit writing from beyond.

Yours ever faithfully | sincerely and affectionately | William Ginty.

Give my kindest and sincerest remembrances to your brother emigrant George Latimer.

Carrick on Shannon | Co. Leitrim. Ireland | Wednesday 19th March 1844

The muse that oft at Tyndall's call
Has bellowed forth her beastly bawl
'Bout 'Ellen's'[9] charms and 'Aubrey's'[10] ball
And soared o'er mountains 'fells' and all
Resumes her 'gray goose quill'![11]

Oh now ye Gods by poets seen
Ye shadow sprites of fairy scene
Ye spectra-lords of fancy's dream
Brace – nerve my soul into my theme
This bath-post sheet to fill!

'Friendship! Mysterious earnest of the soul
Sweetner of life, solder of society
I owe thee much'——— Blair[12]

'Your spirit dares me to forget'[13]
No!– Tyndall – no – too firmly's set
Too deeply graven where 'tis placed
Is Friendship's seal to be effaced
Not Time's dull tarnish e'er shall rest
Upon the mirror in my breast
Nor Time – nor Distance ever dim
Thy portrait pictured there within!
–'The die is cast'[14] – shall we no more
Our kindred souls in fancy pour!
Forefend it Heaven! 'tho' doubly far,
Not e'en that distance could it mar
The broad Atlantic stern and deep
Will soon its billows 'twixt us sweep
Not ten Atlantics e'er shall roar
My Friendship's requiem on this shore
'Tho Fates dark cup of bitter draught
By me perhaps may oft be quaffed
Not grief nor woe shall ever tell
They rung for you my Friendship's knell!
But ah! a pang my spirit sears
And mem'ry weeps o'er faded years
She bears me to those happy scenes
O'er which we laughed in youthful dreams
Those haunts where we so oft have trod
To gaze on Nature and her God!
But all Farewell – in Fancy's birth
They're monuments of absent worth!
–Thy path lies o'er the mountain waves
Thy home no longer – home of slaves!
Thy barque now streams her pennon gay[15]
For Friendship's regions far away
Oh think when there – how in this land
We battled with a miscreant band
How we have grappled side by side
With petty despots in their pride
Have dealt a stout and noble blow
To lay the griping tyrant low
Oh hear the earnest in your heart
That you have played a noble part
In Friendship's cause! o'er thoughts like these
Thou'lt often smile – they'll often please
And conscious Truth – from Heaven a guest
Will revel in thy exiled breast!

And think thee Tyndall – now the toil
Of Tyrants in this perjured soil
There breathes a soul they cant subdue
An honest heart to Friendship true!–
Bear this with thee – beyond the wave
Where Despots find a fitting grave
Within this green and trampled isle
A soul will o'er thy mem'ry smile
In after years – when age shall shed
Her hoary blossoms on his head
Fond thoughts of thee will oft arise
To fire his dim and grave-ward eyes
A dream of thee will cheer the gloom
That presages the chilling tomb!
I've done – Farewell – may fortune pour
On thee and thine her choicest – store
May care or sorrow never rest
One moment in thy manly breast–
May Heaven her holiest blessings shed
In myriads on thy honoured head!
And when thy glass of life hath run
May smiling spirits bear thee home
To realms of happiness on high! –
Goodbye – Dear Jack – a long Goodbye!
William Ginty.

John Tyndall
RI MS JT/1/TYP/11/3628–3630
LT Transcript Only

---

1. *Wednesday 19th March 1844*: 19 March was a Tuesday in 1844; it is unclear whether this was Ginty's mistake or a typo in Louisa Tyndall's transcription.
2. *the one you sent after me*: letter missing.
3. *your departure*: Tyndall was planning to emigrate to America with George Latimer in the spring of 1844; see letters 0237, 0269, 0291, 0295, 0296, 0299, 0304, 0306, 0307, 0313 and 0324 of this volume for further information regarding these plans and their outcome.
4. *your father*: John Tyndall Senior.
5. *that you would not go*: i.e., to America.
6. *George*: George Latimer.
7. *The non-appearance of No. 6*: the expected sixth letter from 'Spectator', Tyndall's pseudonym, to the *Liverpool Mercury*.
8. *you were guilty of the authorship*: i.e., the authorship of the Spectator letters.
9. *Ellen*: see letter 0288, n. 31 of this volume.
10. *Aubrey*: possibly Aubrey O'Brien, who worked with Tyndall on the Ordnance Survey.
11. *'gray goose quill!'*: Lord Byron, 'English Bards, and Scotch Reviewers' (1809), ii.7.
12. *'Friendship! ... I owe thee much ——— Blair'*: R. Burns, 'Epistle to James Smith', ll. 1–3. The word 'earnest' was either a typographical error by Louisa Tyndall or a mistake by Ginty; Blair's poem refers to 'Friendship' as the 'mysterious cement of the soul'.
13. *'Your spirit dares me to forget'*: unknown reference.

14. *'The die is cast'*: an idiom meaning that something has already been determined.
15. *barque now streams her pennon gay*: i.e., the ship's flag flies showily (*OED*).

From George Latimer 20 March 1844 0299

Newtown Gore[1] | March 20th 1844.

My dear John,

I have waited some time patiently to hear from you, and post after post arrives without anything from Leighlin Bridge. I was waiting to hear from you, before I would write, but do now think it better do so that it may prevent any unpleasantness that may occur, and first 'hear me for my cause'.[2] T'was with no peaceable feelings on the part of my friends that I engaged in my American struggle[3] heretofore, being saluted on all sides by this phrase 'I think that you are too hasty, and should wait at least 12 months at home, ere you would engage in such an adventure, for we are persuaded that when you become known to the gentry of the country, you will have perhaps more to do than you will be able to attend to and you know, that you have not been more than 1 year at home for the last 10 years. How then can you expect yourself to be recognized by gentlemen to whom you are a comparative stranger? Again if you go I just give you one twelve months, until you are back again, for with all the experience you have from those who have been in America, and now from those that are in it, in various parts you cannot boast of one solitary exception, where you can say "there is encouragement for me", and it would ill become you to be compelled to set aside all the drawing instruments &c with which you are surrounded, and take up the spade and shovel'. Now I say that I battled with those objections as a man struggling for liberty, and longing to be set free, with this only advantage – that I might go if I would, but mark the consequences. So that after all my fondly cherished hopes of accompanying one who is dear to me, yes dearer I would say than any other beyond my brother, must relinquish my hold, and struggle for liberty amongst this people for another 12 months and then, if I am not satisfied with my advances, I may with all freedom of will spread my sails, and steer my barque[4] for America, for as the whole matter of my opposition is contained in this 'past counsel past grace'.[5] My dear John I know not what may be your opinion of me, but one thing I know, and which I solemnly declare, that I never excited expectations in your breast relative to my going with you that I did not intend to see fully carried out, until transatlantic shores should seal the truth of our long cherished hopes and most faithful intentions. I cannot well describe myself to you, you may be very much displeased with me, but weigh those circumstances and consider that though my father[6] would not prevent me if I were sure of getting anything after I would go, but the thought of not getting employment caused him to calculate for me and to conclude that it is more expedient for me to wait this year, and sound the country well. That there may be no cause of complaint hereafter concerning this gentleman, or that, who may have very considerable work to survey and had I taken my father's advice, I would run a fair chance for it. I am utterly at a loss what to say more seeing that you will be so disappointed in my not going, and would it be inconsistent if I dare ask, for you to come down to Newtown,[7] at the time formerly specified. If not, it is not improbable but I shall see you in Liverpool, as a friend of mine wants me to go there for him on a certain business, but if you would think well of coming you would be as welcome as the flowers of May.

I had a letter from Phil[8] concerning the cash which he said he could not make payable to me sooner than June next. Now if you will oblige me, and honour me so far, in taking that sum which I have no doubt but he can get for you, before your departure, you will confer upon me a great kindness, the sum is £1–10s. I know a little of you! Don't be ruffled at my bequest and till I hear from you which I hope shall be shortly I subscribe myself

As ever your affectionate | George Latimer.

RI MS JT/1/TYP/11/3744–3745
LT Transcript Only

---

1. *Newtown Gore*: a village in County Leitrim, Ireland.
2. *'hear me for my cause'*: W. Shakespeare, *Julius Caesar,* III.ii.13–14.
3. *my American struggle*: Latimer and Tyndall had planned to emigrate to America together in the spring of 1844. See letters 0237, 0269, 0291, 0295, 0296, 0304, 0306, 0307, 0313 and 0324 of this volume for further information regarding these plans and their outcome.
4. *barque*: small ship (*OED*).
5. *'past counsel past grace'*: an Irish proverb meaning that one who cannot listen to advice is beyond help.
6. *my father*: not identified.
7. *Newtown*: a small town in County Leitrim, Ireland.
8. *Phil*: Phillip Evans.

From Archibald McLachlan 31 March 1844 0300

45 Swan Street. York. | 31st March 1844 – | Posted on 3rd April 1844

Dr Mr T.[1]

I have enclosed N°. 2[2] – and on reading it, should you find that it not the one you intended, let me know, and you shall have any one you please, for I have a copy of each – As it is our intention to print all Spec:[3] letters, and as they have been honourably mentioned in several newspapers and as the subject will, before long, cause more enquiry to be made, and as we cannot tell the ultimate effect; would you object to the letters of Spec: being curtailed or condensed? I do not intend to alter any part of them, my intention is to omit some expressions and allusions, which would shorten each letter by about 12 or 14 lines – We cannot tell where you may meet those letters hereafter – Perhaps in America! when you are an old man!! We cannot tell under what circumstances, at what time, or in what place you may have the honor to read them – and as you might wish wish[4] to see them, as originally written – Also – you know some party here might find fault with any liberty taken with them – and say 'this is done without authority' and much more to the same effect – I wish to have your leave, or authority for any omission I may think it necessary to make – This, you know, would relieve me from the charge of meddling with that which I had no right to touch – You may think little of this, but I consider it a ticklish point, and would not like to omit anything without saying 'by your leave' –

Walkem wished to have a Map of the City of Dublin, I promised to send him one – Just before he started from Liverpool I sent for one, to Hayles & Smith Colege Green[5] – the Plan came to Liverpool too late – Mr Tidmarsh called at the Steam-packet office and got the map, he will give it to you, I hope it is a good impression, keep it for auld acquaintance sake, were I a little richer I should send for one that would be of more interest to you – Cork – but I heard that Cork is not published yet –

I have written to Mr Todd – and am writing another to day – I think we understand one another – and will proceed– You once mentioned a wish to have the letters of Mr Ginty – I have a number of them by me now, should you still wish for them you can have them all, send me your Liverpool address, and should you wish it, the whole shall be sent – otherwise I will look over them, pick out the good hints, & then Burn them –

And now, dear Mr T. – our long correspondence has *[closed]*, I look upon you as a man of the other world – I hear, every other day, the question put, 'What has become of Spectator' – it will be a long time before Spectator answers that question! In losing Spectator, every one of us will feel that he has lost a friend – I am no Prophet – that is, no false Prophet – and I predict that the Author of Spectator's letters will meet that reward in America which is denied to him in his own country – Why should you not succeed as well as many others of my 'auld acquaintances', and old O.S.[6] gentlemen too, they had not half your abilities – but they were young, and persevered – And if you show as much 'pluck' in America as you have shown in England, I have no doubt of the results – You will be independent when we are struggling under '– per diem' – – Oh if I were only 20 years younger!! I should not care much whether 'Col. C.[7] is pleased to increase' &c[8] – I would do as you are doing now – leave the country – –

Lest I should forget, they say that if you tie a handkerchief tight round the lower part of the bowels, it will prevent, or at least lessen the distressing effects of sea sickness. If you get sick, try this experiment – And now Mr T. – I must say Good bye – and were it not that we are all scattered, such a great distance between the Divisions, you should not part from us without some Public mark of respect – but you know all our circumstances – and you know that this act of Justice is almost impracticable – but we can do you justice in our pamphlet,[9] and if the authors of the foul injustice have one spark of honourable principle – or one drop of red blood in their veins, that drop shall crimson their face – We will not do it in a passion, but cooly and deliberately – Will you write me a letter when you get to the other side, and say whether a copy of our pamphlet would reach you safe – That you may reach the other world safe, that you may persevere and prosper, & become an independent man, is my sincere & hearty prayer –

A. M^c^Lachlan

To | Mr J. Tyndall – care of | Mr Tidmarsh | Ordnance Survey Office | Liverpool

RI MS JT/1/TYP/11/3831–3832

RI MS JT/8/1/4a

---

1. *Dr Mr T.*: Dear Mr. Tyndall.
2. *No. 2*: the second letter from 'Spectator', Tyndall's *Liverpool Mercury* pseudonym; see letter 0252 of this volume. McLachlan was preparing a pamphlet that was to include the letters; see letter 0296 of this volume.
3. *Spec.*: Spectator, Tyndall's *Liverpool Mercury* pseudonym.
4. *wish wish*: this repetition appears in the handwritten letter.
5. *Hayles and Smith Colege Green*: possibly the name of a bookstore or mapmaker.
6. *O.S.*: Ordnance Survey.
7. *Col. C*: Thomas Frederick Colby.
8. *'Col. C. is pleased to increase' &c*: i.e., increase McLachlan's salary. McLachlan is expressing the wish to not be dependent on the favour of Colby or others like him.
9. *our pamphlet*: see letter 0296 of this volume for more information on McLachlan's plans for this pamphlet.

From Archibald McLachlan [March] 1844 0301

50 Fanshawe S^t^ – Sunday

Dear T.

I have got into such a system of scribbling that I can't break myself off it, the more I try to correct myself the worse I get, have you any secret by which I could correct this confounded scribble hand? –

I wrote you a long and tedious scratch 3 or 4 days ago,[1] enclosing a trace – if you understand the trace tis all that can be expected – it might have been a little finer & plainer, but I had to trace it off in a hurry – or by stealth – I got another since, and would have sent it had I not got a letter yesterday from Mr Kelly[2] saying you were out levelling – You will have it in the course of this week – I saw my learned man,[3] we discoursed of Tides on sea and land; he says there is a book written wherein is described all manner of curves, indeed it is from this book, or the principles laid down there, that the whole system of representing political, religious, and philosophical progress by curve lines – the book is called – 'Quetelets System of Nature'[4] price 1/6 – Chambers's Edition for the people – You could easily get in Manchester – there is a great Socialist bookseller in Manchester, from him you can get any book – up near the Infirmary, or Hospital at the reservoir a good way farther up than the postoffice – when you walk up to the Post office keep it on your right hand, and go on til you come to the Pond or reservoir, keep that on your right also, and go on 100 or 150 yards, turn off to the left – (the infirmary is then behind you) go down a long street, go down about 250 or 300 yards and on your right hand is the Shop – the Name is – Heywood[5] –

My learned man plotted the curves of life and Death taken from the Data furnished by the Irish Census – the Census of Dublin City – and others – one line of curves was plotted for all Ireland – one for the City of Dublin – all these will be published when the Irish Census &c is Complete –

The outlines of mountains – Waves of the Sea – Lives of men – women – Diseases – Deaths – Crime – poverty – ignorance – all can be shown better by Curved lines than any other way – because at one glance you can have all the lines before your eye – just as you have a number of Waves of the sea – you can see how the curves run parallel or cross – or separate – and by simply looking on the lines you can tell the causes – When you see the lines of Ignorance; poverty, and disease running nearly parallel – you can easily fortil[6] the curve of Death before it is even plotted! – All depends on ample and minutely correct registers of Births, Diseases – food – Drink – education – Deaths &c &c – when the list is defective, the results are not satisfactory–

– One general rule applies to the Curves – Generally, you take Time for a base – and Number for a perpendicular – Years – months – or distance for a Base – Quantity – or numbers of individuals for the perpendicular –

Let us suppose that on a certain day, the census of Dublin City was taken – distinguishing between male & female and the Age of each person was taken; & from that a table of those alive between certain ages viz – between 5 & 15 years – between 16 & 25 &c & – on this data a Table of Ages is constructed – and that we wish to lay down the Curve of life – or A curve for the living – We have nothing to do with Death at Present –

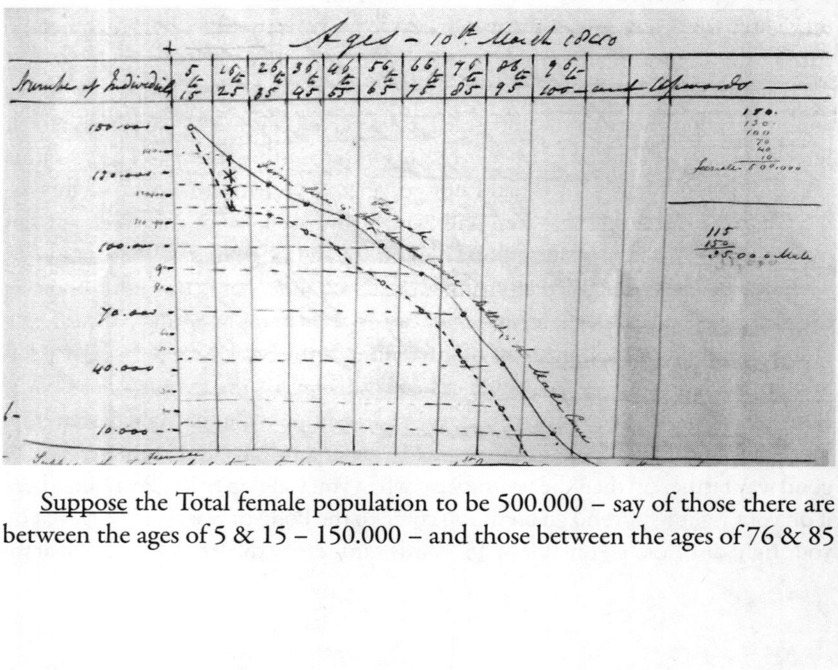

Suppose the Total female population to be 500.000 – say of those there are between the ages of 5 & 15 – 150.000 – and those between the ages of 76 & 85

are = to 70.000 – (It does not signify what scale for the Base & perpendicular you adopt – tis the same in this as Railway Sections, or anything else – the base may be one scale the perpendicular another – The above Table of Ages is made at Random) – You have now 150.000 living, between the ages of 5 & 15 – on your perpendicular scale lay off 150.000 – put a dot in the c^{r}[7] of the proper Column – next you have living between 16 & 25 a number of individuals – 140.000 – lay off on your scale as before 140.000 – in the c^{r} of column $\frac{6}{25}$ – proceed in the same way with all your numbers and ages – the last is $\frac{96}{100}$ and upwards – of those at 100 there are living ~14.000 – lay off fourteen thousand in the proper column – Now connect all your points or dots, this will form the curve of life; Female, – Now for the Male curve – Suppose you have 150.000 males alive between the ages of 5 & 15 (that shows that an equal number of Male & Female are born) – lay off as before 150.000 in the column of $\frac{5}{15}$ – Next, your return or census shows that there are between the ages of $\frac{16}{25}$ a number = 116.000 – lay off 116.000 in the proper column $\frac{16}{25}$ – proceed in the same with all the ages down to 86 | to | 95[8] at this age you find 25.000 alive – Now connect all your points by dots – and this Dotted shows the curve of life for the Males – – Something remarkable in those curves – you see the Male & female are nearly parallel after the age of 16 | to | 25[9] – at $\frac{16}{25}$ an awful difference occurs – We see that at 5 | to | 16[10] there were an equal number of males & females alive – why then should there not be an equal number all through the different ages, & the two lines male & female would be identical – but an awful difference occurs among the population between the ages of 16 & 25 on the Male side – What is the cause of this? Natural death – War – or Emigration? – As yet we cannot tell – But when we come to lay down the curves of Death, those curves will explain all, & show that the difference is owing to death – and the diseases will show the cause of Death – it will show that young men in the city give way at a certain age – give way to intemperance & Debauchery – and that the females escape – hence the difference in the Curves Q.E.D.–![11]

I send a Tracing of The Tide wave,[12] taken at Meades Quay Limerick,[13] the principal difference between this curve & the last I sent, is, that on this tracing the low water is not so rounded as high water, this is perhaps, caused by the narrowing of the channel and rush of the river meeting the Flood Tide – I looked on a map a few days ago, on the place in the C^{o}. Wexford[14] where the low water is sometimes higher than high water, it is a plain open coast – open to the Sea – and I could not, nor could I find anybody who could give any probable reason for this curious fact The different colours represent the different days – the whole thing is not well done, but I did it as well as time and the opportunity would permit –

I send you a Tracing of the Shannon,[15] there were 7 Tide gauges placed along the N. side of the River, beginning at Liscanor[16] and ending at Limerick (Meades

Quay) the tops of all the gauges were on the same level, made so by the spirit Level, – at each gauge a person was placed, and on a certain day & hour (low water) they entered the heights of the rise in a Book & from those books the Sections were plotted. – No 1. is the mean of many observations – and all you can learn from it is the curious difference that exists between the different Stations – if a <u>mean</u> is taken at one place it would not be on the same level with a mean taken at another place – you see the Tide rises more than a foot higher at Meades Quay Limerick, than it does down at the open sea at Liscanor –

N°. 2, is one Tide – the perpendicular Scale represents <u>Time</u> and distance – each marks the <u>time</u> of low water at his station, begin at the right and look at the times of high & low water, look along to the right and at each Station you find the time of high and low water differs – all the way to Limerick – (observe the red lines, I dont understand the blue lines) Look along a red line and it will give you the form of the surface of the Sea – the whole length of the Shannon, at a particular hour – say 50 minutes past one – this line differs considerably from that at 20 minutes past 2 – and so forth, indeed by a simple inspection you will see more than I could tell you Whilst I was making the tracing our trusty man was kept on the look out, I ran my pen along the lines – hurridly – so that the thing is by no means so satisfactory as If I had taken a good Copy – I send it merely as affording an insight into the methods used – as mere hints –

H.W. – is high water –
S.p – Spring Tides
N.P – Neap tides
HW. ST – high water Spring Tides
&c &c –

Since I wrote last, I have been getting lessons on the Electrotype – I begin to understand it and am making my first attempt without the assistance of the Master, I think I will succeed – I sent a note on Saturday to Mr Kelly, enclosing one for you,[17] – I will write in a day or two to the Gentleman in Scotland,[18] I don't know of any person who would Suit him, the Salary is so low – it would not be of any effectual good to write to some one on the O.S. I think they are better as it is – but if you know of one who is out of employment, and would take anything rather than be idle, I would like you to send me his address soon – I think you are in the Country and will not get this letter for some time, therefore I send it to the care of Mr Kelly. I have not heard of you for a long time – write soon after you get this –

Yours truly | <u>A McL.</u>

RI MS JT/1/TYP/11/3826–3829
RI MS JT/8/1/4a

1. *I wrote you a long and tedious scratch 3 or 4 days ago*: possibly letter 0296 or 0300 of this volume, though neither of these had a tracing enclosed.
2. *Mr Kelly*: possibly John Kelly, a Civil Engineer on the Ordnance Survey of Ireland; see letter 0262, n. 15 of this volume.
3. *my learned man*: not identified.
4. *'Quetelets System of Nature'*: unidentified book; probably a translation of something by the French statistician Adolphe Quetelet.
5. *Heywood*: a famous radical bookshop owned by Abel Heywood (1810–93), a socialist and active Chartist.
6. *fortil*: this is probably a misspelling of 'foretell'.
7. *c^r^*: centre.
8. *86 | to | 95*: in McLachlan's text, this phrase is written in a single vertical column in the middle of the sentence.
9. *16 | to | 25*: in McLachlan's text, this phrase is written in a single vertical column in the middle of the sentence.
10. *5 | to | 16*: in McLachlan's text, this phrase is written in a single vertical column in the middle of the sentence.
11. *Q.E.D.*: an abbreviation for *quod erat demonstrandum* (Latin), meaning 'it is demonstrated'.
12. *I send a Tracing of The Tide wave*: tracing missing.
13. *Meads Quay Limerick*: Limerick is a city in Ireland that lies on the Shannon River.
14. *C^o^. Wexford*: a county on the southeast coast of Ireland.
15. *I send you a Tracing of the Shannon*: tracing missing.
16. *Liscanor*: more commonly spelled Liscannor; a town in County Clare, Ireland.
17. *enclosing one for you*: letter or enclosure missing.
18. *the Gentleman in Scotland*: not identified. Possibly the 'Scotch C.E.' mentioned in letter 0290; see letter 0290, n. 3 of this volume.

To Henry Tucker 16 May 1844 0302

Sir

I am induced to trespass upon your attention by circumstances which have placed it in your power to render me an important service. Previous to last March Assizes[1] I had everything arranged for my departure to the United States of America, when I found that the Grand Jury of the County Carlow[2] required an applotment to be made from Mr Griffith's general valuation. I proposed for the job, and after a very close poll I was appointed; through some informality however the proceedings were rendered void and the matter was postponed to the July Assizes, when proposals will be again received. One consideration only renders my proposal doubtful, and that is <u>my youth</u> – the great majority of the Grand Jury are favourable to me personally, but they consider experience indispensable towards the proper execution of the applotment, and measuring my experience by my <u>years</u>, they infer that it must be limited. Now Sir for the last nine months of my engagement under you, I was employed at work which required the constant exercise of steadiness and caution, namely, <u>the final exami-</u>

nation and completion of Tithe Plans;[3] and I trust that nothing has transpired since my removal from the Survey,[4] to show that the confidence reposed in me was misplaced. At all events I can truly say, that while employed under your superintendence, my constant aim has ever been to perform the duties allotted to me so as to merit your approval; your kindness in recommending me twice within the first year after my transfer to the English Survey inclines me to hope that my efforts were in some degree successful; and though my conduct last September was unfortunately contrary to your wishes, still I trust that it furnishes nothing to injure in your estimation, my character as a workman.

Few individuals on the Survey have applied themselves so assiduously as I have in the discharge of its various duties. It was my ambition from the first to render myself as generally useful as possible without being of expense to the public; at present there is no duty connected with the detail Survey[5] which I cannot perform with despatch, though I never was placed for a single day under instructions in any branch whatever. Observation and industry wrought for me what expensive instruction did for others. The accompanying scraps from my old journal which I was permitted to keep will go far to prove this – I joined the Survey on the 1st of April 1839, and on that day I was employed 'writing on fair plan'[6] at which work I continued till the 29th of the same month when I commenced drawing. I joined at 9/– a week and on the 1st of June I was strongly recommended, which obtained for me an increase of 9d a day, or from 9/– to 13/6 a week. From the first I was employed at duties of a high order, such as drawing, examining plans &c. At a time when there were no points in the office I was sent to the field, and my second journal will show, that no man ever lost a day instructing me, nor did I ever lose a day attending on any man, whether examiner, surveyor, or leveller, though I have been engaged in the duties of all of these. While with Lieut. Whittingham[7] in Youghal,[8] the charge of laying down points and of attending to the plotting of lines generally devolved on me; my last year on the Irish Survey was occupied in drawing and examining the plans of the City of Cork;[9] I may further add that up to the 2nd of September a report or complaint concerning me never reached the ears of my Superintendent.[10]

My design then Sir in writing to you is to request that you will have the goodness to certify your opinion of my steadiness and fidelity in the discharge of duties which require the exercise of both. You have always borne creditable testimony both as to my moral character and my qualifications as a draftsman; on these points the testimony of Captain Wynne R.E.[11] with whom I was first employed is even stronger than yours, but I am most anxious to obtain a testimonial which shall contain a marked reference to my uniform steadiness of character; I have taken the liberty of forwarding with this a specimen of my drawing as owing to the peculiar work at which I was engaged in Preston your opportunities of judging of my abilities in this line were limited. I beg also to refer you to the testimonial of

Mr Wynne alluded to above. If you would be kind enough to corroborate the latter and mention further my being engaged at duties which require the attention of a steady and confidential person, you would confer an obligation which should ever be remembered with the warmest gratitude by

Sir, | Your most obdt.,[12] humble servant | John Tyndall | Capt. Tucker | R. Engrs.[13]

RI MS JT/2/13a/34–35
LT Transcript Only

---

1. *Assizes*: a legislative court sitting (*OED*).
2. *the Grand Jury of the County Carlow*: a government body to which Tyndall applied for a surveying position. From Tyndall's journal, 9 July 1844: 'In Carlow all day – had about 10 competitors, very elegant fellows indeed ... At 6 o'clock I was informed that the applotment would not go on, that the Grand Jury had had Counsel's opinion on it, and that that opinion was that the Jury had no power to appoint an applotter; there was an end to all my airbuilt speculations!' (RI MS JT/2/13a/46).
3. *Tithe Plans*: see letter 0231, n. 19 of this volume.
4. *my removal from the Survey*: see letter 0262 of this volume for a copy of the order through which Henry Tucker dismissed Tyndall from the Ordnance Survey.
5. *detail Survey*: the branch of the Survey tasked with surveying and mapping topographical details, rather than boundaries.
6. *'writing on fair plan'*: a surveying term; not identified.
7. *Lieut Whittingham*: Paul Bernard Whittingham (1819–94), biographical details unknown. J. J. Howard and F. A. Crisp (eds), *Visitation of England and Wales* (London: Privately Printed, 1900), vol. 8, p. 45.
8. *Youghal*: a city in southern Ireland (County Cork).
9. *the City of Cork*: a city in southern Ireland (County Cork).
10. *my Superintendent*: the Superintendent of the Ordnance Survey of Ireland was Thomas Frederick Colby.
11. *R.E.*: Royal Engineer.
12. *obdt.*: obedient.
13. *R. Engrs.*: Royal Engineers.

From William Ginty 19 May 1844 0303

Carrick on Shannon | Sunday 19th May 1844.

Mon cher Jack,

Altho' 'remember to keep holy the Sabbath day'[1] has been chiselled on stone by an almighty sculpture – it has not yet been graven on my adamantine heart. The injunctions of the sacred preacher under this test falls upon my ears as 'sounding brass'.[2] 'Six days shall thou labour &c'.[3] But if any right reverend mountebank deduced from that that writing to John Tyndall is a labour then in the teeth of his holy habiliments do I most energetically pronounce him to be a theological humbug and a sacerdotal jackanape.[4] One would imagine from your pious horror and scriptural denunciation of my simple curses that you were some twaddling preacher or tar barrel divine. I however am not that one – oh no! too well do I know how stands the barometer of your righteousness – too perfect is the lithograph – too exact the facsimile which I possess of your very soul to think for a moment that your moralizing was anything but the essence – the quintessence of hypocrisy.

The death thoughts frightened you did they? miraculous! I really thought your looking-glass had braced your nerves pretty well on this score already!!!

Thanks for the valentine[5] – of course I bestowed on it all due admiration (the said admiration being no trifle) but tell me – has any convulsion in the bowels of our little planet changed the geographical position of Leighlin since I honoured it with my presence (ahem!)? Or how came it to be in the 'vicinity' of Preston on the '14th of Feb., last'? – Aye, Preston, you stand connected before the impartial tribunal of my conscience for that not having the fear of God before your eyes did wilfully and corruptly in the said letter stretch the meaning of the word 'vicinity' in defiance of all precedent – thereby bringing the towns of Leighlin and Preston into contact to the utter consternation of the good people of the said towns. The evidence in this case being 'my pretty Jane'[6] short conclusive and incontrovertible. *[Allah]* 'truth is Tyndall's glory'. 'Brutus is an honourable man'![7] 'My first lame effort at a valentine' 'my first' yet 'Brutus is an honourable man' – But quench thy rousing ire take no thought of revenge for here, ever here is my valentine which you 'have yearned to see', forebear! oh forebear.

> 'In mercy spare me if I do my best
> To make as much waste paper as the rest'.[8]

The only thing that decided in favour of this miserable and unutterably shabby production ever coming under your notice was your gigantic oath (quotha 'him that is filthy let him be filthy still')[9] which said oath filled me with perspective horror lest the devil in the plenitude of his benevolence might grant you fortitude to preserve it inviolate.

I really did receive your letter[10] at breakfast in Lakeview (whither it was sent after me) en passant[11] (which being interpreted means sitting by the road side to tell a 'story') that same parenthesis contains a piece of very superfluous information. But, to return, I laughed, long and loud – and uproarious, and two of 'my chief earthly enjoyments' dealt out their respective quantums of bliss in concert – concert? oh la! it was an outrageous discord. The laughter filtered the coffee thro' my nose and teeth and to the unutterable dismay of mine hostess descended in a shower variegating and bedeviling her snowy table-cloth.

Confusion seize thee and thy impertinence – 'my rugged physiognomy'[12] charnel-house[13] deserter! this is unsufferable. But in the plenitude of my contempt and merciless scorn I pardon thee– Yes, I actually laugh at the mere anticipation of the discussion revived. I must procure a cholera morbus belt[14] to avert the dissolution of my ribs. Lord how I will lacerate you if I come within an earshot of you, I have a long account to settle. How is Cupid[15] – is he living or is he dead? As for George Latimer I don't */know/* what he is after. He was in Liverpool lately. My brother[16] is going down to applot the Ph.,[17] he leaves in the course of the ensuing week. I'll write to him and learn particulars. I wrote to Jack[18] this day fortnight, no answer yet – in good truth I have been thinking as you have thought for the last 9 or 10 days with respect to the Belfast gent. If our surmises prove in time to be true I will assuredly wreak my vengeance on him in the shape of a bitter letter. If I can so manage it without creating a feeling to Johnny's prejudice. Have you heard any of the good results of our 'combination' pitiable and shabby as they are – they nevertheless prove how good and efficacious it is to agitate. Hunter[19] writes 'Men on the Survey 20 years are to have half pay while sick. Those from 7 to 14 one-third and none under 7 gets any'. If I had time or space I would shower a volley of malediction on the miscreant herd, but let them pass. A 'Veritas'[20] and a 'Vindex'[21] and some other worthy in the 'Nation'[22] have been walking into them lately.

I will be done in Lakeview[23] in 3 or 4 weeks – if you write within the coming fortnight address to the care of: –

Henry Boyd Esq.[24] | Lakeview | Ballifarudn | Boyle.

for pity's sake write soon for I am wretchedly lonely down there. Give my best respects to your sister[25] and believe me

dear Jack | Yours sans changes | Ginty.

To Evans. A rigmarole.

Child of Lilliput![26] and mine
Ginty greets thee with a rhyme
Dwarfish man with giant soul
As the needle points the pole
As the ivy seeks the tree
Veers my spirit unto thee! –
But say thou 'smallest born of earth' (3)
Has memory sought the 'second birth'
Or why oh Evans has thy pen
Ne'er sought my hapless fate to ken
Is recollection's powers dead
Has every Irish feeling fled
Shades of the past too often wooed
Is this thy base ingratitude!

÷ ÷ ÷ ÷ ÷

Miniature of him of Gath!!!
We have trod a common path
Thro' the sweet and witching dears
Of the southern world for years

O'er 'the green' together pranced
And in Auberys[27] harem danced
Harem! not for lewdness there
Harem for its beauties rare
Such at least 'twas ours to know
Tyndall says <u>we felt</u> it too
And as truth is Tyndall's glory
'I'm <u>half</u> inclined to believe' the story
Now take thy fulsome memory back
To Youghal's[28] oft betrodden track
See what is there! oh ask thy heart
If silence should have been my part
Have I not sought at your behest
To calm Miss Bessie's[29] troubled breast
When your foul fancy for a kiss
Brought infamy instead of bliss
Forsooth 'Miss Sally[30] one – no more
You can't object Bess gave a score'
Oh dearly was that one smack bought
Bethink the ruin that it wrought
Oh think how oft you swore a kiss
Had played the devil with your bliss
Think, think how well I fought your fight
When you were banished from her sight
Aye think of this – ungrateful man
Go – go repent while yet you can

÷ ÷ ÷ ÷ ÷

<u>Giant thro' a microscope</u>! (1)
I have cast a horoscope
Skilled in dark Egyptian lore
I have scanned the heavens o'er
Read thy fate in jumbled stars
Clustered round the bloody mars!
Hear their dismal tale and dread
Vengeance pending o'er thy head
Few short weeks shall fly apace
Then we're fronted face to face
I with fearful vengeance mad
Thou despairing pale and sad
Thou shalt shiver neath my glance –
Nerveless sink in coward trance –
Gasping for thy puny breath –
Looked unto a frightful death
Thus shalt thou die – in terror base
The last of all the Pigmy race!!!

– But jokes avaunt– my day dream is
A wish to view thy ancient phiz[31]
My heavens! I long to have a shake
For quandam 'Cupida' quandam[32] sake
My pent up extacy to shed
Like steam from boiler on thy head
Lord how I long to hear thee whine
Thy sorrows over auld lang syne[33]
As once thou used in Youghal shade
'The rent the envious Bessie made' (2)
To hear thee doat on beauties gone
And hear thee praise the present one
I long for this – write – break the spell
Child of Lilliput[34] farewell.

W.G. Ginty.

(1) That is to say thou art a giant when viewed thro' a microscope– __________! So would a flea.
(2) 'See what a rent the envious Casca made'[35] | Shakespeare
(3) 'Behometh biggest born of earth
Upheaved its wetness'[36] | Milton

Verily I say unto thee Evans thou art the antipodes[37] of the Behometh – the minimum of the animal world – the smallest specimen of genus Homo.
Lakeview | Ballifannon | Boyle | Co. Roscommon.

RI MS JT/1/TYP/11/3630–3631
LT Transcript Only

---

1. *'remember to keep holy the Sabbath day'*: Exodus 20:8. This is one of the Ten Commandments.
2. *'sounding brass'*: 1 Corinthians 13:1.
3. *'Six days shall thou labour &c'*: Exodus 20:9.
4. *a sacerdotal jackanape*: a priestly ape (*OED*).
5. *Thanks for the valentine*: letter missing.
6. *'my pretty Jane'*: possibly Jane Collier; see letter 0320, n. 6 of this volume.
7. *'Brutus is an honourable man!'*: W. Shakespeare, *Julius Caesar*, III.ii.82.
8. *'In mercy ... paper as the rest'*: T. Gent, *Poetic Sketches* (1808), epigraph.
9. *'him that is filthy let him be filthy still'*: Revelation 22:11.
10. *your letter*: letter missing.
11. *en passant*: in passing (French).
12. *'my rugged physiognomy'*: source not identified. Ginty was talking about the shape of his face.
13. *charnel-house*: a burial place or mortuary chapel (*OED*).
14. *a cholera morbus belt*: a belt worn around the waist, made of wool or felt. Thought to protect from cholera.
15. *How is Cupid*: possibly an oblique query about whether Tyndall was interested in anyone romantically.

16. *My brother*: not identified.
17. *Ph.*: possibly an abbreviation for 'parish'.
18. *Jack*: John Tidmarsh.
19. *Hunter*: Richard or William Hunter.
20. *'Veritas'*: a pseudonym under which one of Tyndall's critics wrote to the *Liverpool Mercury*.
21. *'Vindex'*: unidentified pseudonym, presumably for another Ordnance Survey opponent.
22. *the 'Nation'*: not identified.
23. *Lakeview*: probably the name of a house in the town of Boyle in Ireland.
24. *Henry Boyd Esq.*: not identified.
25. *your sister*: Emma Tyndall.
26. *Lilliput*: diminutive size (*OED*).
27. *Auberys*: possibly Aubrey O'Brian.
28. *Youghal*: a seaside town in Cork County, Ireland.
29. *Miss Bessie*: not identified, probably Phillip Evans's love interest; see letter 0259, n. 39.
30. *Miss Sally*: not identified, possibly another love interest of Phillip Evans's; see also letter 0305, n. 6.
31. *phiz*: facial expression, countenance (*OED*).
32. *quandam 'Cupida' quandam*: a desire for a certain Cupid (Latin).
33. *auld lang syne*: old long-since (*OED*), also the title of a poem by Robert Burns.
34. *Child of Lilliput*: a reference to the fictional country of Lilliput, featured in Jonathan Swift's *Gulliver's Travels* (1726).
35. *'See what a rent the envious Casca made'*: W. Shakespeare, *Julius Caesar,* III.ii.175.
36. *'Behometh ... Upheaved its wetness'*: J. Milton, *Paradise Lost,* book 7, ll. 471–2: 'Behemoth, biggest born of earth, upheaved | His vastness, fleeced the flocks and bleating rose'.
37. *antipodes*: the opposite of a given thing or person (*OED*).

From William Wright 7 June 1844 0304

Cincinnati | June 7th 1844

My Dear John

Yours of last month[1] arrived in due time. You will not be surprised when I inform you that it was the most unexpected visitor that ever presented itself to my view. For a week previous to its reception I had been in daily expectation of your arrival. I had all things in readiness for your comfort and had even put off a fishing excursion until you had arrived in order, as I expected that you would participate in the enjoyment even your seat at the oar had been preserved. But at the arrival of your letter all my castles vanished and for a moment I indulged in selfish thoughts, but it was only for an instant. Before I got half through your letter I was satisfied that you had acted right in deferring your anticipated voyage. Nay was rather gratified than otherwise. The principal cause of my gratification was that it would contribute a great deal to your father's[2] happiness to have you near him, that alone compensated me for my disappointment, for I learned from my sister[3] that he was taking your departure very much to heart and I was thinking he would not be at all obliged to me for my advice to you. Oh John we little know how the aged parent's heart is rent when he looks as he supposes for the last time upon the features and form of his beloved one. It's a sad struggle which takes place in his lonely bosom all in this life worth living for seems to depart from his grasp. If it's your lot to be thus separated may you be both spared to meet again is my ardent prayer. I little thought when I last looked upon my honoured Father[4] that it was the last I should see of him in this life. I feel deeply deeply feel what I have been deprived of myself and my fate seemed to interest him more than that of any other of the family. May he rest in peace. I know not why I have indulged in this strain. But these thoughts seemed uppermost in my mind and I was obliged to give them utterance. I expect you would rather have this letter filled with something more instructive. Another source of pleasure your absence was that it would give you an opportunity of signalizing yourself in the Co. Carlow,[5] for I must confess it would be my highest ambition to get big in my own country. It is said in the good Book that a prophet hath no honour in his own country.[6] I hope it may not be so with surveyors, or whatever you style yourself. I forgot to state in the commencement of this that upon the first perusal of your letter that I intended to inform the Grand Jury of Carlow[7] of the black sheep you were, I would soon have you out here then, but upon second thoughts I thought I would let you take your fling, for like a true Irishman I despise and detest a Pirater. Well Jack I wish you a hearty success, its speedy completion, and a quick voyage across the Atlantic as soon as you have got through with it, until then your bed shall be kept aired and a vacant seat at my hearth. Believe me I am very happy that you were so agreeably disappointed in my not being aboard of the illfated Shepherdess.[8] I assure I have no fancy to be put under water or frozen into a statue as some of the poor participators in that catastrophe were. Do not let the recent riots in Philadelphia[9] have any weight with you as the sober intelligent part of the community denounce them in no measured terms. I am sorry to add that the Irishmen were the first aggressors but nothing can justify the awful retribution with which they were visited. I sent you a newspaper[10] giving an account of it, I hope you may receive it. It also contained several wood cuts giving a description of the several places during and after the conflagration. The excitement seems to have entirely subsided. Here in Cincinnati I have not heard it once mentioned for 3 weeks at least.

John keep that meet again forever before your eyes, never for a moment lose sight of it, as soon as you have your job completed I hope you will still determine if possible to visit the Queen City of the West.[11] Ever remember that within its limits beats a heart whose highest gratification will be accomplished when its possessor beholds you face to face. We are well here. By this time I suppose you will have learned that my sisters[12] arrived safe in about seven weeks. Business has been pretty good with me this Spring and we are getting along pretty smooth. Sarah[13] is in the State of Indiana on a visit and the 2 children are with her also. So you see I am a bachelor once more. Give my best wishes to all the old stock about Leighlin. I am much obliged to William Murray[14] for his kind enquiry I hope he is doing well. Now Jack you must keep up a regular and frequent correspondence with me, and give me all the news Civil Military Foreign and Domestic about Leighlin. If you could inform my mother[15] of our welfare I will be much obliged. I will now bid you farewell

I remain yours forever | William Wright.

RI MS JT/1/TYP/5/1820–1821
LT Transcript Only

---

1. *Yours of last month*: letter missing.
2. *your father*: John Tyndall, Snr.
3. *my sister*: possibly Mary Ann Wright.
4. *my honoured Father*: not identified; see letters 0187, 0193, and 0216, *Correspondence of John Tyndall, Volume 1* for further information on Wright's father and his passing.
5. *Co. Carlow*: County Carlow.
6. *a prophet hath no honour in his own country*: John 4:44.
7. *the Grand Jury of Carlow*: see letter 0302, n. 2 of this volume.
8. *my not being aboard the illfated Shepherdess*: the Shepherdess was wrecked on 11 February 1844 carrying passengers from Cincinnati to St. Louis. Approximately 70 people perished in the wreck. Tyndall had read an account of the shipwreck and had worried Wright might have been on board. From Tyndall's journal, 4 March 1844: 'reading a portion of a letter from Wm. dated Feb. 1st. This relieves my mind from the sickening state of doubt in which it has been plunged since I saw the account of that fatal shipwreck. Wm. is safe and I am happy' (RI MS JT/2/13a/22).
9. *the recent riots in Philadelphia*: a reference to the Philadelphia Nativist Riots of May and June 1844. The riots were the result of anti-Catholic sentiments and persistent rumours that Irish Catholic immigrants were attempting to remove the Bible from public schools.
10. *I sent you a newspaper*: clipping missing.
11. *the Queen City of the West*: a nickname for Cincinnati.
12. *my sisters*: possibly Mary Ann Wright. The name of Wright's other sister is unknown.
13. *Sarah*: Sarah Wright.
14. *William Murray*: a friend of Tyndall's and Wright's from Leighlin Bridge, who worked as a civil assistant in the 5th Division, C District of the Irish Ordnance Survey between May and October 1840. He was then transferred to the English Survey.
15. *my mother*: not identified.

From William Ginty [June or July] 1844[1] 0305

Lakeview | Ballifarnon | Boyle.[2] Sunday

Dear Tyndall,

Undismayed by the terrible magnitude of your oath, unmoved by the high wrought frenzy of your invective I have ventured not to write so soon as you demanded, because I could not help it. Another proof to the doctrine of circumstances. Yes 'my pretty Jane'[3] I respect it. And I'll venture to say that the said 'pretty Jane' will not discern where exists the affinity between me and a pismire.[4] Alas, poor devil, you were lost for summat to advocate, in one line you make her an angel and in the next the said angel is filtering her _______.[5] I don't think Miss Jane would 'vindicate your taste' neither would 'Miss Sally'[6] be quite pleased at your obscene hint about the 'damaged meshes of her net'. 'Value my skin'.[7] Pshaw – I'm able to leather the angel _____ out of you!

I will be here until the last day that you want me. I would feign be in Leighlin a few days before we commence – for sports' sake. Are you sure of Latimer? I heard a few days ago since that he had engaged with the Co. Surveyor of Leitrim.[8] I will go to Kingstown to see the Misses Tidmarsh and their mama.[9] I had a note from Jack, his is still in Dervock.[10] I wrote to him this evening. I wrote my last to him in phonography[11] he took a month to read it. If I had written to McLaughlin[12] 5 or 6 months ago I would have a snug situation now. That opportunity does not now offer – he is on the alert for me. I was damned unfortunate in not writing to him. Here's the lines for 'Jim,'[13] I bungled them together in a hurry as you will see altho' Jim may not. I cannot definitely answer your question about the applotment. It chiefly depends on the quality of the work. Mr. B.[14] and I have not been regularly at it, we have done about 20,000 acres in the last 2 months, exclusive of the fair booths which remain yet to be done. Some townlands are exceedingly easy some are very difficult, according to the number of tenements or holdings and lots of the Ordnance valuation.

I'm in haste boy o. Write write write! | Yours | Ginty.

Kindest remembrances to my Jane.

Write soon. Address | Lakeview | Ballifannon | Boyle | Co Roscommon

I have a letter of yours from McGrath[15] – Jack[16] sent it me. McGrath is a great fellow – a genius!

RI MS JT/1/TYP/11/3832
LT Transcript Only

---

1. *[June or July] 1844*: Louisa Tyndall annotation: '1844. | probably after June 30 | before July 13'.
2. *Lakeview | Ballifarnon | Boyle*: probably the name of a house in the town of Boyle in Ireland; Ballifarnon is commonly spelled Ballyfarnon.
3. *'my pretty Jane'*: unknown person, possibly Jane Collier; see letter 0320, n. 6 of this volume.
4. *a pismire*: i.e., an ant (*OED*).
5. *filtering her* _______: unknown reference. Louisa Tyndall has omitted one word from this phrase.
6. *'Miss Sally'*: not identified, possibly a love interest of Phillip Evans's; see also letter 0303, n. 30 of this volume.
7. *'damaged meshes of her net'. 'Value my skin'*: Ginty appears to be quoting a missing letter from Tyndall.
8. *the Co. Surveyor of Leitrim*: the County Surveyor of Leitrim, full name not identified.
9. *Misses Tidmarsh and their mama*: relatives of John Tidmarsh, full names not identified.
10. *Dervock*: a village in Country Antrim, Northern Ireland.

11. *in phonography*: i.e., in shorthand.
12. *McLaughlin*: Archibald McLachlan.
13. *Jim*: probably Phillip Evans.
14. *Mr. B.*: not identified.
15. *McGrath*: not identified.
16. *Jack*: probably Jack Tidmarsh.

From John C. Chadwick [first half of 1844][1] 0306

You get from New York to Philadelphia in about a day – the City has many large squares the largest is Franklins Square which has a splendid jet de eau[2] in its centre. I have now given you all the news which my scanty means can afford and trust it will be of some use to you in your sojournings.[3] May God bless and prosper you in all your undertakings is my earnest prayer and I trust that the time will arrive when we will meet in the land of promise and live over old times with new zest where we may live and have our being free and untrammelled and lift our heads into the day with as lordly an air as the best in the land mix in the best society. It is the only place for white men but they make the niggers jump Jim Crow.[4] John F.[5] will be glad to see you. Remember me to Latimer. I was sorry to hear that McGowen and others had to offer their necks to the yoke at the footstool of Tyranny,[6] but some are born to cringe and others to soar – You will have stirring times of it – If you call on my brother[7] in Liverpool he might assist you in your choice of a vessel. Write to me before you leave, if you have time, if not farewell and blessing on thy way where'er thou goest. You and Latimer will be pleasant company. I should not like to go alone. All here are well. Mary[8] joins me in best wishes for your success. You ought indeed to be glad to get out of this wretched country which is becoming the mere sink of humanity. I must now conclude, if you write to me you will hear from me again, if not I may bid you farewell forever or the opposite according as fortunes wheel revolves; but under every phase of our acquaintance I shall cherish your memory as that of one whom I will always consider the most recherché[9] acquaintance I ever possessed and to whom I would now bid farewell with more regret did not hope whisper to me in a far land and far time that atom of humanity John Tyndall will again become your associate through good and ill. So now my dear fellow I must conclude, reiterating my wishes for your good fortune (le bon temps viendra)[10] take that for your motto and charge boldly.

I am most sincerely | your old Chum | J.Chadwick

RI MS JT/1/TYP/11–3507
LT Transcript Only

1. *[1844]*: Louisa Tyndall annotation: 'Must be early in 1844 | First half missing'.
2. *jet de eau*: jet of water (French).
3. *your sojournings*: Tyndall had planned to emigrate to America with George Latimer in the

spring of 1844; see letters 0237, 0269, 0291, 0295, 0296, 0299, 0304, 0307, 0313 and 0324 of this volume for further information regarding these plans and their outcome.

4. *jump Jim Crow*: a song and dance from 1828 performed in blackface.
5. *John F.*: John Fudge, an acquaintance of Tyndall and John Chadwick who lived in Cincinnati. See also letters 0268 and 0307 of this volume.
6. *McGowen and others had to offer their necks to the yoke at the footstool of Tyranny*: Louisa Tyndall annotation: 'McGowan returned to the Survey on [old] terms'.
7. *my brother*: not identified.
8. *Mary*: probably Chadwick's wife.
9. *recherché*: sought after (French).
10. *le bon temps viendra*: good times will come (French).

From John C. Chadwick 9 July 1844 0307

Kinsale,[1] | July 9th 1844.[2]

Dear Tyndall,

I need hardly say that the receipt of your last epistle[3] gave me a most uncommon deal of surprise as in verity I never expected to have heard from you again during the remaining term of this mortal life. I thought that you had departed long since[4] for another and a better world, and that the green isle[5] had lost one of its greenest sprouts. I wrote to you[6] at some length at the time that you expected to leave the country, giving you any little information I could respecting your route. As you did not mention it I suppose it did not go to hand. Perhaps as you played so distinguished a part in the memorable revolt of the sappers,[7] that Sir James Graham considers you worthy of official espionage by taking a peep at your letters. So mind your eye my old chap – but there was very little necessity for inditing such a lengthy monologue of repentance to excuse your epistolary hiatus. I am indeed a weak and erring mortal in that respect myself, and I should be the last to throw a shell towards any man's condemnation. I am satisfied like Lazarus of old[8] to pick up the crumbs of your correspondence whenever they may fall, well knowing that I am unworthy of more. So please yourself my dear fellow, and do not again bend your proud spirit to make excuses for an imaginary fault to a poor retailer of tape and thread.

(how are the mighty fallen).[9]

I quite conform to your creed of circumstances as indeed I do to your thoughts and theories in general. Indeed I have bowed so much before the shrine of your memory and genius that I think I will make you the head of a new sect called Tyndalites and in this age of dogmas it will be as worthy of followers as many of them. You can advise me in your next what you think of it, and what form of worship you would as a prophet, saint, or devil, (which you like) – prefer – You can make the power of circumstances a fundamental doctrine and mix it with a spice of fatalism and then we can shift the blame of our errors on our predestination and sit down blameless. I hope that the Spinning Jenny[10] of your fate may weave the £300 you mention[11] into your web. Your reminiscences of the past impressed my spirit with a tender melancholy. They came like the sweet south breathing o'er a bank of violets, and alas I often think that the spring and elasticity of life has departed and left me like a defunct bottle of ginger beer, or like a frog after spawning vapourising in an eternal *[cogsfundity]*[12] of cogitation, the fertility of life's morning having changed into a mental morass. Your memories came over my mind like beams of sunlight breaking from the west after the departure of day's orb only to make the long night appear more gloomy. Cus you you tarnal devil could not you as well paint the future in the glowing colours of heaven's bow as to rake up the ashes of the past merely to give me up a prey to the attacks of blue devils and ennui.[13] I hope that your next may be more exhilarating, giving an account of judge Doherty's[14] repentance and the fruition of your £300 sterling desires. Such a length of time has elapsed since I indited any but business letters that you may deem my present scribble weary flat stale and unprofitable. Indeed I could not attempt to compare anything in your *[1 word illeg]* style as poetic as the wreath of Nourmahal[15] so that you must be content with a plain dish of Kinsale cod and provide your own butter &c. You speak in anticipation of again seeing me. That would indeed be a pleasure to me. Oh speak again, and be more explicit. How I should like to give you a call in the event of your getting this job when Latimer[16] and Ginty would be with you and to board you for at least two meals of Carlow potatoes. I have spoken your desired remembrance to old friends who were happy to hear of your welfare. Excuse me for not sooner answering your last as I did not receive it for a few days after its arrival having been from home. I was at Clonakilty[17] bidding farewell to the Abbots[18] who will be leaving about the first of August, Mrs Fudge[19] and George[20] will go with them, and I may deem them my precursors, for I feel that this is not my abiding place. This place remains pretty much as you left it. There is generally an interval of twelve years between every event in this place. The word Kinsale seldom intrudes on the newspapers. Do you ever hear anything about the old survey? I hope you will write soon, as I should wish to hear how you get on about the applotment.[21] If you do not get it I suppose it will be hey over to America with you once more. Indeed I should not wonder if you were disappointed, as every job is hunted up with the ferocity of a pack of wolves by the many solicitors in this miserably overpopulated land. One has hardly room to breathe freely in it. I suppose that America will eventually bring you up and I still fondly hope that our fates will bring us together there. Perhaps next spring you might be going with me. I have had a good deal of boating this summer. I often wish you were with me, as all I want is a companion of a congenial spirit to make the time fly happily. But we cannot look for too much happiness in this life. Mary[22] desires to be particularly remembered to you. Miss Donoclift[23] is not yet married but will be in time to Barry the tutor[24] at Mr Perdue's.[25] All friends send a length of love to you. My daughters[26] are growing up stout and strong and handsome as myself.

I was thinking of rearing one of them for you if you will honour me with your alliance. It will be just the thing in your old age, that is if some fair Leighlinite does not forestall me. And now having spun my yarn to the last strand I must give up for the want of more wool, and wishing you every luck I am here

your steadfast friend | J.C. Chadwick.

I am in mourning for my sister Mrs Harrison[27] of Liverpool who departed lately so let not the black frighten you.[28]

RI MS JT/1/TYP/11/3508
LT Transcript Only

---

1. *Kinsale*: a village in County Cork, Ireland.
2. *July 9th 1844*: Louisa Tyndall annotation: 'See Envelope July 12/44'.
3. *your last epistle*: letter missing.
4. *I thought that you had departed long since*: Tyndall had planned to emigrate to America with George Latimer in the spring of 1844; see letters 0237, 0269, 0291, 0295, 0296, 0299, 0304, 0306, 0307, 0313 and 0324 of this volume for further information regarding these plans and their outcome.
5. *the green isle*: a nickname for Ireland.
6. *I wrote to you*: see letter 0306 of this volume.
7. *memorable revolt of the sappers*: Chadwick is referring to Tyndall's role in the protest of the Ordnance Survey employees against their wages and working conditions, most notably in the letter of protest sent by the workers of the Ordnance Survey of England to George Murray on 23 September 1843; see letter 0236 of this volume. On sappers, see letter 0232, n. 10 of this volume.
8. *like Lazarus of old*: a reference to the Biblical parable of Lazarus and the rich man, which appears in Luke 16:19–31. In the story, Lazarus was content to eat what fell from the rich man's table. Not to be confused with Lazarus of Bethany, the Biblical figure whom Jesus resurrected from the dead.
9. *how are the mighty fallen*: 2 Samuel 1:19.
10. *the Spinning Jenny*: a spinning frame with multiple spindles, designed to spin multiple balls of yarn or thread at once.
11. *the £300 you mention*: appears to be a reference to a previous letter from Tyndall which is missing.
12. [*cogsfundity*]: not identified.
13. *ennui*: boredom (French).
14. *judge Doherty*: not identified.
15. *the wreath of Nourmahal*: possibly a reference to M. J. Quin, *Nourmahal: An Oriental Romance* (London: Henry Colburn, 1838).
16. *Latimer*: George Latimer.
17. *Clonakilty*: a town in Cork County, Ireland.
18. *the Abbots*: probably the family of John Abbot. Louisa Tyndall annotation: 'Abbots going to America'.
19. *Mrs. Fudge*: wife of John Fudge, an acquaintance of Tyndall's who lives in Cincinnati; see also letters 0268 and 0306 of this volume.
20. *George*: probably George Latimer.

21. *the applotment*: see letter 0302, n. 2 of this volume.
22. *Mary*: Mary is probably Chadwick's wife, although it might possibly be Mary Hanlon, a friend of Tyndall's; see also letter 0291, n. 23.
23. *Miss Donoclift*: probably Sophy Donaclift; see letter 0268, n. 20 of this volume.
24. *Barry the tutor*: not identified.
25. *Mr Perdue*: not identified.
26. *My daughters*: Elizabeth and Sarah Kate Chadwick; see also letter 0268, n. 17 and n. 19 of this volume.
27. *my sister Mrs. Harrison*: not identified.
28. *let not the black frighten you*: this statement implies that Chadwick wrote this letter on paper with black borders out of respect for the passing of Mrs. Harrison.

From William Ginty 13 July 1844 0308

Lakeview, Ballifarnon.[1] | Saturday & Sunday | July 13th & 14th 1844.

My very dear Tyndall,

And is it so? Are all our dreams of happiness dispelled? All our hopes of social enjoyment destroyed? – All our anticipations of a delectable winter vanished into thin air?– All our glorious visions of six months uninterrupted bliss faded, all our dreams, hopes, anticipations and wishes consigned to a dull untimely and ignominious grave! Alas 'tis far – far too true? All, all are gone! Gone, gone forever ---! How heartily, how sincerely, how emphatically could I shower a fierce and bitter torrent of imprecations on their heads – A flame is kindled, it burns, it blazes in my boiling and contemptuous breast, stretching its withering influence to my very tongue, prompting me to curse them in all the intense bitterness of my soul – But no I will not, henceforth and forever, the Grand Jury of Carlow,[2] be it honour or be it disgrace have looked for and have found – a resting place – a black hole in my memory, in which they never can be viewed by me again only when their 'darkness is made visible'[3] by the torch of impassioned scorn and contempt. Yes! you did kindle hopes, not of lucre nor of gain, not of pension nor of place, but hopes preeminently elevated above the truckling avarice and sensualities of these days, hopes of a purer kind, hopes of meeting with friendship and with friends, hopes of happiness and of pleasures, hopes, heaven-born hopes too pure to enter the conception of the despotic aristocrat, too noble to fill the breast of the willing and submissive slave. These hopes are trampled on and crushed but not by Tyndall. Hope again asserts her power, again she rules the wide extended empire of my breast and points to a time when we may meet free from the trammels of tyrants and together hail their memory with scorn and their actions with contempt.

I must as this is the Sabbath curb, nay quell, my ire which burns as furious as when on yesterday I penned the above malignant lines. If as they say it is a characteristic of <u>charity</u> to 'endure all things'[4] I have I fear little of it. What of the Lancashire job[5] – What is the nature of it? Is it surveying? Do think further on it. I shall expect to hear from you soon and hope you will not be altogether so sore at heart as you were when you last wrote. I am just as much disappointed as you are and between you and I – I will now tell you you would not have made £1 by the job – We are now nearly 3 months applotting 20 thousand acres and we can and could do as much and have done as much as any other two sons of Adam.

Yours in haste | very faithfully | Ginty.

RI MS JT/1/TYP/11/3633
LT Transcript Only

1. *Lakeview, Ballifarnon*: probably the name of a house in the town of Boyle in Ireland; Ballifarnon is commonly spelled Ballyfarnon.
2. *the Grand Jury of Carlow*: see letter 0302, n. 2 of this volume.
3. *'darkness is made visible'*: while the specific source of this quotation is unknown, this phrase appears frequently in missionary literature when discussing the effects of spreading Christianity, for example, British and Foreign Bible Society, *The Twenty-Third Report of the British and Foreign Bible Society, Volume 12* (London: R. Clay, 1837), p. cxlii.
4. *'endure all things'*: 2 Timothy 2:10.
5. *the Lancashire job*: possibly the job for which Tyndall applied in letter 0311 of this volume.

From George Latimer 27 July 1844 0309

Newtown Gore[1] | July 27th 1844.

My dear Tyndall,

If it be time to answer your last letter[2] I now snatch from the world these few moments which I hope shall be uninterrupted. When I did not receive your reply the first or second day after the important decision[3] took place, I entertained but little hope of your victory, though in days gone by I was always or more generally recognised as the sanguine Latimer, now I am led to trace you in your epistle and conclude that you were a little sanguine on that occasion but I dare not blame you my dear fellow, for the importance of the case no doubt was well calculated to excite that spirit, if there was but one spark existing in the being of cool and calculating Tyndall. Yes, you certainly would have got 'big in your own country'[4] as your estimable friend (Wright)[5] terms it, and need I say that there would have been an additional spark of that temperament found rankling in me which hitherto characterized my existence. Well John 'tis past, irrevocably, sorry though I am and must be that our correspondence will not be so frequent as usual once you take your departure for that (I trust) more friendly clime. Yes and I believe that, to be more worthy of the man than this, for you must but pine away in this tainted land, polluted by every species of dishonour and distrust. But I am of opinion and see no doubt existing that there is to be found reserved for you a nitch in the society of those, better calculated to do honour to merit, and to recognize such with all that respect of which it is worthy. I often endeavour to dream in my waking moments beholding Tyndall occupying that station in society which shall move me to rejoice in ever knowing you, and in being able to say for my own pleasure and gratification, that we oft have lain together and talked over the necessity of the times of which we were not thought worthy, and though vanquished by Aristocratic power yet we were found capable of argument and remonstrance which may be thought of when other men and other days shall be found to do justice to our claim as individuals worthy of note and respect and attention. What about the Toryism of your friends or my friends? Are they not all, Whig and Tory, endeavouring to elevate self on the ruins of others or at the expense of more honest characters, though I never was much prejudiced as to party politics I must be happy in the thought of my lot been cast in Albion[6] for any length of time, by which I have been enabled to learn the state of the Times. Well then I should apologize for not writing before this, but my father[7] went to the water[8] this season, and a good deal of his business devolves on me in his absence, so that I have not much time on hand that I can call leisure. Even since I commenced this I was obliged to desist and attend to other business. What are your thoughts like at present? How far in the perspective does America lie now? and are your thoughts fully bent on that tour? If so, may every prosperity attend you and may every blessing of a Kind Providence be your portion both in time and for Eternity. My dear Tyndall I would fondly hope that you have considered those things so as to cleave to all that is good and all that is your duty as well as your privilege, but I won't preach more, and will conclude by saying that I do think you have a friend in Wright that will not desert you but who will with all the warmth of a brother prove to be your port in case of your embarkation for that land of liberty and freedom and above everything promise. I won't say that I shall be always bound up in my native soil. If you go, and think I would do any good there, my time will be short here in all probability. And now with every good wish and desire in sincerity for your welfare

Believe me to be | Your ever affectionate | George Latimer.

RI MS JT/1/TYP/11/3745
LT Transcript Only

---

1. *Newtown Gore*: a village in County Leitrim, Ireland.
2. *your last letter*: letter missing.
3. *the important decision*: probably the decision made by the Grand Jury of Carlow regarding Tyndall's possible applotment job; see letter 0302, n. 2 of this volume.
4. *you certainly would have got 'big in your own country'*: see letter 0304 of this volume.
5. *Wright*: William Wright.
6. *Albion*: an archaic name for the island of England, Scotland, and Wales.
7. *my father*: not identified.
8. *went to the water*: i.e., went to the seaside, possibly to treat ill health.

To William Ginty 29 July 1844[1] 0310

My dear Bill

For the last few days my thoughts have had rather a vagrant kind of life, and I now lift the pen in the hope that you will prove a rallying point – a kind of temporary bivouac[2] for the gipsey tribe. Your last arrived[3] in due time – your remarks on my style are just. I do write from impulse, or at least the feeling exists anterior to the writing. Why should I write from invention? I am not yet a hireling scribe – I write to please myself, cheered by the hope that what pleases me will please you – for the present I neither seek nor care for the approbation of anyone else. On this subject however I would say a few words: men, Bill, are inheritors of common feelings – these may differ in degree, but there are the same in kind and though external circumstances may conspire to create diversity in this respect, still man is man, and retains much of his originality unto the end of his existence – thus, the poet who is skilful enough to rouse the passion or win the tear of an individual, will in all probability find that his power extends to the multitude – someone sings:–

> 'Poetic fire, like vesta's pure and bright,
> Should draw from Nature's sun its holy light
> With Nature should the musing poet roam,
> And steal instruction from her classic tome;
> When 'neath her guidance, least inclined to err –
> The ablest painter when he copies her'.[4]

Now Bill, if men had not a common nature this advice would be a mere 'tinkling cymbal'[5] if each individual stood isolated from the rest of mankind and possessed of feelings peculiar to himself, what would be natural to me might be un-natural to the rest of the world – how are we to copy Nature but by studying our individual relation to her? How are we to know the feelings of mankind in general but by a reference to our own? – this then is my guide. When I write as I feel, I expect that you will feel what I write. But I don't think you ever intended to dispute this. You say writing from feeling might injure the 'professional literati' and I agree with you. Southey[6] and Scott[7] published each a work at the same time[8] – Southey's feelings lay in favour of an Epic. Scott produced 'A Lay' – a year after their publication Southey in writing to a friend mentions that up to that time the profits of his poem amounted to the sum of three pounds fifteen shillings, while Scott's Lay had realized upwards of one thousand pounds! Southey may have copied Nature truly, but unfortunately for his exchequer, he struck a chord which found no response in the heart of the Community.

It would not however be just to judge of a man's merit by the reception of his works. Witness Milton.[9] Ten pounds for the copyright of John Milton's works! Is not 'the physiognomy of a world verging towards dissolution, reduced now to spasms and death-throes pictured in that one fact'![10] Who will judge of Milton by such a standard? Not I by Heaven!

RI MS JT/2/13a/22
LT Transcript Only

1. *29 July 1844*: from Tyndall's journal, 29 July 1844: 'Wrote to Ginty' (RI MS JT/2/13a/22). The text of the letter was then transcribed in the journal.
2. *bivouac*: a night watch by troops (*OED*).
3. *Your last*: letter missing; letter 0308 of this volume is from Ginty to Tyndall but does not mention his style.
4. *'Poetic fire ... when he copies* <u>her</u>*'*: E. B. Browning, *An Essay on Mind* (1826), verses 69–70.
5. *'tinkling cymbal'*: 1 Corinthians 13:1.
6. *Southey*: probably Robert Southey (1774–1843), English Romantic poet best known for works such as *Joan of Arc: An Epic Poem* (1796) and *Roderick the Last of the Goths* (1814). Southey was the British Poet Laureate from 1813–43.
7. *Scott*: Sir Walter Scott.
8. *published each a work at the same time*: probably a reference to 1805, when Scott published the narrative poem *The Lay of the Last Minstrel* and Southey published the epic poem *Madoc*.
9. *Milton*: John Milton (1608–74), English poet and civil servant under Oliver Cromwell, best known for the two epic poems *Paradise Lost* (1667) and *Paradise Regained* (1671).
10. *'the physiognomy ... in that one fact!'*: T. Carlyle, *Past and Present* (1843; London: Chapman and Hall, 1872), p. 75.

To Mr. Wren 30 July 1844[1] 0311

Sir

I have to acknowledge the receipt of yours[2] of the 27th and am sorry to find that I entertained a misconception as to the nature of the services you require.

I am aware that I would have a better sphere of action in America than in these Kingdoms, but my feelings are strongly in favour of an engagement nearer home. This would induce me to forego considerable pecuniary advantages if the sacrifice of them would ensure me employment in England.

You want a person who could draw up advertisements, catalogues, books, pictures &c and keep your books '<u>correctly</u>'. In each and all of these capacities I may briefly say that <u>I'm your man</u>. I am willing to bend to circumstances, and where no compromise of principle is involved, to mould myself to suit the peculiarities of time and place. It is all equal to me whether the subject of an Advertisement be an estate or a Cruet-stand,[3] whether I have to stretch my imagination over green fields, waving woods, and silvery streams, or concentrate description on a pocket edition of the 'Beauties of Byron'.[4] I know something of pictures too – I have made the acquaintance of nearly all the old Masters and may, I trust, descant on their peculiar excellences without being guilty of profanation.

You have in your note laid an emphasis on the word 'correctly'. Man is fallible. But if steadiness and fidelity avail anything towards counteracting the effect of man's inheritance I can assure you that neither shall be wanting in me.

I follow the course of your note seriatim.[5] You wish your son[6] to learn Surveying – here also I'm your man. The remarks under this head which my last letter contained, had reference to my qualifications as a Surveyor exclusively. I may now add that I shall feel most happy to take your son under instructions and to give him a knowledge, not only of Surveying in the common acceptance of the term but also of the concomitant branches of Engineering. These include trigonometrical observations, constructing maps, of cities &c having references to sanitary purposes, to drainage, sewerage &c., traversing, levelling, contouring, and constructing sections, laying out lines of road &c &c. all according to the most approved systems – systems recommended by men who unite the greatest professional experience to the highest scientific attainments.

I may, I think, with safety assert, that if the young gentlemen you intend for this profession lay his shoulder to the wheel and face every difficulty with a determination to overcome it, at the end of one year he will find himself possessed of more solid information than he would acquire in any Engineer's office in the kingdom during the same time.

From your position in Lancashire I have little doubt of your ability to render the Surveying worth something – If we get a fair trial, I flatter myself that the execution of a single job will give you a strong claim to further patronage.

Now comes 'the rub'.[7] What salary will I require? Here I am thrown out of my reckonings. As a Surveyor I would know what to say without hesitation, but for the present I must consider myself an Accountant &c. Well I shall at all events endeavour to render further correspondence on this subject needless. I propose then to engage with you on trial for one, two, or three months at a salary of 30 shillings a week. Your knowledge of me at the end of this time will enable you to judge of my efficiency, and I hope to find you willing to act on that judgement.

I am Sir | Yours very truly | John Tyndall

Mr Wren | 122 Fishergate.

RI MS JT/2/13a/51
LT Transcript Only

---

1. *30 July 1844*: from Tyndall's journal, 30 July 1844: 'Had a letter from Mr Wren, he is willing to engage me and is only fearful that his establishment is too limited for me – wrote to him in the evening' (RI MS JT/2/13a/51).
2. *receipt of yours*: letter missing. According to Tyndall's journal he received the letter on 30 July 1844.
3. *a Cruet-stand*: a small container for cruets containing condiments for food (*OED*).
4. *the 'Beauties of Byron'*: probably A. Howard (ed.), *The Beauties of Byron, Consisting of*

Selections from his Works (London: T. Davison, 1824).
5. *seriatim*: in the order in which they appeared (*OED*).
6. *your son*: not identified.
7. *'the rub'*: W. Shakespeare, *Hamlet*, III.i.65.

To Mr Wren 9 August 1844[1] 0312

Dear Sir

This day's post brought me yours of the 5th Inst.[2] It gives me sincere pleasure to learn that we have arranged matters without much expense of words – well then, I'm at your service any time you wish to call upon me – though I am anxious to get into the harness at once, still I could not for an instant harbour the wish that any person at present in your employment should suffer inconvenience to accommodate me.

I have been informed that it is your intention to open a Picture Gallery[3] – and your note proves that you possess every facility for doing so. Doubtless such a thing would be duly appreciated by the Preston people. I have never seen their patronage wanting when a picture was in question.

I shall act in strict accordance with your directions as to secrecy. I may however remark that previous to the receipt of your note, my friends Martin and Hill[4] who introduced me to your notice were acquainted with my proposal, I have written to them to prevent it going further.

And now I shall close with a cordial response to the hope that when we have seen each other and understand our relative duties as Governor and Subject we shall be in no hurry to separate.

I am Dear Sir | Yours very faithfully | John Tyndall

Mr. Wren.

RI MS JT/2/13a/53
LT Transcript Only

1. *9 August 1844*: from Tyndall's journal, 9 August 1844: 'Copy of Note to Mr Wren' (RI MS JT/2/13a/53).
2. *yours of the 5th Inst.*: letter missing.
3. *a Picture Gallery*: a room or building for the display of paintings.
4. *Hill*: not identified.

From William Wright 9 September 1844 0313

Cincinnati | September 9th 1844.

My dear Jack

What in the name of all creation keeps you from writing for the last two months? I have been expecting a letter from you on the arrival of every steamer which conveys the Mail from England to the States. It is with regret I witness their arrival and departure without their conveying either a letter to or from me. I always put off writing until the monthly steamer's arrival is announced and then I am on tip-toe for a line from some friendly pen. But her arrival in the most instances only brings disappointment to me. I do not like to give up the hope of receiving a letter at once and so put off expecting that it may have been mislaid. But when I find that it is all useless I then begin to storm and swear that I will not write home again, it takes some time before I am cooled down, against then the steamer has departed on her voyage and I suppose only to convey the same annoyance to you. In good earnest Jack what has hindered you to send me an answer or reply to that letter which I forwarded in May[1] to you immediately after I received yours which contained your determination of not emigrating for some time. I suppose you are so much engaged in your undertaking that you have not a few minutes to devote to your old acquaintance and friend. Well time changes all things and I suppose there is little use in finding fault with her ravages. I did not expect she would have so soon made impressions on you. If you ever put your foot on fair Columbia's shore I would be just as anxious to hear of your welfare. This morning about ½ past 2 o'clock by the cry of fire and being a member of No 3 company I was obliged to turn out, we got it under about 9 o'clock, since that time I have been busy in the store till now ½ past 9 p.m. I inform you of this merely as a set off against all the errors which this sheet will contain. I suppose you are hard at work prosecuting your survey, I hope when you have got it finished your mind will be more intent on coming out than it has ever been. I regret very much that you were not here this summer as a vacancy occurred which would have suited you exactly. A brother-in-law[2] of my uncle John's[3] wanted a book keeper and I am certain that 3 weeks' study would make you perfect master of it you would have no difficulty in obtaining a situation. I am not aware whether it is agreeable to your friends that I should be thus advising you, still when I address you I cannot refrain from encouraging you to come out. And I would say to you if you ever do arrive make up your mind to view the brightest side of every landscape for the first 2 months. After that time you will be thoroughly Americanised and every little difference and annoyance (which a foreigner meets with on his arrival) will be unheeded by you. There are a great many little differences and oddities which will beset you and if you are at all inclined to find fault you will have plenty of opportunity to do so. I am led into these remarks by the conduct of my sisters[4] since their arrival – every little matter which a person of good sense would laugh at seems to annoy them most astonishingly. I was unfortunate enough to advise them to come out and I must confess that I am very, very sorry for it. For instead of being an assistance to me it is the reverse and for the life *[of me]*, I cannot fathom their excentricities. Whether it is our free air that has made them more than independent or that their bump of self-esteem has grown so large that it has left no room for those of a more admired nature I cannot tell, not being anything of a phrenologist. Now Jack I know that you have got too much good sense and information to be annoyed by trifles. When I first came to this country, as yourself knows I was much inclined to find fault. But after 3 years experience through rough and smooth I would not exchange or leave it on any consideration. I feel that I and all around me are free men, each on a part of the mighty Sovereignty of this widespread republic, whose sword has more than once been measured with the mighty nations of the earth and has always come off victorious. Oh that I could say so for the land of my birth. But it will and must be said be the day far or near. I have merely written this letter to stir you up to write to me. I have only to add that we are getting along pretty comfortably – business very fair for the season, and I find prospect of a good fall business, and to renew my invitation to you and your messmate.[5] I have been contemplating a trip to Santa Fe in New Mexico next March. There is a large trade carried on between the cities of the West and that country, it lies West of the Rocky Mountains. The merchandise is conveyed in wagons over the boundless prairies of the West. The traders go in large companies of 3 or 4 hundred to protect themselves and wares from the bands of hostile Indians who occupy the country through which they pass. They occasionally have a smart brush with them and also rare sport hunting the buffalo which they meet with in large numbers on their route. I have not quite made up my mind about it yet, I am getting acquainted with the traders and it is more than probable I will go. Give my respects to your Father and Mother[6] Caleb[7] and William[8] and all the rest of the old stock, may they be as happy and as well off as I would wish them. If you have any opportunity of informing Mother[9] of all our welfare I would be much obliged. A note addressed to her to No 12 Nicholas Street would find her out. You might say that the girls are in good health and wish for a letter from her very much. And now old fellow good bye and let me soon hear from you and know all the old folk.

Your old friend and shipmate | William Wright.

RI MS JT/1/TYP/5/1822–1823
LT Transcript Only

---

1. *that letter which I forwarded in May*: letter 0304 of this volume.
2. *A brother-in-law*: not identified.
3. *my uncle John*: not identified.
4. *my sisters*: one sister may be Mary Ann Wright; other sisters not identified.
5. *your messmate*: a reference to George Latimer, who also intended to emigrate.
6. *your Father and Mother*: John Tyndall, Snr, and Sarah McAssey Tyndall.
7. *Caleb*: Caleb Tyndall (1798–1879), John Tyndall's uncle.
8. *William*: William Tyndall (1804–?), John Tyndall's uncle. In 1844 William Tyndall was living in Philadelphia.
9. *Mother*: not identified.

From E. D. Steuart 27 September 1844 0314

1 Burdete Avenue | September 27th 1844

John,

If I had required any assurance of your kind anxiety to fulfill your promise by the gratification of my wishes, it would have been afforded to me by your letter,[1] received this morning. But I felt quite certain that your time had been fully and, I trusted, much more profitably employed than in finishing the sketch of Steuart's Lodge,[2] to which I merely expected you to devote a leisure hour whenever such should occur, and I must now request that you will not think of even touching it, until your present hurry of business is quite over, and your time and thoughts entirely disengaged. If you could give me an idea of the size of a frame for it, I should be glad to take one home with me, as I yesterday left the etching you gave me in Dublin, to have a frame made for it, and would wish the other to be the same. It has given Mr Steuart[3] and me great pleasure to hear you have been able to procure employment,[4] and we hope there is now a sphere open for the full display of your talent, which we have no doubt, combined with the steady perseverance you have hitherto exhibited, will yet raise you to the highest rank in your profession, and I can assure you, such is our earnest and most sincere wish and desire. I am thankful to say Mr Steuart is much better, and I trust we shall be able to return home early next month.

Believe me | Your sincere friend | E.D. Steuart.

RI MS JT/1/TYP/10/3322
LT Transcript Only

---

1. *your letter*: letter missing.
2. *the sketch of Steuart's Lodge*: Mrs. Steuart had evidently asked Tyndall to sketch her family's home.
3. *Mr Steuart*: William Richard Steuart.
4. *you have been able to procure employment*: see letters 0311 and 0312 of this volume.

1845

To Phillip Evans 30 January 1845 0315

My dear Phil,

Here I sit weary and jaded, every muscle of my body quivering thro' pure fatigue. I have just returned from a day's work sufficient to expunge any ordinary mortal from the catalogue of human existence. I'm completely fagged – these York hills are sufficient to wring the energy out of the body of a Hercules and to reduce his strength to that of childhood. On this day week I started from Leeds to set out and test the survey of one of the lines I levelled. Some serious differences exist between me and the West Junction levellers, and the matter has been referred to the decision of a third party;[1] he has already tried the point where the most important discrepancy exists and, Heaven be praised! he agrees almost to an hairsbreadth with your old messmate.[2] I made Leeds my quarters until Monday, when I started to my present locality.[3] We had a slight sprinkle of snow on the Sunday night, and towards 3 o'clock on Monday it fell so heavily that I was compelled to 'retire from the field'; from that time to the present each succeeding night has contributed its 'fleecy store'; the sum total of the deposits up to the present forms a very beautiful counterpane about 15 inches in thickness. Charles Swain[4] in one of his poetic vagaries has rung a stanza from his lyre in praise of

> 'The snow, the snow, the silvery snow'–[5]

had he been with me this morning I fancy he would touch another string; it strikes me that he has little idea of the pleasure of tugging through it knee deep 'from morn till frosty eve'. I was forced to step out of my warm bed at half past seven and immerse my physiognomy[6] in a basin of water so cold that when I raised my head you might mistake my nasal promontory for a huge misshapen root of mangel warzel[7] instead of a stout orthodox structure of skin and gristle! This ordeal ended, I sat down to a most comfortable breakfast, beside a fire which sparkled like the joy of boyhood, but what is comfort under such circumstances? Why it merely serves to render your subsequent misery more distressing. Breakfast over, I buttoned on my leggings and sallied forth warming my cheek with the last half inch of an expiring Havannah.[8] The first object of interest I met was a thick wood thro' which the line ran and through which I must set it out. I stood for a while gazing with a feeling distinct from admiration on the 'Silvery snow' which loaded the branches; there it clung in all its vestal purity – fit type of the constancy of woman, it required but a breeze to annihilate its embrace. It seemed a species of desecration to disturb it; but it was useless to hesitate – forward! was the word, and forward I went; every movement of course brought a most interesting shower about my ears. From my soul I wished Mr Swain in my situation, stooping and stretching his neck to get through the tangled oak and hazle, ever and anon greeted by the descent of his 'bright beautiful' commodity, soft and cool as the dew of Hermon[9] and exciting a thousand pleasurable sensations as it thawed and trickled down his back! Byron also sings:

> 'There is a pleasure in the pathless woods
> There is a rapture on a lonely shore'.[10]

I wonder could his lordship form an idea of the delight which a surveyor extracts from the conjunction of a 'pathless wood' with a spirit level![11]

RI MS JT/2/13a/73–74
LT Transcript Only

---

1. *a third party*: probably a Mr. Cooper mentioned in Tyndall's journal. Tyndall's 24–5 January 1845 journal entry noted: 'Mr Cooper commenced levelling and I measured. This gentleman was a kind of umpire between myself and the Junction levellers. I had made the crossing of the Leeds and Whitehall Turnpike Road some 8 or 9 feet different from them. To determine which was correct Mr Cooper was employed. I may just add that he and I agree throughout. My levels were also corroborated by Mr Crowther' (RI MS JT/2/13a/73).
2. *your old messmate*: i.e. Tyndall, as he and Evans had worked together on the Ordnance Survey.
3. *my present locality*: Tyndall's 27 January 1845 journal entry noted: 'Put up at my old quarters at Fulneck' (RI MS JT/2/13a/73). Fulneck is a village in Leeds, England.
4. *Charles Swain*: (1801–74), an English poet and bookseller born in Manchester, best known for *Beauties of the Mind: A Poetical Sketch with Lays Historical and Romantic* (London: Simpkin and Marshall, 1831).
5. *'The snow, the snow, the silvery snow'*: C. Swain, 'The Snow', line 1: 'The silvery snow! – The silvery snow!'
6. *my physiognomy*: my face; physiognomy was the practice of studying the features of one's face as a means of determining their character and qualities (*OED*).
7. *mangel warzel*: mangalwurzel, also called 'field beet'; a type of root vegetable.
8. *Havannah*: a cigar from Havana, Cuba or a cigar made with Cuban tobacco (*OED*).
9. *the dew of Hermon*: Psalm 133:3: 'As the dew of Hermon, and as the dew that descended upon the mountains of Zion: for there the LORD commanded the blessing, even life for evermore'.
10. *'There is a pleasure ... on a lonely shore'*: Lord Byron, 'Childe Harold's Pilgrimage', IV.clxxviii.1–2.
11. *a spirit level!*: letter ends here.

From William Ginty 1 February 1845 0316

Langford Arms | W. Meath | Saturday 1st. Feb./45

Dear Tyndall,

Can your memory paint Ginty in one of his ancient sneezing fits – with his proboscis[1] dyed a deep crimson and swelled into a German sausage – his eyeballs inflamed and blinking in the sunbeam – and a well saturated 'wipe' stuck in his fist which finds its way noseward at least 40 times in the minute – if it can – you have a true portrait of his present state – I have been very ill for the past week and now half-blind and with my nether end in close communion with the sweets of a roaring coal fire I am about replying to your philosophic rantings[2] – damn that acrostic[3] I must have been (as Jack would say) 'labouring under the influence' when I submitted it to your review – you laid it on terrible thick – you lashed most unmercifully – oh dam that acrostic or rather the pen that copied it for Tyndall's perusal. I sent Tid[4] your address at his own request yesterday, his father is very ill – and Jack has assumed the important office of amanuensis.[5] Evans is in Carlow, he has not yet given up the idea of coming here – but prefers waiting for a longer day – now for a stave – the mountain is in labour – a mouse shall come forth a mere Brutum fulmen[6] – Dijecta membra poetae[7] – The scattered remains of the Poet (save the mark)

Free from fetter curb or chain
My spirit Jack is still the same–
Unchanged – unaltered still it glows
As fearless as when first it rose
And hobbled side by side with thine
By 'moonlit shone' on 'Ellen's'[8] shrine

Here I'm stuck in the mud – I can't budge a peg – The Bird can not sing Tyndall – tis all up the spout – I'll try it again – but I think tis useless – labour may make an orator – but a poet must be born a poet – I'll at it again however –

It was but hobbling Jack you know
Where'er the silly thing would go
You propped her up – you cheered her on
And pumped her out in blubbering song
Disjointed – unconnected stuff
Rough as the muse: herself was rough
The prop is lost – the guide is gone
The creature now sits sick alone
A mule – a cripple and a drone

Sleep on then dull and heavy muse
Thoust trod too long in others shoes
Sleep on thou ill-begotten strap
And take thy all-inglorious nap

The grass above thee's not more green
Than thy poor simpering e'er has been
Sleep on – while neighbouring spirits soared
You dirty witch you only snored –

far enough in that strain – I ween[9] for your edification – and now I shall gallop thro' the conclusion of the snotty production in double quick time – So you see the quondam.[10] Mary[11] – is she much the worse of the wear? I am most infernal lonely here. I have not a single being to open my mouth to. Jim[12] tells me you are about visiting the House of Commons on business[13] – don't say you are a radical or you'll be kicked out – How is that little celestial being in Kingston[14] and you getting on? I was very much surprised when Jack[15] asked for your address – for I really did think you and her kept up a 'constant fire' – I was in Tyrrellspass[16] lately and was enquiring for 'Blackthorn'[17] he has not yet made his appearance there all I could hear was that he was 'laying out a railway in England'. How is our friend of the monster nose[18] getting on? Why forget all in your letters but your self.

In my next I hope to be able to give you a more lengthened stave – even with your terrible volcano of abuse – staring me in the face – I am not terrified – blast you and your 'Defunct' – Just as you get in to it – you get out of it. Do not be so long in writing as you usually are.

ever faithfully yours | W. Ginty.

I sent you 'The Dublin Evening Mail'[19] yesterday.

RI MS JT/1/TYP/11/3634–3635
LT Transcript Only

---

1. *proboscis*: elongated nose (*OED*).
2. *your philosophic rantings*: an earlier letter from Tyndall to Ginty; letter missing.
3. *acrostic*: a poem in which the first letter of each line spells out a word or name.
4. *Tid*: John Tidmarsh.
5. *amanuensis*: employed at writing based upon the dictation of another person (*OED*).
6. *Brutum fulmen*: an empty threat (*OED*).
7. *Dijecta membra poetae*: scattered poetic fragments (Latin).
8. *Ellen*: see letter 0288, n. 31 of this volume.
9. *ween*: to think or surmise (*OED*).
10. *quondam*: former (*OED*).
11. *Mary*: possibly Mary Edwards, a former romantic interest of Ginty's who married in June 1843.
12. *Jim*: probably Phillip 'Jim' Evans, previously mentioned in this letter.
13. *visiting the House of Commons on business*: Tyndall was in London helping to prepare Parliamentary testimony in favor of the West Yorkshire Railway.
14. *that little celestial being in Kingston*: not identified; presumably a romantic interest of Tyndall's.
15. *Jack*: John Tidmarsh.
16. *Tyrrellspass*: a village in central Ireland.

17. *Blackthorn*: a nickname for Martin Cuddy; see also letter 0274, n. 33 of this volume.
18. *our friend of the monster nose*: Tyndall; see letter 0315 of this volume for further discussion of Tyndall's nose.
19. *The Dublin Evening Mail*: an evening newspaper in Dublin. It was published under the name *Dublin Evening Mail* from 1823–1928, when it was renamed the *Evening Mail*.

From E. D. Steuart 5 February 1845 0317

Steuart's Lodge | February 5th 1845.

John,

It has given Mr Steuart and me very great pleasure to hear that you are getting on so well, and we sincerely hope you will continue to prosper and meet with every encouragement in your professional career.

I now write to say that I believe it is in your power to do a little matter which would both oblige and gratify Mr Steuart: he wishes to have a porch built to this house, and the architect being in London, requires a sketch (he does not care how rough) of the front elevation: as you have the etching of it with you, I hope you will not think me unreasonable in asking you to let me have just a rough pencil sketch from it, one that will not take you more than an hour to draw: I can give the heights required by Mr Robertson,[1] and forward it to him: he must know the length of front, height of walls, string course &c. &c. all of which I can add to the sketch: but on no account do we wish that you should devote time to it, or take the least trouble about finishing it, all of which is quite needless. Mr Steuart is anxious to get a plan, in order to have everything provided as soon as possible, and have it commenced the moment the weather will allow.

I feel most anxious to hear if you have succeeded in detecting the errors in your rival's[2] line of rail road, and should like much to receive a long letter from you, telling me all particulars. With every good wish, in which Mr Steuart unites I remain

Your sincere friend | E.D. Steuart

John Tyndall | Post office | Halifax | England[3]

RI MS JT/1/TYP/10/3323

LT Transcript Only

---

1. *Mr Robertson*: unidentified person; possibly a carpenter whom the Steuarts had hired to build their porch.
2. *your rival*: Tyndall's then-employer Richard Carter was the Surveyor of the West Yorkshire Railway. In 1845, the West Yorkshire Railway was competing with a rival company for Parliamentary approval of its railway plans.
3. *John Tyndall | Post office | Halifax | England*: address given by Louisa Tyndall.

From E. D. Steuart 17 February 1845 0318

Steuart Lodge | February 17th 1845.

John,

Yesterday morning the drawing arrived[1] – and I lose no time in offering you our best thanks for it, and also for your prompt compliance with our wishes: We are however very sorry that you took so much trouble with the sketch, as a rough pencil one would have answered the purpose equally well

I received the Newspaper[2] you so kindly sent, and read with great interest the account of the meeting held on the rail road question. I feel quite anxious for the success of the 'West Yorkshire'[3] and shall watch the Newspapers for information on the subject. Your letter of the 10th[4] was very satisfactory, as it explained much I was wishing to learn, though I would rather you had been more explicit with respect to your own affairs and prospects – I ventured a hint through your Father[5] a short time since relative to any savings you have been able to lay aside, and I expect you will agree with me in thinking it would be desirable to secure such sum, in such a way that you might call upon it at any time you pleased, and receive interest for it until then, I mentioned the Loan Bank here, as offering a safe deposit, and 5 per cent interest, but there are many other ways of securing your money, which you perhaps might prefer.

With very kind wishes | Your sincere friend | E.D. Steuart

RI MS JT/1/TYP/10/3324
LT Transcript Only

1. *the drawing arrived*: a drawing of a front porch to be built on the Steuarts' home that Mrs. Steuart had asked Tyndall to make; see letter 0317 of this volume, drawing missing.
2. *the Newspaper*: unidentified reference.
3. *the 'West Yorkshire'*: the West Yorkshire Railway, a small railroad company. Tyndall's then-employer Richard Carter was the Surveyor of the West Yorkshire Railway. In 1845, the West Yorkshire Railway was competing with a rival company for Parliamentary approval of its railway plans.
4. *Your letter of the 10th*: letter missing.
5. *your Father*: John Tyndall, Snr.

From John C. Chadwick 11 April 1845 0319

April 11th

Dear Tyndall,

In the hurry of a busy evening and after having written 6 letters before this I send you a few lines for fear you might think me unkind I have but little to say only that I may now sit down under my vine and fig tree being a settled old curmudgeon the father of a family and so forth quite patriarchal but of course you heard this before there is nothing now but that this proverbial pig's pot of a town[1] has been as dry as a bone for the last week or so I think that the clouds must have got the gravel and cannot continue their aqueousness there is a whopper for you so get seven dictionaries I hope please providence to be able to see you before you leave Cork even were I to walk there like a Loughderg pilgrim[2] upon my bare bones. Write to me soon you quintessence[3] and I will send you an answer that might as the Arabian nights[4] say be copied in characters of gold and tied up in the archives of the ordinance survey, which in its meanness made a descent upon me for 4 and a half sterling for the purpose of building new barracks at Kinsale[5] but which demand I will resist as a White boy would resist the tithes.[6] I will visit Cork before the end of the month, so hoping you will excuse haste

I am yours &c &c. &c. &c. &c. | Corny[7]

Mrs C.[8] is getting quite well and the young un[9] is thriving and desires to be remembered to you amen.

RI MS JT/1/TYP/11/3509
LT Transcript Only

---

1. *this proverbial pig's pot of a town*: not identified. Chadwick moved to Cork after leaving the Survey, but this letter indicates that he and his family no longer lived in Cork as of 1845.
2. *like a Loughderg pilgrim*: a pilgrim to Saint Patrick's Purgatory, a site on Station Island in the middle of the Irish lake Lough Derg. According to legend, in the fifth century AD Christ showed Saint Patrick a cave or well on Station Island that contained an entrance to Hell. Saint Patrick's Purgatory has been a popular pilgrimage destination since the fifteenth century. Pilgrims approach Saint Patrick's Purgatory on their knees.
3. *quintessence*: the 'fifth essence' of which celestial bodies were composed, according to classical and medieval philosophy (*OED*).
4. *the Arabian Nights*: a reference to *One Thousand and One Nights,* a famed collection of stories written in Arabic.
5. *Kinsale*: a village on the southern coast of Ireland.
6. *resist as a White boy would resist the tithes*: the Whiteboys were members of a secret Irish Catholic organization that used extreme, occasionally violent tactics to protect tenant rights. The group was named for the white smocks they used when conducting nightly raids, and was active for various periods in the eighteenth century.
7. *Corny*: a nickname for John Chadwick.
8. *Mrs C.*: Mary Chadwick (dates unknown), John Chadwick's wife; see also letter 0306 of this volume.
9. *the young un*: probably Elizabeth, Chadwick's elder daughter.

From James Byrn 16 April 1845 0320

Leighlin Bridge | April 16th 1845.

Dear John,

You may thank the Lord's prayer which teaches me to forgive as I expect to be forgiven all trespassers[1] who will not immediately thrust a letter in to stop a gap in friendly correspondence and you may perceive I act according to it by sending you this. Now, you thing of the world, begin on the receipt of this to fence the breach, for fear that any unpleasant stray ideas might get in. I wrote about six weeks or a month ago[2] and received no tidings of you I am afraid you have gone where you sent my letters handy to blazes if you have not gone or is about going to take a survey for a railway in his sulphurous or dark tartarean[3] Majesty's dominions, upon my soul I am almost afraid you have gone thither. If you have not write. If you would only say Dear James I am alive and kicking still. Faith if you were not in the above mentioned place or next door to it you would let me have some tidings of you before this time. How soon you would write to a fellow if he said you are a damned scoundrel aye would you in a minute. – well then you ought to write me in half a minute that would think nothing of doing my best to knock the brains out of him that would dare say the like to you. Your long silence has made patience and I part company however we will be friends again on receipt of your letter which will be welcome as light and a calm to the weary and tempest tost mariner after toiling out a whole dark night without the ray of a single star to guide him in an unknown sea and along a dangerous shore. I hope this will not interrupt any pleasing thoughts of your adored one.[4] I had a full gazette of my angel's charms to send you ready made, only I feared to make this more tedious than it is. Now as I have come to a conclusion, 'tis hoped by me you will deem it no intrusion.

I beg <to> remain ever | Your affectionate friend | James Byrn[5]

I am thinking you are married for it makes people forget their old friends.

P.S. I am in a very bad state of health these two or three months. Your old charmer Jane Collier[6] and her brother William[7] has set up a milliners, trimmers, and haberdashery shop in Dublin Street, Carlow, and I think they are doing well. In a letter previous to this I sent a few verses namely Sir Walter Scott's study[8] and the Spaniard of 1830.[9] I now send you a beautiful but unpublished song by Thomas Moore Esq.

Memory

Oh, wilt thou remember that song of our youth,
    That brings back its lustre to light our decline
Or weepest thou often when memory's truth
    Comes mingling with music to make it divine?
For 'Where are those voices', my heart will inquire,
    'That fondly and faithfully whispered that lay?
Oh where the affection that kindled its fire
    On the altar of friendship? are all passed away?

If fortune will fade, and affection with years
    Go sleep in the grave of oblivion at last
Still music shall live 'mid the friendship of tears
    And fondly survive the beloved of the past.

Like waters that roll from these rock covered hills,
The muse's sweet tears shall undying remain;
And if lonely me linger the last amid ills,
Our own native music shall soothe with its strain.

Like roses we've loved in our infancy's hour,
Which oft the dull pathway of manhood adorn;
Thus memory brings back to her own vesper bower
Those glimpses of melody taught her at morn.

Like Israel's harp as by Babylon's stream
It wept on the willow for lands it loved best
Thus dear to old Erin,[10] in music's soft dream,
Sighs faithfully and fondly the Harp of the West. [11]

Beautiful.

RI MS JT/1/TYP/11/3489
LT Transcript Only

---

1. *the Lord's prayer ... forgive all trespassers*: a quotation from the Lord's Prayer, which reads 'Forgive us our trespasses / As we forgive those who trespass against us'.
2. *I wrote about six weeks or a month ago*: letter missing.
3. *tartarean*: of Tartarus, pertaining to hell (*OED*).
4. *your adored one*: not identified.
5. *James Byrn*: a friend of Tyndall's from Leighlin Bridge. Possibly related to the 'Monk Byrn' mentioned in letter 0066, *Correspondence of John Tyndall, Volume 1* or the 'Mickey Byrn' mentioned in letter 0124, *Correspondence of John Tyndall, Volume 1*.
6. *Jane Collier*: probably a former romantic interest of Tyndall's; biographical details not identified; see also letter 0303, n. 6 of this volume.
7. *William*: not identified.
8. *Sir Walter Scott's study*: R. Thomas, 'Sir Walter Scott's Study', a poem published in *The Guide to Knowledge; An Interesting Literary Repository, and popular Scientific Instructor*, (London: Orlando Hodgson, 1837), vol. 1, p. 212.
9. *the Spaniard of 1830*: unidentified poem.
10. *Erin*: a modern derivative of Éirinn, the Irish-language name for Ireland.
11. *Oh, wilt thou ... Harp of the West*: T. Moore, 'Memory' (1844). On Moore, see Biographical Register.

To [Robert Allen] 20 May 1845 0321

Mrs Wright | Ferguson Street, Halifax | Monday May 20th 1845.

My dear Bob,

Here I am once more, having ended my pilgrimage in London. Your kind and welcome favour reached me at Covent Garden, you must have written to Sayers at the same time as on the following evening I received a note from him requesting me to call to see him. I paid him a visit on Saturday week and found him the same harum scarum[1] devil as usual. I spent a pleasant day with him, he lodges at 47 King street, has a nice garden to walk in and employ his botanical propensities, he brought me through the Dockyard and showed me everything worth seeing; on my way I paid the Great Britain Steam Ship[2] a visit, inspected her from stem to stern and trod the same planks that our gracious Queen stepped upon three days before. I spent Whitsunday[3] at Richmond;[4] 4 of us including governor[5] hired a coach and went out there in the morning; oh! it is a paradise! Upon our arrival we called at the Star and Garter Hotel and ordered dinner to be ready at 4 o'clock, then set out through the park, scoured it all and returned along the banks of the Thames. I am quite unable to describe the beauties of Richmond as seen from the river side, it stands upon a hill or rather upon a terrace ground, the houses receding one above another in beautiful order – perhaps disorder would be a better expression; the houses are not continuous but isolated while the intervening spaces are filled with trees of the most beautiful foliage, evergreens and flowers – these basking under the influence of the May sunlight looked charming. We arrived at the hotel at the time appointed and after a preparatory wash entered the dining room. This is a superb concern; the fact of its requiring not less than twenty waiters to attend the guests may convey to you an idea of its extent – At one end there are three large windows which can be raised so as to admit visitors out upon a circular balcony which commands a view of nearly 30 miles. The waiter stood our friend, tho' there were 50 gentlemen and a sprinkle of ladies in the room when we entered it, the very best table was kept in reserve for us – it commanded a complete view of the Thames and scenery outside – Here we pitched into a sumptuous dinner, salmon cutlets, stewed eels, roast mutton, maids of honour, rhubarb and apple pie and half a dozen etceteras – After the cloth was removed we sat for two hours over our wine. Champagne, port, and sherry sparkled in abundance of course we all became elevated and could I only catch a gleam of the spirit which bounded within me at the time I would make the dull page sparkle under my pen. At one moment I was up to the ears in discussion with governor respecting the reflection and refraction of light, the topic being suggested by the rainbow hues dispersed through the room from the 10 thousand and one lustres which hung pendent from the ceiling. Sometimes we amused ourselves watching the transient shadows of a May afternoon shedding their ever-varying effects upon the wide woodland before us – now the horizon would appear in deep shadow while the foreground came forward in the richest sunshine – anon a patch in the centre would become shaded while the back and foreground were dashed with a flood of light and glory from the descending sun. Oh! Bob it was lovely I felt more than once influenced by the genius of the spot and unconsciously murmured –

> 'On Richmond Hill there lives a lass
> More bright than May-day morn'[6]

Surely if ever external nature was calculated to refine the heart of man, to subdue the brute within him, and to cherish his softer sympathies the spot of which I write must have that effect. And now in conclusion, let me give you the amount of my estimation for this scene in contrast with a position alluded to in your letter.[7] The smile of nature is fair, but being common to all loses its personal interest. Give me the smile of my heart's pure planet; my own ÷------÷[8] before heaven's sunlight – Give me the music of her tongue before the melodies of the murmuring Thames – Give me the hue of her lip before the rose of Sharon, and the assurance of her love before an amalgamation of them all together! Without it happiness cannot be mine – with it the desert becomes a fruitful vine.

RI MS JT 1/TYP/11/3485
LT Transcript Only

1. *harum scarum*: reckless (*OED*).
2. *the Great Britain Steam Ship*: the SS *Great Britain*, designed by Isambard Kingdom Brunel, was the first oceangoing steamship to cross the Atlantic on 26 July 1845.
3. *Whitsunday*: a feast day observed seven weeks after Easter (*OED*).
4. *Richmond*: a market town in North Yorkshire.
5. *governor*: unidentified person.
6. *On Richmond Hill ... May-day morn*: J. Upton, 'The Lass of Richmond Hill', ll. 1–2.
7. *your letter*: letter missing.
8. ÷------÷: This may be an instance of Louisa Tyndall omitting the name of one of Tyndall's former love interests.

From James Byrn [20] June 1845[1] 0322

Mon cher Jean,

<*Unknown number of lines missing*>[2]

before this only your father[3] desired me not to write when I asked him for your direction, as he did not know where to direct his letter also that he had sent you two to the old place where your letters were re-directed and received no answer, now you may perceive I was determined not to let my instructive or rather entertaining communication go astray (Ahem hem take breath) and now you have Byrn's true and genuine reasons all else being counterfeit for his not sooner acknowledging the receipt of your very entertaining new year's gift, alias journal though last not least what shall I say about the valentine nothing save that there is something indescribably beautiful in the four lines it contains and that I consider myself blessed in having a friend such an one as you to call on in time of need do not conclude by my saying this that I am bespeaking another for Feb. 1846. No such thing I regret to say it was not in my power to send it away thro' my being very ill and remain not much better still, with a bad cough and pain in the side I was so sick when I attempted to direct <*1-2 lines missing*>[4] I will in my next letter mention her name I am serious indeed ('Tis true tis a pity tis true')[5] Verily verily I say unto you that it is easier for a dromedary to go through the eye of my mother's[6] cambric[7] needle than for me ever to forget her. Another never shall usurp her place in my memory or heart. When I see her and cannot speak to her without a third person being by I feel in spite of myself a kind of joyful melancholy stealing over me. Excuse this nonsensical preamble. I know you will when you recollect that I am in love and sick besides indeed John I am getting very bad health.

I remain ever | Yours sincere and affectionate | James Byrn[8]

P.S. I send you a scrap of poetry called Sir Walter Scott's Study[9] I consider it beautiful let me know how you like the Spaniard of 1836,[10] to Cupid, adieu | J.B.

Sir Walter Scott's Study. By R. Thomas.[11]

Tread lightly! 'tis a land of spells!
'Tis 'fairy Fiction's' throne!
And here her great magician dwells,
Unrivall'd and alone!
Glance round in silence and in awe,
Unlike to him who rashly saw
Agrippa's[12] chamber shown;
For who would dare in lore to face
This master spirit of this place?
The relics of a distant age,
The helmet and the spear,
The blazon'd shield, the <missal> page,
And Gothic towers are here:
Like fragments gathered from the tomb,
By Wizard in the midnight gloom,
To make the dead appear!
So these, with art as great constrain
Long buried years to rise again!
And they were mighty spells of yore,
<For knight or priest to wield;
Which chain'd the soul in mysic lore>
Or fired it for the field:
But, though they oft might wrested be,
Of old, to craft and cruelty,
Yet now those arts they yield;
And like dead giants, most avail
To weave the magic of a tale.

RI MS JT/1/TYP/11/3490–3491
LT Transcript Only

1. *[20] June 1845*: Louisa Tyndall annotation: '1 Jas Byrn re-addressed to: 13 Princess St. [Manchester] | Postmark: Leighlin Bdg, June 18 to Halifax June 20 | 1845'.
2. *Unknown number of lines missing*: Louisa Tyndall did not transcribe the opening of the letter, probably because the letter was damaged; see n. 4.

3. *your father*: John Tyndall, Snr.
4. *1–2 lines missing*: Louisa Tyndall annotation: 'bit torn out'.
5. *('Tis true tis a pity tis true')*: W. Shakespeare, *Hamlet,* II.ii.100: 'That he is mad, 'tis true; 'tis true, 'tis pity'.
6. *my mother's*: not identified.
7. *cambric*: fine linen (*OED*).
8. *James Byrn*: see letter 0320, n. 5 of this volume.
9. *Sir Walter Scott's Study*: see letter 0320, n. 8 of this volume.
10. *the Spaniard of 1836*: unidentified poem, possibly never published.
11. *R. Thomas*: unidentified poet, author of 'Sir Walter Scott's Study'.
12. *Agrippa*: probably not the Roman statesman, but rather Heinrich Cornelius Agrippa (1486–1535), an occult writer, astrologer, and theologian best known as the author of *De occulta philosophia libri tres* (1531).

To Deborah McAssey 22 June 1845 0323

Covent Garden, London | June 22, 1845

My Dear Debby[1]

I declare I don't know how to commence a letter to you – You must consider me a worthless, ungrateful fellow, and indeed I regret to say that I have afforded you too much grounds for that opinion – but I beg of you not to judge of my feelings by my apparent neglect – I often think of Bally Cromwell[2] and never without remembering the kindness which I have always met there. Will you act the part of mediator between me and my aunt and uncle[3] – tell them that I am not so bad as appearance indicates and that I shall ever remember them with sincere affection – And Rachel[4] – What must I say to soften her anger – Will she not forgive her repentant cousin? Let her think of the time I asked to feed her turkeys and mend her stools and let her reflect that I am the same John Tyndall now with the exception of a trifle of beard and whiskers – let her think on this and she cannot steel herself against me. – She will pronounce my pardon; – I might forge an excuse from the excitement in which I am continually involved but I shall do no such thing I admit that I am guilty – very guilty and I throw myself unreservedly on your mercy –

This is my second trip to London this season – My present business is to give audience before the Lords Committee[5] in one of the Northern lands of Railway. I am sworn in the House of Lords on Thursday but don't expect to be called upon until Wednesday – I may expect a due laceration from the opposing counsel, however I am endeavouring to brace my nerves for the contest and meet the terrors of a *[1 word illeg]* – with the fortitude of man –

As I don't know the reception which this will receive I must cut it short and assure you in conclusion that I am with feeling unchanged by time or circumstances,

Your affectionate cousin | John Tyndall

My country address is at

Mrs Wright | Ferguson Street | Halifax, Yorkshire

where I expect to return in a week or ten days. Hope to hear from you then I am in a <*1 word missing*> JT

From Martin Nevin, Private Collection.
Copy in an unknown hand.

---

1. *My Dear Debby*: Deborah McAssey (*b.* 1805), the daughter of John McAssey and Mary McAssey (née Smith/Smyth), was Tyndall's cousin on his mother's side of the family, although the precise relationship is not known. In 1845 she lived at Ballybromwell, a village in the parish of Fenagh, County Carlow. She emigrated to Wisconsin in April 1847.
2. *Bally Cromwell*: possibly a mistranscription, or a play on Ballybromwell, the town in Ireland where Deborah McAssey lived.
3. *my aunt and uncle*: possibly John and Mary McAssey, Deborah's parents.
4. *Rachel*: Rachel McAssey, later Rachel Harris (1822–1908). Rachel was Deborah's younger sister. She would also emigrate to Wisconsin in 1847, where she married her fellow Irishman Richard Harris (1822–63) in October 1847.
5. *give audience before the Lords Committee*: Tyndall's then-employer Richard Carter was the Surveyor of the West Yorkshire Railway. In 1845, the West Yorkshire Railway was competing with a rival company for Parliamentary approval of its railway plans. Tyndall was sent to London to help prepare testimony in favour of the West Yorkshire Railway's plans.

From William Wright 9 July 1845 0324

July 9th. 1845.

My dear John

I thank you a thousand times for your ever welcome letter.[1] I am sorry I was not able to send a reply by the same steamer which brought it out, if I sent one by the return Mail, it would arrive in Boston 2 days after the departure of the steamer for England. You never penned a line which received as hearty or as grateful a reception as it met with. The pure spirit of youthful friendship and forbearance which pervaded its every line gave me untold satisfaction and was heartily responded to by my grateful heart. The forbearance I must admit was exclusively yours. I am glad you did not accuse me of indifference un-kindness or forgetfulness for to all the charges I would plead not guilty although you have not received a line from me since September often has 'My dear Jack' appeared upon the heading of a sheet intended for him. But the fact is Jack, visions of notes coming due and a dull business prevented their completion, for while so impressed I could not write to my satisfaction (nor do I expect I will do so now). Never Jack while memory keeps her seat will you be forgotten, and although at times I may seem neglectful you will always remain the green spot on memory's waste,[2] just now a dark cloud seems to envelope the sunshine of my prosperity and heaven only knows when it will be dispelled. Last fall I laid in a heavy stock of winter goods in expectation of doing an increased trade, the winter was an unusually mild one and the consequence was that a depreciation was obliged to be submitted to in connection with small consumption. The Spring has done likewise both together have crippled me considerably and I am in rather unpleasant circumstances. If I am obliged to give way before them, I intend to remove to Upper California[3] which will ere long be the theatre of a mighty struggle between the U.S. and England. As you are aware it lies immediately South of Oregon[4] and as that country is not Texas annexating that will disturb the friendly relations which exists between the two countries,[5] which ever power retains her hold on Oregon will also be soon master of California, although the country naturally belongs to Mexico. Yet as sure as Texas was wrested from her grasp[6] so will the fertile valley of California. By the way Jack, you mistake Mr Polk's[7] situation in the affair of annexation if you suppose he can create or control public opinion on any question of importance – certainly not on the Oregon or Texas question.[8] He may direct or oppose it but when the fiat has gone forth from a majority of people he is bound to submit to its dictates and as a public servant carry out its will. Annexation as far as the United States are concerned is completed, and news *[which]* has been received by the very mail which has just arrived from South Texas[9] has consummated it. Notwithstanding the subtle interference of France and England, they succeeded in obtaining the consent of miserable and distracted Mexico to acknowledge the independence of Texas forsooth, as if they had not done so for the last 9 years in defiance of all her actions, ah and more than once defeated her legions and made her greatest chieftain Santa Ana[10] beg his life like a dog at the foot of Old Sam Huston[11] the victorious leader of the Texans at the battle of San Jacinto.[12] By the Lord if Mexico does not keep quiet and not be revolutionising itself every 2 or 3 years we will have to annex it and unfurl the glorious stars and stripes upon the wall of the Montezumas[13] as the principles of self-government as developed in the United States is and will attract the attention of all who are oppressed and I believe the day is not far distant when this whole North American Continent will be under one confederated government. The surrounding States cannot but behold with admiration the progress which the United States are making in all the arts and sciences useful to civilized life and in the arts they have taught the old world something, at the present moment you have a Cincinnati man[14] learning you something about the electric magnetic light. It is much doubted here whether it will ever be productive of much good. If he should succeed according to his expectations there will be no more use for the sun as a luminary, however as I see you do not take much notice of it in the public prints you would seem to be satisfied with your present mode of lighting. I must leave these grounds and come to something more akin to both of us. I was surprised to find you were not aware of the return of my sisters[15] although they must be in Ireland some 3 weeks before you wrote. I was not in the best humour imaginable when they left and consequently did not write home by them, this is the first letter I have written I believe since January, so that you have no reason to be jealous. I must also inform you that I have had another youngster added to my stock since then, a tolerably good looking chap, you will perceive how slow my imagination is getting since the arrival of the first. We have called him Samuel Watson Wright,[16] after a brother of Sarah's.[17] Well Jack, I suppose we must give up the idea of seeing you in Cincinnati, your present employment must be pleasant and lucrative. I have often thought if you had come out that time perhaps you might be disappointed, and that it was quite as well things turned out as they did. It is a hard thing to scan the future even in this world, I often think if I had staid at home I might have done as well. But no, if I had lived at home I never could feel as I do now even under disappointment. Should I be obliged to give up business I must only turn in and work a little harder or as is most likely on the far off shores of the Pacific. I wish I had an opportunity of sending you one of the numerous works which are written on the country. Forbes[18] an English Consul has written a good work, I wish you would peruse it although it is not as favourable an account as other writers have given of it. But if one half that is written of it is true it must be the most delightful country in the world. All agree that the climate is the most uniform they have ever visited. The productions of its soil surpass that of every other country known, in their variety and numbers almost all the grains are indigenous productions. Oats and wheat grow wild. I suppose it must be a bad place for a man to sow his wild oats in. Nearly 8 thousand stout hearts and rifles have left the West this Spring and I have no doubt but that twice the number will leave for there the following Spring. In a very short time they will be able and will turn out all the minions of Mexico, in 2 years the concentrated power of all Mexico could not dislodge them. The journey there is exceedingly interesting and exciting, attended with a great deal of privation, severe toil, and suffering. 4 months is about the time spent in crossing the country – the distance 2,100 miles, from Independence on the Missouri river to San Francisco on the Bay of S. Francisco. The emigrants principally subsist on the buffalo which is met with in large numbers on the prairie. The emigrants generally go in companies of 200 to 500 so that they are always sufficient to repel any attack that may be made by hostile Indians that infect the route and also to supply a number of hunters who forage for the camp. The expense I suppose of the whole outfit would be about 20 pounds sterling. I have no doubt but numbers from the old world will find their way out there next Spring. The sea voyage is very long and dangerous Perhaps we may meet there yet, if so what joy will be in store for us. I feel if I was single I would set out there now; as it is Spring, if I am alive, will behold me up and doing. So now Jack I must bid you farewell and

believe I remain yours for ever | William Wright.

I must not close without enquiring after your Father[19] and Mother's[20] health I hope they enjoy a good state of it, Emma[21] I hope is also well. You from your migratory occupation must be hale and hearty – that you may long enjoy it along with all the other blessings of this life is my most sincere wish. Give my best regards to your father mother and sister do not forget remembering me to Billy Tyndall[22] when you are writing next which must be very soon, write to Leighlin and tell father[23] to give you all news which has accumulated there for years in order that you may transmit it to me, whenever opportunity offers by letter or otherwise. You would make my best respects and kind wishes to the old folks in Leighlin. Tell father I will be very much obliged to him if he would transmit the intelligence which I have requested of you, and not to be particular, you cannot be aware how slight a thing which occurs there interests me, excuse me for the trouble it may occasion and once more I bid you farewell.

William Wright

We are all in good health here I hope you will not lose much time before you write again. I suppose you have noticed in the newspaper the arrival at Liverpool of the /Murkingam/ Bank from Cincinnati,[24] it must have been received with interest. The idea alone of a vessel being built and freighted 2000 miles from the sea board is rather novel, at present there is lying at the wharf a revenue cutter ship rigged with screw propeller attached 'Jim Hall',[25] that would do credit to any shipbuilders. She belongs to uncle Sam.[26] I wish you would send me a few Irish newspapers say the Carlow Sentinel. I like to see the local news.

Direct your letter to the post office here as probably I may not be in 5th Street when you write. I send you some newspapers, if you know anything of Joe[27] or Tom Payne[28] I wish you would let me know and oblige.

yours | W. Wright.

RI MS JT/1/TYP/5/1824–1827
LT Transcript Only

---

1. *your ever welcome letter*: letter missing.
2. *a green spot on memory's waste*: a saying meant to suggest a particularly vivid or pleasant memory.
3. *Upper California*: the northern part of California, immediately south of present-day Oregon.
4. *Oregon*: a reference to Oregon County, which at this time encompassed present-day Oregon State, Washington State, Vancouver Island, and part of British Colombia. In 1845 it was a disputed territory with both the United States and Britain claiming ownership.
5. *the two countries*: i.e., the US and Great Britain, which both laid claim to Oregon County.
6. *as sure as Texas was wrested from her grasp*: the United States had annexed Texas in March of 1845.
7. *Mr Polk*: President James K. Polk (1795–1849), the President of the United States from 1845–9. Polk, a Democrat, finalized the US annexation of Texas in 1845 and threatened war with Great Britain over the ownership of Oregon County in 1846.
8. *the Oregon or Texas question*: the question over whether the United States would incorporate Oregon County and Texas into its territory. President Polk moved to annex Texas later that year and the 1846 Oregon Treaty between the United States and Great Britain divided Oregon County along the 49th parallel, with the British keeping the northern part and the Americans retaining the southern part.
9. *news which has been received by the very mail which just arrived from South Texas*: news that Mexico had conceded the American annexation of Texas.
10. *Santa Ana*: Antonio López de Santa Anna (1794–1876), a famed Mexican general and

politician. He was the President of Mexico for twenty-two non-consecutive years.

11. *Old Sam Huston*: Samuel Houston (1793–1863), Commander-in-Chief of the Texas forces during the war for independence from Mexico, and President of the Republic of Texas from 1836–1838 and again from 1841–4. He negotiated with US President John Tyler for the American annexation of Texas.
12. *the victorious leader of the Texans at the battle of San Jacinto*: Samuel Houston was the commander of the Texan forces at the Battle of San Jacinto on April 21, 1836. Houston famously surprised Mexican General Antonio López de Santa Anna and won a resounding victory at San Jacinto, forcing the Mexican government to sign the Treaty of Velasco granting Texas independence from Mexico.
13. *wall of the Montezumas*: i.e., the former residences of the Aztec emperors of Mexico.
14. *a Cincinnati man*: possibly John Wellington Starr, inventor of an incandescent light bulb.
15. *my sisters*: Wright had two sisters, one with an unknown name. The known sister was Mary Ann Wright.
16. *Samuel Watson Wright*: William and Sarah Wright's son, born in 1845.
17. *Sarah's*: Sarah Wright.
18. *Forbes*: unknown English naturalist who wrote about California; almost certainly not the famous naturalist Edward Forbes (1815–54), who did not publish any books about the Pacific coast of the United States.
19. *your Father*: John Tyndall, Snr.
20. *your ... Mother's health*: Sarah McAssey Tyndall.
21. *Emma*: Emma Tyndall.
22. *Billy Tyndall*: William Tyndall.
23. *Wright to Leighlin and tell father*: Wright's father, whose name is unknown, still lived in Leighlin Bridge.
24. *[Murkingam] Bank from Cincinnati*: a ship for transatlantic travel built in Cincinnati.
25. *screw propeller attached 'Jim Hall'*: unidentified description of a shipbuilding technique.
26. *uncle Sam*: a personification of the United States of America.
27. *Joe Payne*: Joseph Payne, a civil assistant in the 5th Division, C District of the Irish Ordinance Survey. He joined the Survey on 1 April 1839, the same day as Tyndall, and worked closely with him. He was initially employed in plotting but subsequently in drawing and correcting. On 8 April 1842 he was transferred to the English Survey. He is possibly the Joseph Payne, son of William and Jane Payne, who was born on 5 November 1821 in Dunleckny, County Carlow (Irish Church Records).
28. *Tom Payne*: Thomas Payne, a civil assistant in the 5th Division, C District of the Irish Ordnance Survey, where he worked alongside John Tyndall. He joined the Survey in May 1839 and his principal duties were registering and calculating. He left the Survey later in 1839, only to return in May 1840. He was transferred to the 3rd Division in April 1841 and then to the English Survey on 8 April 1842.

To Jack Tidmarsh 9 November 1845 0325

Halifax | Monday evening July 14th 1845.

My dear Jack,

There was a great big god once in Egypt who was christened Memnon.[1] The people of that country say that in the far gone ancient time this god used to sing like a blackbird. Every morning when the rising sun took his first squint at the pyramids and climbing higher still cooled his hot lips with the exhaled dew from the gardens of Thebes, and laughed and danced and gambolled on the river Nile for all the world as he plays upon the burnished surface of a new tin saucepan. There the sweetest sounds that ever god or mortal manufactured used to proceed from the lips of Memnon! There was no settled measure in the notes, but it seemed as if the summer breeze was wandering over the strings of an Eolian harp,[2] sometimes rising loudly, anon dying into the most touching melody, as if soft showers of celestial honey were falling from the lips of the big god! Oh! it was grand to listen to him, with his wild, beautiful measureless cadences, enough to make the Sphynx neglect her riddle, or old Father Nile forget the law of gravitation and ask whether the rules of politeness required him to flow up or down!!

I had my head once examined by a phrenologist and he said that I had the bump of comparison largely developed; be that as it may, I cannot help comparing the feelings awakened in my heart on the receipt of your letter with the music of Memnon. 'And Parker[3] has written to me at last' said I; and immediately a host of little vagabond ideas which I fondly fancied were sleeping the sleep that knows no waking, rushed from the dark crannies of the past, that I really thought the mantle of the big god had been anatomised for their especial inspiration!

There was once an infernal young scamp who stole the sum of fourpence halfpenny from a poor butterwoman in Paris, as he was leaving the shop the old woman twigged him and seizing her poker pursued him as fast as her dilapidated energies would permit. Some men in the street sympathising with the poor old creature joined the hue and cry – 'Arretez le voleur!'[4] was shouted from a dozen throats – to scan the strength and swiftness of his pursuers was the work of a moment with the urchin, and seeing that he had nothing to fear from that quarter, the mustered force of his lungs burst forth in a wild echo to the cry behind – 'Arretez le voleur!' The people in front of course fancied from this that he was one of the pursuers, and not the pursued, and he was therefore allowed to pass unmolested.

Must I explain the moral Jack? 'Your silence Tyndall has proved your professions of friendship insincere' quoth Felix.[5] 'Stop thief' cried the guilty boy, at the same time saying in a whisper 'I know that I myself am the thief but this is my only chance'. In one respect you and the boy differ, he escaped, you shall not. I have a little Yorkshire salt from your tail[6] my boy. I'll nail you with the very pin on which you hang your charges against me!

Of course I heard of Allen's marriage; I have had letters from him since, I intend to pay him a visit shortly. But the most wonderful part of Allen's performance is cited in the following quotation from a letter to Bob Martin received yesterday – 'on the beginning of this week the lady of Mr Robert Allen presented him with a daughter!' There's an exhibition of 'manly vigour' for you wasn't it speedily manufactured? But your revelation respecting Lizzie[7] has almost unmanned me; I compressed my lips with fearful energy when I read it. The blood rushed to my temples, and it required no small effort to make the rebellious torrent retire to its proper channels. I never had the slightest intimation of this before. Well, one fond dream has exploded. Though I had never any 'intentions' towards Lizzy still I cannot resign her without a pang. I had promised myself three days pleasuring and Goosnargh[8] was set down among the places I was to visit, but the light of the village is gone and a visit under present circumstances would be a mournful affair. I'm glad you told me, it will preserve me from keener disappointment. What a time Marques[9] and I spent there! My recollection of it forms a kind of intellectual banquet, where the 'dear departed shades' of roses and ringlets and lips and kisses sweetly mingle – alas! alas! 'My peerless Lizzy'! 'Spoiled for ever! :–

> 'My peace is gone
> My heart is broken!'[10]

I can never forget the evening to which you allude.[11] And I shall long remember a remark you made at the same time 'By God Jack, he should not have all the fun to himself!'

Do I remember Miss Wilding?[12] Oh! don't I just? But man you're a fool to say you would go through 17 hells for her. Your hair would be burnt off in the passage, and you know that like Samson of old you lose your power when you lose your hair![13] You remember Jack — eh? You know I was in love with Miss Wilding, and had I seen you kissing her, in all human probability you would have experienced, in a certain nameless region, the momentum of my boot toe!!

> Johnny my dear
> I am no engineer
> From which you will see
> That the letters 'C.E.'[14]
> Apply not to me.
>
> Hope whispers a day
> Will come when you may
> To the letters J. T.
> Attach the 'C.E.'!

I don't know any thing that would give me greater pleasure than seeing you here – if ever you do come you may calculate on as cordial a welcome as an Irish heart can give. I have never seen Marquis since we parted in Preston which will be two years next November. I wrote to him a few days ago, having obtained his address by mere chance. There is a sweet gal here named Lizzy, I must endeavour to fall in love with her and forget the former – that cussed apothecary![15] I should like to kick him! Now have I not sent you a monster scroll, perhaps I have inflicted too much on your patience – finis!

Good bye for the present
My dear dear Jack
and believe me
in consonance with old professions

Your attached
Tyndall.

I am sincerely sorry for Margaret Mulhall tho' not at all surprised, I knew she was frail, and it required all my virtuous attention to preserve her from Latimer![16] Jokes apart, I regret her destiny – She was very attentive.

RI MS JT/1/TYP/11/3870–3871
LT Transcript Only

1. *Memnon*: an Ethiopian king from Greek mythology, said to have fought in the Trojan War.
2. *Eolian harp*: a harp that is played by the winds, named for the Greek God of the winds, Aeolus (*OED*).
3. *Parker*: possibly a nickname for Tidmarsh. There is also a Mr. Parker occasionally mentioned in the letters; see letters 0279, 0282, 0289 and 0297 of this volume.
4. *Arretez le voleur*: stop the thief (French).
5. *Felix*: possibly a nickname for Tidmarsh.
6. *I have a little Yorkshire salt from your tail*: unidentified idiom or reference; possibly an allusion to the folk belief that a bird can be caught by putting salt on its tail.
7. *Lizzie*: unidentified nickname, evidently a woman of whom Tyndall had been enamoured.
8. *Goosnargh*: a small village in Lancashire, England.
9. *Marques*: probably William Marquis.
10. *'My peace is gone / My heart is broken!'*: possibly a reference to J. W. von Goethe, *Faust*, 3136–3137: 'My peace is gone / My heart is heavy' (in German: 'Meine Ruh ist hin, / Mein Herz ist schwer').
11. *the evening to which you allude*: previous letter from Tidmarsh is missing.
12. *Miss Wilding*: a woman Tyndall knew in Preston. Tyndall seems to have been romantically interested in Miss Wilding. His journal for 14 November 1843 reads: 'At 9 1/2 went to the dance ... [I saw] Miss W. ~~who by some strange spell has absorbed all my affection~~' (RI MS JT/2/13a/2-3). According to his journal entry for 16 November 1843, he wrote Miss Wilding an acrostic poem (RI MS JT/2/13a/3). Miss Wilding may be the woman Tyndall and Ginty referred to as 'The Northern Briton'; see letter 0235, n. 31 of this volume.
13. *Samson of old ... lose your hair*: the Biblical Hebrew warrior Samson had incredible strength until his duplicitous lover Delilah cut his hair while he slept. See Judges 13:1–16:31.
14. *the letters 'C.E.'*: Civil Engineer.
15. *that cussed apothecary*: unidentified person, probably the new husband of Lizzie (see n. 7).
16. *Latimer*: probably George Latimer, or possibly his brother William.

From William Ginty September 1845 0326

Post office Thirsk[1] | Yorkshire, Sep. 1845.

A selection of verses written by the poet Ginty during his several amours[2] both in England and Ireland. These verses bespoke an honest soul and fully confirms the old saying that 'true lovers are always the most liberal men', and so it is with the poet Ginty, it is well known that when he does chance to get entrapped by the fair ones he loves with that degree of ardour so peculiar to the Irish character but more particularly so to the inhabitants of an oft spoken of province I mean Connaught![3] as the richness of the subjoined lines will show.

To Miss S. Aubury[4] on leaving Youghal[5]

We've met when heavenly morn was bright
When twilight shadows fell
When festal halls were bathed in light
And dreamt not of farewell

We've wandered thro' the shady grove
The bright and flowery dell
The eve has heard our sighs of love
But never our farewell

But now the fatal word must come
It strikes me like a knell
From love and Sally[6] I must roam
Farewell! sweet girl farewell!!

These are beautiful lines but they unavoidably show in so few words the bewildered state of the poet's mind. Here followeth an acrostic.[7]

'Acrostic to my beloved on leaving her for ever'

Must I at length believe my life a dream
And clasp the pang which withered hopes impart
Rend every tie as flits the radiate beam
Your presence but to cheer my drooping heart

Even as the mantling ivy fondly flings
Despite the storm its tendrils round the oak
Wreathed round my soul thy loved idea clings
Amid the fury of each worldly shock
Rest crushed affection in thy silent cell
Deep cherished shrouded there for ever dwell
Sweet idol of my bosom, fare thee well!

Surely after the perusal of this a little more in the same strain will be wished for, so one more shall be given, to them –

To Miss Maria Hopkins[8]

May heaven its choicest holiest blessings strew
And throw its beautiful beam on you
Rest my sweet love devoid of care or dread
And sylphs[9] stand sentinels about thy bed.

Serenade
'Oh! Mary[10] pon my soul my love
I love you very dear
The midnight moon now rolls above
And yet you are not here
'Amid the far blue ether love
My thoughts in wildness roam
Can lovely Mary faithless prove
Oh! why am I alone!!

When last I pressed your cheek my love
And trembled in your eye
The world was half asleep my love
And full of stars the sky
You said you'd meet me here my love
Where I my vigil keep
And while I dream of blighted love
I blow my nose and weep

'Then come oh! come my only love
And cheer thy poet's heart
Else Mary Hill the window sill
Is warmer than thou art
The moon is high and cloudless love
And yet thou art not here
With healing on thy wings my dove
Appear! appear!! appear!!!

Finis.[11]

RI MS JT/1/TYP/11/3636–3637
LT Transcript Only

---

1. *Thirsk*: a small market town in Yorkshire.
2. *amours*: loves (French).
3. *Connaught*: a province in western Ireland.
4. *Miss S. Aubury*: not identified; the end of the poem suggests that the woman's first name is Sally.
5. *Youghal*: a seaside town on the southeastern coast of Ireland.
6. *Sally*: Sally Aubury (see n. 4), not identified; possibly another love interest of Phillip Evans's. See also letters 0303 and 0305.
7. *acrostic*: a poem in which the first letter of each line spells out a word or name. In this case, the acrostic spells out 'Mary Edwards'. Mary (*c.* 1822–?) lived with her uncle and aunt in Kirkby Lonsdale, and, despite having a putative romance with William Ginty, married William Henry Owen, an Irish solicitor, in June 1843.
8. *Miss Maria Hopkins*: not identified.
9. *sylphs*: a creature that lives in the air; a beautiful, graceful woman (*OED*).
10. *Oh! Mary*: possibly a nickname or variation on the name 'Maria'.
11. *Finis*: Finished (Latin).

1847

To Margaret Ginty 29 January 1847 0327

Halifax, | (Jan. 29.1847)[1]

Dear Peb,[2]

You might have spared yourself the trouble of asking me to go to Manchester, as I intended, weather willing, to bundle off to you on Sunday morning whether you like it or not – I would have been with you last Sunday, but I saw the consequence of going – a long day's discussion with Ginty by the fireside – and I had no wish to split your ears with our noisy logic. We all expected you this week – our fair friends[3] opposite were dancing with delight at the bare thought of seeing you once more. Mrs Wright had your bed prepared – and I intended to inflame four pipes of tobacco as a domestic bonfire to welcome your arrival – I now stop with the promise to be with you tomorrow night – Good luck to you dear Peb and believe me

your affectionate old | John Purybingle,[4] | Mr Hall desires me to present his warmest regards.

To M[rs] W Ginty 34 Gt *[Dalie]* St | *[Strangways]* Manchester[5]

RI MS JT/1/TYP/11/3637
LT Transcript Only

---

1. *Jan. 29. 1847*: Louisa Tyndall annotation: 'Postmark'.
2. *Peb*: a nickname for Margaret Roberts Ginty.
3. *our fair friends*: not identified; probably female friends in Halifax.
4. *John Purybingle*: a nickname Tyndall occasionally assumed when writing to the Gintys.
5. *To M[rs] W Ginty 34 Gt [Dalie] St | [Strangways] Manchester*: address given by Louisa Tyndall.

To the Editor of the *Sligo Champion* 11 March 1847 0328

Gentlemen[1]

You have lately attained an unenviable notoriety. The wide world of newspaper readers is aware of your physical entity. Scraps from the Sligo Champion transferred to the columns of Jerrold[2] and the Times[3] have carried your fame further than many of your mothers ever anticipated! Thus have I made your acquaintance.

I have not been in Ireland many days, still have I seen sufficient to sadden my heart[4] and relax the fastening of my breeches pockets. Inconvenient stirrings of humanity, unworthy perhaps of your participation! You are no ordinary men. Any fellow with an average impregnation of the Devil might by an effort button up his sympathies against the pleadings of Woe everywhere existent. By a stern act of volition I might do the same myself. But you, Sligo officials, stop not here. You push your heroic principles beyond the boundaries of passive indifference. Joyfully death-defiant you stand forth a rollicking antithesis to the scenes which surround you. Ireland's year of mourning is your year of jubilee! Gallant souls! Jolly sparks from that primeval spirit which sung tenor to Nero's cat-gut when he fiddled over burning Rome![5]

Sligo Officials, there are two laws to which I may appeal – The first and holiest is God's law written in the hearts of all though oft illegible through the disfigurements of selfishness, snobbism and folly – that law which binds your sympathies with humanity and claims soft pity for the wretched. This to your own consciences Gentlemen. If you heed it not then I tell you there is a law to control perforce your Baccanalian exhibitions. Potent as your gastronomic coterie may deem itself, it is less than a maggot beneath the heel of Public Opinion. Provoke this agency and

> 'The joys and the wishes
> The loaves and the fishes[6]
> On which you so wantonly revel
> Will go – and quite right –
> To the people's delight
> Most exceedingly quick to the Devil!'[7]

Wat Ripton[8] | Ex. Gov. official

RI MS JT/2/13a/185
JT Journal Transcript Only

1. *Gentlemen*: Tyndall's journal entry for 11 March 1847, in which this letter was copied, recorded that Tyndall wrote this letter 'to the Editor of the Sligo Champion concerning the Government officials' (RI MS JT/2/13a/185). *The Sligo Champion* was an Irish newspaper founded in 1836. Tyndall appears to be responding to recent accusations that government officials in Ireland's County Sligo were profiting from the Great Famine by accepting government aid and not passing the funds or food to the neediest citizens in their county. An untitled editorial for the *Times of London* on 25 February 1847, for example, mentioned that "a Sligo paper ... distinctly charges the gentry and magistracy of the county with trafficking in the people's misfortune for their own lucre" (*The Times,* 25 February 1847, p. 4).
2. *columns of Jerrold*: probably either *Jerrold's Shilling Magazine*, which flourished in the mid-1840s, or *Douglas Jerrold's Weekly Newspaper*, which lasted from 1846 until 1848.

M. Fryckstedt, 'Douglas Jerrold's Shilling Magazine', *Victorian Periodicals Review*, 19:1 (Spring 1986), pp. 2–27, on p. 2.

3. *the Times*: a widely read daily British newspaper published in London.
4. *still have I seen sufficient to sadden my heart*: probably a reference to the Great Famine in Ireland.
5. *Nero's cat-gut when he fiddled over burning Rome*: Roman Emperor Nero allegedly fiddled while the Great Fire of Rome burnt his city in AD 64. Cat-gut was occasionally used to string violins and fiddles.
6. *The loaves and the fishes*: a reference to two Biblical miracles, in each of which Jesus was able to feed thousands of people using just a few loaves of bread and fish.
7. *'The joys and wishes ... to the Devil!'* : unidentified poem, possibly written by Tyndall himself.
8. *Wat Ripton*: one of Tyndall's most frequent pseudonyms.

To Thomas Hirst 29 May 1847 0329

Written at Railway speed – | May 29th 1847.

My Dear Tom

En passant[1] as the French say I may observe, that the frigid 'dear Sir' is unnecessary when you write to me – my name is Tyndall – I have about 10 minutes to respond to your epistle it is now twilight and candles are scarce here – Richardson informed me of his victory. I was sincerely glad to learn that Mr Carter settled it without resorting to extremities – I will look sharply after my compatriots in Ferguson Street[2] when I return – I am astonished and confounded – and am almost disposed to chime in with the thread-bare cry of the depravity of human nature –

Geology is a most interesting study, but a hammer and chisel and a bona fide ramble amid crags and quarries are indispensable to a thorough knowledge of it – Hall must read Past and present[3] once more – to deduce Carlyle's ideas of religion from the work is impossible – he purposely avoids touching upon religious theories, and insists on the general principles of earnestness of soul and energy of action – Your question regarding Bulwers statements[4] is by no means foolish – It gratifies me much to find that you dont allow such passages to slip through your intellect like sand through an hour glass – The subject is purely speculative, and I have for some time shut my eyes against such matters comforting myself with the sentiment of Saint Paul[5] 'Shall not the Judge of all the earth do right?'[6] your opinion of the happiness of heaven seems to be that it is infinite – tested by this Bulwers theory falls – but Bulwer will grant you no such postulate – his idea of happiness has a relation to man's capability for enjoying it – take for instance the happiness of sense – many animals have but one or two senses – gratify these to their utmost power of enjoyment and you make the animal perfectly happy; but the gratification of these two senses would not be perfect happiness to an animal possessing five senses such as man – If we apply the illustration to the immaterial world we can readily conceive a capacity for enjoyment increasing ad infinitum as if sense after sense were added still supposing the being in each stage of its existence to be perfectly happy, that is enjoying happiness to the full extent of his power of enjoying it – this I think was Bulwers idea and certainly it accords with my views more nearly than the heavenly Chartism[7] which you seem to advocate – you are a leveller and would contract the enjoyments of a Newton in the world of spirits to the span of Cowper's[8] peasant[9]

> 'Who knew and knew more her bible true
> A truth the brilliant Frenchman[10] never knew'[11]

I write hastily and may therefore be obscure but I know I write to a man who can brush away a cloud –

I'm not a fellow who would deliberately catch a bull by the horns – I shall most certainly avoid a collision with that Skircoat gentleman[12] – if he be all you describe him I shall be happy to be permitted to call him 'friend'. Good bye my dear Tom

& believe me | Most sincerely yours, Tyndall

Kindest regards to your mother[13] and brother[14] – present my sincere thanks in return for their kind wishes | J.T.

RI MS JT/1/HTYP/1
RI MS JT/1/T/510

1. *En passant*: in passing (French).
2. *compatriots in Ferguson Street*: Ferguson Street was the location of Mr and Mrs Wright's boarding house in Halifax, where Tyndall lodged.
3. *must read Past and present*: T. Carlyle, *Past and Present* (London: Chapman and Hall, Ltd., 1843).
4. *question regarding Bulwer's statement*: possibly a reference to the writings of Edward Bulwer-Lytton (1803–73), an English writer and politician (*ODNB*). Bulwer-Lytton was the author of several novels, including *Paul Clifford* (1830), which famously began 'It was a dark and stormy night'. It is unclear which of Bulwer's works Tyndall might be referring to.
5. *Saint Paul*: also called the Apostle Paul, to whom fourteen of the twenty-seven books of the New Testament are attributed.
6. *'Shall not the Judge of all the earth do right?'*: Genesis 18:25. Tyndall was mistaken when he referred to this quote as a statement from the writings of St Paul.
7. *the heavenly Chartism*: Chartism was a national British movement in support of the People's Charter of 1838. It was highly active from 1838–48. The People's Charter, whose proponents became known as Chartists, demanded democratic reform of the political system but had been rejected by Parliament.
8. *Cowper's*: William Cowper (1731–1800), English poet and hymn-writer (*ODNB*).
9. *Cowper's peasant*: William Cowper's poem 'Truth' focuses on a female peasant.
10. *the brilliant Frenchman*: a reference to Voltaire; Cowper's poem compares Voltaire to an English peasant.
11. *'Who knew...never knew'*: W. Cowper, 'Truth', ll. 327–8: 'Just knows, and knows no more, her Bible true – / A truth the brilliant Frenchman never knew'.

12. *that Skircoat gentleman*: not identified; possibly someone from the Skircoat Green district of Calderdale, near Halifax.
13. *regards to your mother*: Hannah Oates Hirst.
14. *regards to your ... brother*: Thomas Hirst had three living elder brothers (*ODNB*); it is unclear which brother Tyndall was referring to in this letter.

From E. D. Steuart 1 June 1847 0330

Steuart's Lodge | June 1, 1847

John,

I fear you must have thought me negligent in not having written before now to thank you for the beautiful engraving you so kindly sent me,[1] and I should have done so but my entire time has been taken up with Mr Steuart who has had a most severe attack of gout for the last ten or twelve days: he is still confined to bed, and suffering great pain, but Doctor Roche[2] hopes it will soon subside, as it has moved from the knee to the ankle: he is very much exhausted for want of rest, and much weakened of course, still I am thankful that his complaint is not one of those which are raging around us,[3] and I trust when it is over his health may be improved by it. 'The Heart's Misgivings'[4] arrived yesterday, quite safely, and I feel greatly obliged to you for it, and shall doubly value it as your gift. I hope you will let me hear sometimes how you are, for believe me your welfare is always a subject of warm interest to Mr Steuart and me.

I remain, | Your sincere friend, | E.D. Steuart.

RI MS JT/1/TYP/10/3325
LT Transcript Only

---

1. *beautiful engraving you so kindly sent me*: letter missing.
2. *Doctor Roche*: not identified.
3. *his complaint is not one of those which are raging around us*: possibly a reference to the Great Famine.
4. *'The Heart's Misgivings' arrived yesterday*: possibly an engraving of the 1844 painting 'The Heart's Misgivings' by Frank Stone, and this could be the engraving referred to at the beginning of the letter. Alternatively, Tyndall may have sent Mrs. Steuart a copy of Cuthbert Bede's 1846 poem, 'The Heart's Misgivings'.

To Margaret Roberts Ginty [12 or 13 June 1847][1] 0331

Halifax | Saturday, My dear Mrs Ginty,

You have indeed been the medium of 'good news' I don't know any intelligence that could have afforded me more sincere satisfaction than what you have just communicated – Poor Higginson![2] – I don't know any one on earth, ladies excepted, that I like better – He has come nearest to my ideal of a true man – we were bound together by the warmest friendship, and when half the world lay stretched between us the more I thought on his character the more I loved him – But how did you come to know of his appearance in Ireland? I wrote to him about two months ago, but my letter had not time to reach him. He belonged to the 78th. I missed his name in the Army list for two or three months successively – I thought he was dead and wrote to the Secretary at war to learn the worst, I was informed in reply that he had exchanged to the 62nd – and your note places the fact of his existence beyond doubt – I owe you many letters! How many? Now don't be hard on me – let us balance our books and begin a new account. Accept my sincere congratulations on the late happy event – I trust little Kathleen[3] is doing well. Bill[4] has returned to England with 6 inches added to his stature – there is no standing the chap's pride. I saw him on Wednesday evening. I go to Manchester tomorrow and may possibly push on to Preston. To whom are you going to introduce me? It will do no good – when the dry bones in the valley of Hinnon drill themselves into skeletons and clothe themselves with flesh, the spring time of my heart will return.

'Their eyes have scarce a charm for me'[5]
Good bye for the present and believe me always

most sincerely yours | Perybingle[6]

RI MS JT/1/TYP/11/3705
LT Transcript Only

1. *12 or 13 June 1847*: From Tyndall's Journal, 12 June 1847: 'A note from Mrs. Ginty containing an account of Higginson's return from India!' (RI MS JT/2/13a/208)
2. *Poor Higginson!*: Thomas C. Higginson; see letter 0251, n. 26 of this volume.
3. *little Kathleen*: Kathleen Ginty (1847–?), the oldest daughter of Margaret and William Ginty.
4. *Bill*: William Ginty.
5. *'Their eyes have scarce a charm for me'*: Lord Byron, 'Personal, Lyric, and Elegiac: To Inez', l. 16.
6. *Perybingle*: also spelled Purybingle, a nickname Tyndall occasionally assumed when writing to the Gintys.

From Fanny Smith 24 June 1847 0332

Preston, | June 24th, 1847[1]

Dear Mr Tyndall,

You anticipated my feelings of surprise, when I received your letter;[2] it is quite impossible for me to write as my heart would dictate, but I trust you will excuse any incapacity I may evince in my letter to write. I find myself placed in a very difficult position, indeed I am afraid of trusting my pen. I wish to inform you that you have received a very wrong idea about H.M.[3] I did not give up his friendship on your account, considering such an act as mean and unjust. I have thus far told you the exact truth, and trust it will give you satisfaction. The perusal of your letter gave me pleasure because of the liberality of your views; and the candour of your expressions give me to hope that I am resting on a firm foundation of honour and integrity. There is one matter on which I may ask an explanation – you remember the conversation we had at Sister Allen's,[4] when we were alone at tea-time. Your remarks may flash into your mind. I should never consider that I was doing an act of justice, if I continued an intercourse or an attachment when I did not feel an increasing degree of respect and love. I <u>never</u> would bind myself for life on such conditions.

At present I do not consider that you are in a situation to render life very comfortable, but I have no doubt that as time rolls on by the power of your 'own right arm' you will be so strengthened that happiness and comfort will not dwell in name alone. Will you be kind enough to direct any of your epistles to Mr or Sister Allen without my name being outside at all, as I never receive a letter without Papa[5] opens it, unless it is directed to somewhere besides Stanley Terrace.[6] I beg to be excused writing any more at present. Hoping you will excuse any imperfections my letter may contain, I remain, dear Mr T.,

Your affectionate friend, | Fanny Smith.[7] | Mr Tyndall.

RI MS JT/1/TYP/11/3866
LT Transcript Only

---

1. *June 24th, 1847*: Louisa Tyndall annotation: 'See envelope'. It is possible that this is the postmark date, not the date on which the letter was written.
2. *when I received your letter*: letter missing. From Tyndall's journal, 21 June 1847: 'While in Preston Allen m'a dit que le gage d'amour entre Mlle F. et M.M. était cassé à cause de moi! Je sentais beaucoup agité, quoique je ne le montraîs pas. Je suis comme un esprit mauvais sans en avoir l'intention. J'écrivais à Mlle F. une letter très longue dans la quelle je l'ai informé touchant ma position incertaine. Je lui ai dit que je suis beaucoup plus agé qu'elle; en un mot, j'écrivais comme je ne pouvais pas écrire en Francais. Je lui ai dit toute la vérite et j'ai laissé la conclusion à elle même' (RI MS JT/2/13a/213–214). Translation: 'While in Preston Allen told me that the pledge of love between Miss F. and Mr. M. was broken because of me! I felt very agitated, though I did not show it. I am like a bad spirit without having the intention. I wrote a very long letter to Miss F. in which I informed her about my uncertain position. I told her that I am much older than she is; in a word, I wrote that I could not write in French. I told her all of the truth and I left the conclusion to her.' Tyndall's choice to write about these events in French suggests that he felt the episode was of a private nature.
3. *wrong idea about* <u>H.M.</u>: Henry Merryweather. From Tyndall's journal, 24 May 1847: 'Tea at Mrs. Allen's, where I had the pleasure of meeting Miss Fanny ... elle est [une] bonne fille ... Elle est engagée quoiqu'elle n'a que 15 ans. Son amant est un surveyor .... du l'appelle Henry Merryweather. Hereux homme!' Translation: 'Tea at Mrs. Allen's, where I had the pleasure of meeting Miss Fanny ... she is [a] pretty girl ... She is engaged even

though she is not but fifteen years old. Her lover is a surveyor ... by the name of Henry Merryweather. Happy man!' (RI MS JT/2/13a/200).

4. *conversation we had at Sister Allen's*: probably the wife of Robert (Bob) Allen, who may also have been a sibling or cousin of Fanny Smith's.
5. *Papa*: not identified.
6. *Stanley Terrace*: the Smith family home in Preston.
7. *Fanny Smith*: (*c.* 1831–?), a Preston resident and friend of Robert Allen's; possibly a young relative of Allen's wife's. From Tyndall's journal, July 28, 1845: 'Allen came and actually dragged me from the house. He told me that I was expected to dinner at Stanley Terrace [the Smith family home] ... On my left sat "Miss Fanny Smith", the "sweet" young lady who had asked a blessing at dinner.' Tyndall thought she was around 14 at the time (RI MS JT/2/13a/13a/92), making her around 16 when this letter was written.

To Robert Allen 2 July 1847 0333

July 2 1847

Mon cher Bob

Votre lettre a jeté une triste clarté sur cette petite transaction et *[l'invests]* avec une gravité que je ne pensais y appartenir. Quant à mes pensées, ils sont tranquilles à present comme si toute la nature était en harmonie. Je ne croyais pas jusqu'à maintenant qu'il y a aucune chose approchant la dans l'affaire et je ne puis m'informer pourquoi il serait ainsi. Pour moi, je n'exciterais pas sans raison le mépris du plus chétif des créatures de Dieu; et surement vous vous trompez de supposer que Mrs S. & Me. F. voulaient murement agir dans une manière calculée expressément de m'exciter ce sentiment là.

Sans doute vous m'avez souvent ecouté m'opposer contre la doctrine de la dépravation de la nature humaine. Cette semaine même j'écrivais une lettre de dix neuf pages soutenant mes opinions sur ce sujet. Quoique je tiens ces opinions un contact avec le monde me force n'en esperer trop. J'attend la malhonneteté et le vice et contre elles je dirigerais les meilleures facultés des hommes, mais dans cette instance je me confesse franchement que j'ai rencontré la dissimulation oû je ne l'attendais point.

Mes sentiments vers M. F. étaient tout-à-fait fondé sur la conviction de sa verité et de sa bonté – cette fondation brisée, le batiment restant sur elle partège nécessairement de sa destruction, il me cause un vive regret qu'une si jeune aurait appris d'etre fausse. Je n'étais pas preparé pour cette revelation et desormais je procederai sur ma route un homme plus pensif et plus sage.

Je la pardonne de bon coeur je crois qu'elle a péché sans reflection, mais Bob, qu'elle s'assure qu'elle a entré dans un chemin perilleux. La fin de la duplicité est presque toujours défaite et misère.

Le dimanche passé à neuf heures au matin je n'avais point d'expectation que cette correspondance prendrait lieu, quand nous marchions à la chapelle. Votre première intimation éclatait sur moi. Ma première impulsion était condamnation de moi-même, pour ce que je croyais ma propre imprudence. Je resolvais immédiatement de comprendre le tout, et après diner vous me disiez tout ce vous saviez. A ce temps-là je vous disais que j'étais resolu d'agir dans une manière que personne ne pouvait blâmer, et j'appelle à votre jugement si j'ai tenu parole.

Il n'est pas probable que des sentiments si courtes dans leur duration étaient profondement énraciné. Le cas était ceci. Je croyais M. F. d'être habile et vraie et je pensais que celui-ci serait un sol favorable pour l'infusion de mes propres principes. Je croyais qu'elle n'avait été embrouilli dans les folies du monde et c'était mon intention d'en la preserver. Mes sentiments étaient purement raisonnés. Sans doute le temps les aurait fait plein de dévotion, mais comme de temps n'était pas donné il n'y a pas de violence fait. Je plaçais sa dernière lettre dans ma poche aussi froidement qu'elle était la mémoire de mon tailleur, et depuis, je n'ai point eprouvé de malheur touchant la.

Ne croyez pas, mon cher Bob, que cette transaction puit pour un moment affecter la rélation qui a subsisté pour longtemps entre nous. Notre amitié était formée independemment de telles transactions et elle continuera en dépit d'elles. Je suis satisfié que M. A. a agi partout comme j'étais son frère. Loin d'être irrité contre elle, ce sentiment n'a pas d'existence vers celle qui a si follement taché de me trompé.

Un dernier mot d'avis à cette belle demoiselle. Si elle value le respect et l'estime du sage et bon, je voulais la prier ne permettre pas la vanité la tempter de la vérité. La fraude est une <*1 word missing*> qui recule inévitablement sur ceux qu'elle nourrit.

W.R.

My dear Bob,

Your letter had cast a dim light on this little transaction and invested it with a gravity that I did not think belonged there. As for my thoughts, they are peaceful now as if all of nature was in harmony. I did not believe until now that anything was happening in this matter, and I could not tell why it would happen. As for me, I would not *[exercise]* contempt for any reason towards the weakest creatures of God; and surely you are wrong to assume that Mrs. S[1] & Miss F.[2] wanted to act carefully in a calculated manner specifically to awaken such a feeling in me. No doubt you have often heard me arguing against the doctrine of depravity of the human nature. This very week I wrote a nineteen page letter in support of my opinion on this subject. Although I hold these opinions, when I am in contact with the world I force myself not to hope too much. I expect dishonesty and vice, and I would set the best powers of men against them. But in this case, I frankly confess that I met a dissimulation where I did not expect it.

My feelings towards M. F. were in fact based on the conviction of her sincerity and kindness – this is a broken foundation, and the building that remains on top will surely meet its destruction. It causes me great regret that such a young one would have learnt to be false. I was not prepared for this revelation and now I will continue my journey as a more thoughtful and wiser man.

I forgive her with all my heart; I believe she sinned without thinking, but Bob, *[that ensures that she went down a perilous path]*. The end of deception is almost always defeat and misery.

Last Sunday at nine o'clock in the morning I did not expect this correspondence to take place, as we walked to the chapel. Your first insinuation fell upon me.[3] My first impulse was to condemn myself for what I believed to be my own carelessness. I immediately decided to understand it all, and after dinner you told me all that you knew. At that time I told you that I was determined to act in a way that no one could blame, and I call to your judgement whether I kept my word.

It is unlikely that feelings of such short duration were deeply rooted. The case was this. I believed M. F. to be clever and truthful and I thought that this would be a favourable ground for the infusion of my own principles.[4] I believed that she was not yet embroiled in the madness of the world and it was my intention to keep her from it. My feelings were of a pure reasoning. Without doubt, time would have made them full of devotion, but since time was not given, there was no harm done. I put her last letter[5] in my pocket as coldly as if it was *[my tailor's record]*, and since then I did not experience misfortune in regards to it.

Do not think, my dear Bob, that this transaction could affect even for a moment the relationship that persisted for a long time between us. Our friendship formed independently of such transactions and will continue despite of them. I am gratified that M. A.[6] acted throughout as if I was her brother. Far from being angry at her, this feeling was nonexistent towards the one who crazily tried to deceive me.

A final word of advice to this beautiful lady. If she values the respect and esteem of the wise and good, I would want to ask her to not allow vanity to stain truth. Fraud is a <*1 word missing*> that inevitably falls back on those it nurtures.

W.R.[7]

RI MS JT/2/13a/224–225
LT Transcript Only

---

1. *Mrs. S*: Mrs Smith, Fanny Smith's mother and possibly a relative of Robert Allen's wife.
2. *Miss F.*: Fanny Smith; see letter 0332, n. 7 of this volume.
3. *Your first insinuation fell upon me*: Allen told Tyndall that Fanny Smith had broken off her engagement on Tyndall's account; see letter 0332, n. 2 of this volume.
4. *favourable ground for the infusion of my own principles*: Tyndall had offered to tutor Fanny Smith. From Tyndall's journal, 26 June 1847: 'Posted letter to the Witch [Fanny Smith] and have thus made the first step in a direction which a week ago I never dreamt of or at least if the thought arose it was only to be discarded as absurd and unrealizable. Je la enseignerai. Il faut qu'elle pense comme moi et j'avais beaucoup de temps pour faire des opinions comme les miennes – et il le sera!' (RI MS JT/2/13a/215). Translation: 'I will teach her. It is necessary that she thinks as I do and I have much time to make her opinions like mine – and it will be so!'
5. *her last letter*: probably letter 0332 of this volume.

6. *M.A.*: probably Allen's wife, first name unknown. Fanny Smith may have been a relative of hers.
7. *W.R.*: probably an abbreviation for 'Wat Ripton', one of Tyndall's most frequent pseudonyms.

To Thomas Hirst 25 August 1847 0334

Queenwood College near Stockbridge | Hampshire | Aug. 25th 1847.

My dear children –

Though my ideas are sadly out of gear this afternoon, nevertheless as I have half an hour that I can call my own I devote it to writing to you – I started from Manchester on Thursday morning, dined at Birmingham, and reached London at half past 8 the same evening – met Mr Edmondson according to appointment and spent thursday night at a private and Commercial boarding house kept by a quaker – It is one of the most comfortable places that I have met, stuck right in the nucleus of the city yet perfectly quiet and retired – on Friday morning we all – thirty perhaps, mostly quakers – sat down to breakfast together – our host[1] occupied the head of the table – his daughter a rich and *[*lovely*]* brunette sat at the foot – the former read a chapter from the bible after which we all paused in silent meditation for a minute or two, and then fell voraciously upon a capital breakfast – If ever I have occasion to go to London I shall most certainly stop there –

After breakfast Mr Edmondson and myself made the tour of London and purchased books and instruments – during our peregrination we visited *[*Delarue*]*'s paper and fancy stationary manufactory,[2] and a more wonderful sight I never beheld; one machine an envelope folder performed the function of the human hand with perfect exactitude had a savage looked upon the instrument his first impulse would doubtless be to fall down and worship it.

We reached Queenwood[3] at 10 o'C on Friday night when I was introduced to Mrs Edmondson and her fair daughter[4] – From the limited premises I now possess I am inclined to hope that I shall like this situation exceedingly – Mr Edmondson is a man of most comprehensive attainments. I like him very much, he is almost paternally kind to me – on Saturday he placed two superb works on Natural Philosophy in my hand and requested me to compile a series of lectures – 'Thou seest John Tyndall' he remarked 'what an atmosphere thou wilt breathe in'

The geologic features of this County are all Chalk, Chalk, Chalk – In the evening you fancy you see a gentle mist hanging over the fields – you are deceived it is nothing but vapoury looking chalk powder – you see a white patch here and there and imagine yourself in the vicinity of a wash house – wrong again, it is all chalk – the sole stones to be found are the associated flint nodules and many of the houses, for want of better material are built *[*of*]* clay and marle.[5]

We purpose making an excursion to Stonehenge[6] tomorrow and there surrounded by the ghosts of the Druids and shaded by their gray alters will we demolish our ham sandwiches and pop – In a few days we shall make the tour of the Isle of Wight – I shall be at liberty during three months of the year – I might possibly take a look in at you next Christmas – And now I have not written so long a letter to any one else so I shall conclude – Remember me to all who care any thing about me

Goodbye and believe me | most sincerely and affectionately yours | J Tyndall | Is the level[7] safe?

RI MS JT/1/HTYP/2
RI MS JT/1/T511

---

1. *our host*: not identified.
2. *[Delarue]'s paper and fancy stationary manufactory*: Thomas de la Rue, paper manufacturer, stationer and printer, invented an envelope machine in 1845.
3. *Queenwood*: the school where Tyndall taught from 1847–8. See Introduction to Volume 2.
4. *and her fair daughter*: Jane Edmondson Benson (1823–?). Jane wrote two books about her father's work: *From the Lune to the Neva* (1879) and *Quaker Pioneers in Russia* (1902). C. A. Russell, *Edward Frankland: Chemistry Controversy and Conspiracy in Victorian England* (Cambridge: Cambridge University Press, 1996), p. 62. In 1859 she married Davis Benson (1820–?). S. B. Foster, *The pedigree of Birkbeck of Mallerstang and Settle, Braithwaite of Kendal, Benson of Stang End* (London: Printed for private circulation, 1890), p. 111.
5. *clay and marle*: 'marle' probably refers to marl, a mudstone composed of clay and silt.
6. *Stonehenge*: the oldest archaeological site in the United Kingdom.
7. *the level*: possibly a piece of surveying or scientific equipment.

From Conrad Kupper 16 September 1847 0335

Halifax | 16 Sept. 1847.

My Dear Mr Tyndall,

I had the pleasure of receiving your very kind letter[1] from the 22 *[inst]* and thank you very much that you kept me so well in memory and honoured so very soon after your arrival with your agreeable news. I am very glad that you are content with your new situation.[2] Of course it cannot feel so agreeable as if your brothers[3] there had views and principles like yours, but nevertheless it is nothing what does make unhappy considering the great multitude which is walking yet in darkness and blindness and whose oppositions minds like ours have continually to meet. If the result of our trouble has not been rewarded in proportion to those of our opposers,[4] who keep with little labour and wrong principles their inexperienced and not thinking fellow-men on their side, we have that advantage that men like ours value 50% more than theirs and that each of them has so much more influence upon his fellow-men because we act by principle after being perfectly convinced of our sake. Our opposers, on the contrary, are not able to work upon men by reason and principle, but merely by what they have heard and what they are told, and in the same time a great deal does not believe in what they pretend to be. The painful and responsible situation of trust is very easy to imagine, also those who leave themselves at their care, but there is a good time coming. It cannot remain long time in that way. I should feel very much obliged if you were so kind as to send me an extract of philosophical delivery.

Mr Hall has returned. He felt very sorry when he got to know you had left. There lived an independent preacher[5] for about a fortnight in the room which you formerly occupied. I got very soon acquainted with him, and also in a short time very familiar in changing our opinions on different subjects. Some very interesting conversations we had. You will imagine I was very careful in speaking to him about religion. He was not able to contradict one of my views which I clearly expressed several times. He was a more liberal thinking man than most probably we had not agreed with each other without him using reason.

I had yesterday very agreeable letters from home; my brother[6] is quite well and charged me to remember him to you – his wife[7] is also very well, she got delivered from a daughter[8] during the time her husband was in England. ¼ of an hour before he left I got a letter with that agreeable news, so that I was just able to make him acquainted with it. He wept for joy, deeply regretting of his being from home, but it was impossible to alter – what he only could do and what he did also was to hasten home without any delay. His happiness was in the highest degree when he found his wife and his little daughter perfectly well. In a fortnight he is going to Berlin, Breslau, Leipzig and Russia.

You will be so kind as to excuse it that I have waited such a long time before I answered your very kind letter. Next time I will be more punctually in that respect. I shall be most happy in hearing very soon from you.

Mrs Ginty has been here some time ago. By George, she is one of the most agreeable women I ever saw in my life. Had she not been married I would have fallen in love with her, there is almost no doubt in it.

I was on that ball which was given in honour to Mr Edwards[9] and enjoyed it very much. I have ....[10] but excuse in telling it not.

Halifax is always the good old town, as a certain gentleman[11] whom you know very well foolishly says – nothing essential has happened.

I was several times obliged to put Mr Wright on the right way – he was behaving worse than ever. Yesterday I had him and his wife until about midnight in my room, and after his wife had gone I remained with him till 2 o'clock when in the meanwhile I gave him some earnest instructions as well as some well deserved reproaches. I like always to hope there might be a possibility of a reconciliation.

Believe me yours truly, | Conrad Kupper.

RI MS JT/5/15c
LT Transcript Only

1. *your very kind letter*: letter missing.
2. *your new situation*: as of August 1847, Tyndall had begun teaching at Queenwood College in Hampshire.
3. *your brothers*: i.e., Tyndall's fellow teachers at Queenwood. It is unclear why Kupper believed the other teachers would not have 'views and principles like' Tyndall's; Kupper may be referring to the fact that George Edmondson, the school's founder, was a Quaker.
4. *our opposers*: not identified; presumably those who did not share Kupper and Tyndall's views about religion. This letter suggests that Tyndall was beginning to reject religious belief by 1847.
5. *an independent preacher*: not identified.
6. *my brother*: not identified.
7. *his wife*: not identified.
8. *a daughter*: not identified.
9. *Mr Edwards*: not identified.
10. *I have ....*: Louisa Tyndall annotation: 'sic'.
11. *a certain gentleman*: not identified.

To Jack Tidmarsh 26 September 1847 0336

Queenwood College | near Stockbridge | Hampshire | 26th Sep. 1847.

My dear Jack

Not wishing to put you off with a fragment of an epistle I have waited day after day until I should have time to send you a long one. Contrary to numerous predictions and to my own expectation I left Halifax on terms of amity with our late governor.[1] On Sunday morning the 15th of August he and I found ourselves tête a tête at Wards End. I had made half a dozen appointments with him previously but was unable to see him and I began to fear that I should be compelled to break my engagement with Mr Edmondson. In half an hour we settled matters, and though the settlement did not reach what under other circumstances would be considered compensatory, still all things taken into account I think I have reason to be satisfied – he allowed me £60 extra for over time – had he got his money peaceably from the company I should have expected more than twice the sum. I will send you a prospectus of this establishment[2] in a few days and from it you can learn its nature – the building is immense, it was constructed seven years ago by the Socialists and was by them called Harmony Hall[3] – the adjective however misnomer and the harmony of the establishment was in three or four years converted into utter confusion.[4] On one of the gables were the letters R.O. set in flint – on the other still stand the

1841

letters C of M – The first you will recognise as the initials of Robert Owen[5] the founder of Socialism in England the latter means 'Commencement of Millennium' poor Owen imagined he could mould humanity as he could mould bricks and the foundation of this edifice was to have been the dawn of a new era. We have an estate of 800 acres to practice upon and I take my band out at 3 ½ o'clock and work until 6. I also deliver occasional lectures – I have already given three on physiology and have been fortunate enough to acquit myself with credit. When my course of this subject is ended I shall take up engineering and I promise myself some substantial advantage from the comprehensive reading previously necessary. We have a most able young fellow a chemist,[6] he was formerly demonstrator of chemistry at Putney College,[7] he and I make an exchange of knowledge – he teaches me the mysteries of his art while I teach him mathematics. We meet together every night and read for a certain time: afterwards each questions the other, and thus the matter is planted in our minds. I have also commenced German and promise myself a trip up the Rhine next summer when I shall want to use it. We paid a visit to Stonehenge a few days ago and visited Salisbury Cathedral[8] upon the way. The site of Stonehenge must have been peculiarly suitable for the exercise of druidical rites – it stands in the centre of a vast plain boundless as the sea and unbroken by fence or human habitation – fancy tens of thousands wild hairy Britons collected here with the moon shining down upon them while their demoniac priest plunged his sacrificial knife into his victims heart! I now pause – let me hear from you, and present my very best regards to all at home

Most faithfully yours | Tyndall | I saw Archy McLachen[9] in Southampton a few days ago

RI MS JT/1/TYP/11/3871
LT Transcript Only

---

1. *our late governor*: Richard Carter.
2. *this establishment*: Queenwood College in Hampshire, the school where Tyndall taught from 1847–8. See Introduction to Volume 2.
3. *Harmony Hall*: The building which came to house Queenwood College had previously been used by Robert Owen as the home of a utopian community for equalitarian, communal living. E. Royle, *Robert Owen and the Commencement of the Millennium: A Study of the Harmony Community* (Manchester: Manchester University Press), pp. 122, 181.
4. *converted into utter confusion*: Harmony Hall was financially unsuccessful and closed in 1845. Royle, *Robert Owen and the Commencement of the Millennium*, pp. 205–7.
5. *initials of Robert Owen*: Robert Owen (1771–1858), a Welsh utopian socialist best known for works including *A New View of Society* (1813) and *Observations on the Effect of the Manufacturing System* (1817) (*ODNB*). The Owenite movement had once owned the building and grounds of Queenwood College, but Owen's lavish spending on this and other buildings nearly bankrupted the Owenite movement. See Introduction.
6. *most able young fellow a chemist*: Edward Frankland.
7. *Putney College*: Putney College of Civil Engineering, in London.
8. *Salisbury Cathedral*: one of the earliest built and most beautiful of England's cathedrals, founded in 1220. It is about 10 miles from Stonehenge.
9. *Archy McLachen*: Archibald McLachlan.

To Robert Martin 13 November 1847 0337

Queenwood. | 13th Nov. 1847

My dear Bob,

In a postscript to a letter received a few days ago from Bob Allen he informed me that you were looking very anxiously to the coming year, that you dreaded being thrown out of employment. Now Bob to a man in your position anxiety about this subject is quite natural, but it is possible to dwell too much upon gloomy contingencies, this pondering upon dark possibilities often disarms us of energy. He said wisely who said 'sufficient to the day is the evil thereof',[1] and though matters at the present time may and do look disheartening still the world has bravely weathered far gloomier seasons. You must cheer up Bob. It will never do to suffer your spirits to sink – I have written to Halifax[2] about you, if there is anything doing there I shall be informed of it in a few days. I confess I am not sanguine but if the application fails we must try elsewhere and supposing the worst to come, supposing you are thrown out of employment for a month or two, what then? Just look the difficulty in the teeth and answer me. Here I sit, a young fellow quite disencumbered, with strength enough to assist a friend who has often assisted me, and shall it not be done? I would not give a brass button for friendship which did not lead to this result. Cheer up then, you 'depraved' biped and look world-defiant into the eyes of futurity. Believe me all will yet be well.

Yours most faithfully, | J. Tyndall

R. Martin | Reford Street | Preston | Lancashire[3]

RI MS JT/1/TYP/11/3758
LT Transcript Only

1. *'sufficient to the day is the evil thereof'*: Matthew 6:34.
2. *I have written to Halifax*: i.e., to Tyndall's friends in Halifax, possibly including Thomas Hirst and Tyndall's former employer Richard Carter.
3. *R. Martin | Reford Street | Preston | Lancashire*: address given by Louisa Tyndall.

From Robert Savage 14 November 1847 0338

Achill Sound.[1] | 14th. Nov./47

Dear Sir,

I received this day two post office orders from you, one for £5 and the other for two shillings – for Mr Roberts,[2] and would feel grateful to you for his address – as I would wish to mention to him some of the movements here. I am sorry to inform you that 28 of the Achill fishermen partially under my charge were drowned on the night of the 8th inst.[3]

Yours faithfully, | Robt. Row^{d} Savage.[4]

RI MS JT/1/TYP/11/3857
LT Transcript Only

1. *Achill Sound*: a village on the western coast of Ireland.
2. *Mr Roberts*: not identified. Possibly a relative of Margaret Roberts Ginty. It is not clear why Tyndall sent post office orders for Mr Roberts through Savage.
3. *drowned on the night of the 8th inst.*: the fishermen drowned on the night of 8 November 1847. From 'The Storm and Loss of Life on the West Coast', *Freeman's Journal and Daily Commercial Advertiser,* 18 November 1847, p. 1: 'Monday the 8th November, will be a memorable day to the inhabitants of the sea coast in the county of Mayo ... A company of fishermen on the island of Achill, at Bowford, were preparing to go out with their nets, and two boats and three curraghs had got upon the waters, when the tempest rushed upon them ... Nineteen persons were lost ... The fishermen belonging to Mr. R. Savage, who was from home were making some repairs in their nets, and had the storm delayed another half hour, they too would have been sharers in these dreadful perils'. The source of the discrepancy between the newspaper's death count and Savage's account is unclear.
4. *Robt. Row^{d} Savage*: Robert R. Savage, a fisherman and member of the coastguard at Achill Sound in County Mayo, Ireland. H. E. Hatton, *The Largest Amount of Good: Quaker Relief in Ireland* (Kingston: McGill-Queen's University Press, 1993), p. 135.

To Thomas Hirst 28 November 1847 0339

Queenwood 28th Nov [1847][1]

Dear Tom.

I rose at 6 oC this morning and have worked away ever since paying off my epistolary debts – your account being high in the list demands my attention now. I can well appreciate the treat which Dawson[2] afforded you, he had a splendid array of characters – We had some time since an admirable discourse in our lecture room on the history of one of Dawson's group Pestalozzi[3] the lecturer had the good man's bones in his arms some short time ago. They were removed from an obscure resting place and placed beneath a monument characteristic of the labours of the dead. This method of lecturing has become very prevalent of late years – Carlyle was perhaps the first that introduced it into this country. instead of dealing with the influences which operate upon humanity in their metaphysical and abstract character, the great man is now chosen as the representative of a class, his thoughts, power, and influence form tangible illustrations of general principles and infuse an intense personal interest into discourses of this nature – Switzerland has long been the theatre of great events – at least the play has been often rehearsed there previous to being acted upon a wider stage.

I should be sorry to find you and Mr Carter differing and still more sorry to find you both closing with his present views – his opinions are the summation of all anarchy, the refuge of men who see no hope of human progress or are too apathetic to assist in it. There are men who prefer seeing a smooth orderly exterior, a meek submission to authority however abused, a quiet assent to the established order of things however absurd, to seeing a resolute protest against falsehood and injustice no matter how heavy they may be. M[r] Carter thinks humanity has taken the stereotype form, that the same nest which contained man's chickenhood and afforded shelter to his unfledged intellect must suit him for ever – but M[r] Carter may draw his deductions from too limited a premises – whether he can understand them or not men will arise to shake asunder such respectable fallacies as he has propounded, men who will not stoop to falsehood for the sake of order, well knowing that this is semblance and not substance that in fact as I said before it is merely the sum total of <u>dis</u>order. – I should be sorry to see you an infidel in the bad sense of the term but the taunt is the refuge of weakness against strength, of error against truth and too often of envy against worth – you must be an infidel to many things that you see around you, you must shake away a precious lot of the tag rag and bobtail of opinion before you can expect to lay hold of anything better. there is danger in your position. You will go forward but you <u>may</u> go in the wrong direction. Shall I therefore stop you altogether? God forbid – if so where would be the hope for humanity? –

I read M[r] Hutchinson's letter[4] with much satisfaction he has a strong dash of Germanism in him –

Now Tom after all this will you allow me a word of admonition – dont allow your discussions to interfere with your studies, all speculative things must give way before the practicalities of life – mind this & goodbye

Your sincere friend | John Tyndall

RI MS JT/1/HTYP/3
RI MS JT/1/T/941

1. *1847*: year given by Louisa Tyndall.
2. *Dawson*: George Dawson (1821–76), a preacher and political activist who gave a series of lectures in Manchester in 1846 and 1847, some of which Tyndall attended (see Tyndall's journal, RI MS JT/2/13a/168-9). See also letter 0342. From Tyndall's journal, 26 November 1847: 'A letter from Tom Hirst containing an account of Dawson's lecture on the Characters of Switzerland' (RI MS JT/2/13a/270).
3. *Pestalozzi*: Johann Heinrich Pestalozzi (1746–1827), a Swiss educational reformer.
4. *Mr Hutchinson's letter*: letter missing.

To Thomas Hirst and James Craven 11 December 1847 0340

Queenwood, Saturday night[1]

My Dear Boys–

I send you a number of our reporter which appeared today – the mechanical portion of it will perhaps interest you – the class to which the questions were proposed consists of boys from 12 to 15 years of age on this account the matter is purely elementary – might I request Jemmy[2] to make himself master of these 12 questions – in subsequent numbers we shall advance to more elaborate calculations – It is a long time since I heard from Halifax, I dont know what is going on there – have you any Chartist[3] meetings in the Cattle Market? that affair in London passed off very peaceably – what a waspish little fellow Ernest Jones[4] is – I have little faith in a movement whose materials are held in such red hot crucibles – I read some time ago a magnificent lecture from Mialè[5] on State Education the general principles he laid down were excellent but I think he shot himself wide of his mark – all his arguments might be readily turned upon himself – – I shall be much obliged if you send me a long letter containing an abstract of your local history for the last 2 months – how is Lord?[6] how is McArthur?[7] – Tom Perkinton[8] wrote me such a brief note[9] some time ago that he actually froze the fluid on my steelnib – it will gush out upon him some of these fine days –

believe me dear | Tom and Jim[10] | most truly yours | J. Tyndall | How are all my friends at Carlton Street?

RI MS JT/1/HTYP/4
RI MS JT/1/T/984

1. *Saturday night*: Louisa Tyndall annotation: 'probly Dec 11/47'.
2. *Jemmy*: James Craven.
3. *have you any Chartist*: The People's Charter of 1838, whose proponents became known as Chartists, demanded democratic reform of the political system. Chartism was a national British movement in support of the People's Charter that was highly active from 1838–48.
4. *Ernest Jones*: probably Ernest Charles Jones (1819–69), a writer and leader of the Chartist movement (*ODNB*).
5. *Mialè*: not identified.
6. *how is Lord?*: not identified.
7. *McArthur?*: not identified.
8. *Tom Perkinton*: probably the same 'Perkinton' who surveyed with Tyndall in Halifax; see, for example, Tyndall's journal entry for 29 June 1846 (RI MS JT/2/13a/131).
9. *Perkinton wrote me such a brief note*: letter missing.
10. *Jim*: another nickname for James Craven.

From Henry McClintock 28 December 1847 0341

Macloneigh Glebe, | Macroom.[1] | Dec. 28 1847

Dear Sir,

I hope you will excuse me for not having replied to your kind letter[2] of the 20th inst.[3] before now, but I have been so engaged by Christmas duties that I really had not time to do so.

The misery and destitution of my large and impoverished parish is beyond description great, and I would gladly and thankfully avail myself of any help that I could procure for the poor people. But, before sending you any statement respecting the distress around me, I feel that it would be but right that I should be aware whether you are acting in this work of mercy as a benevolent individual, or on the part of a charitable Society. I trust that you will not blame me for asking this question, which makes me feel quite ungrateful after your kind letter, but I should not like to appeal to the public without knowing exactly whether it was to be in a private manner through you, or through some of the many charitable Associations.

I recollect the time you allude to, when you were one of my class at Castle Bellingham,[4] and wishing you many happy new years, I remain,

Dear Sir, faithfully yours, | Henry F. McClintock.[5]

John Tyndall Esq.

(Have the goodness to enclose this when you next favour me with a letter. | J.T.)[6]

RI MS JT/1/TYP/11/3761
LT Transcript Only

---

1. *Macroom*: a market town in southern Ireland.
2. *your kind letter*: letter missing. From Tyndall's journal, 20 December 1847: 'Wrote to Mr McClintock requesting some information relative to the poor of his district' (RI MS JT/2/13a/285). Tyndall was visiting Ireland at the time and was dismayed by the poverty and malnutrition he observed.
3. *letter of the 20th inst.*: the letter was from the 20th of December.
4. *Castle Bellingham*: also spelled Castlebellingham, a town in County Louth, Ireland where the Tyndall family lived from 1833–6. From Louisa Tyndall's unpublished biography: 'In 1833 the Tyndalls were transferred to Castle Bellingham, Co. Louth, where they remained until the autumn of 1836. The teaching in the school which Tyndall attended was in many ways defective, and was not of a quality to attract' (Louisa Tyndall, 'Biography', vol. 1, p. 12).
5. *Henry F. McClintock*: (?–1879). An Irish cleric whom Tyndall evidently knew in Castle Bellingham. McClintock died in Macroom in October 1879.
6. *(Have the goodness to enclose this when you next favour me with a letter. | J.T.)*: this postscript is included in Louisa Tyndall's transcript; it is unclear who made this request, or exactly what is meant by it.

1848

To Thomas Hirst 1 February 1848 0342

Tuesday night[1] 1848

Dear Tom

I handed your question over to a youth[2] who commenced geometry with me about 3 months ago – I send you his demonstration without addition or diminution – I trust you will understand it – Well, Kuppers[3] has gone, and the probability is that we shall see his face no more – I was very much attached to him and though at a distance from Halifax, his leaving it affects me as if I had just bidden him good bye –

– In Emerson you behold one of the noblest souls that ever was stuck in clay – every time I rise from his book[4] I find a new *[vigour]* in my heart – he teaches one to be so independent that you almost feel disposed to quarrel with himself just to shew how little you cared about even him – There are many parts of his writings very difficult, especially some portions of the Transcendentalist, and Idealism[5] – the rule he lays down will I believe make all clear – let us by enacting our best insight by doing that which we feel to be right strengthen our powers and purify our vision, and all will be understandable – there is a world of meaning in those two words he uses so emphatically I ought[6] – you remember Dawson's[7] reference to a principle of duty[8] – and see its identity with the views of Emerson – In his 'Christian Teacher' a passage relative to Jesus Christ occurs which surpasses any thing I ever read on the same subject – It is difficult to deal with such a class as yours. I see no way open to you but to allow each to push forward to the best of his ability – it would be a pity to harness a racer to a slow coach – keep them in sections of threes and fours if possible & thus enable yourself to address this number at the same time. I may add that the difficulty lies at your own feet and you alone must meet it – I shall however be happy at all times to render you any assistance in my power – I would recommend you to accompany Dr. Inglis[9] throughout his lectures – purchase a few things and repeat his experiments yourself – I see no better way of fastening the thing in your mind – the apparatus will be very cheap, bits of chalk, iron filings, clippings of copper, and such mean things are the chemists principal materials – Could you contrive to send me tracings of those small sections which Tidmarsh reduced? If so I should be much obliged – Send me the Guardian[10] by all means –

Good bye & bel[11] me dear Tom | Sincerely yours | J. Tyndall | Is Mr C.[12] still in London?

<Handwritten enclosure missing; LT Transcript only henceforth>

Given the chord[13] of a circle and its versed sine to find the diameter.[14]

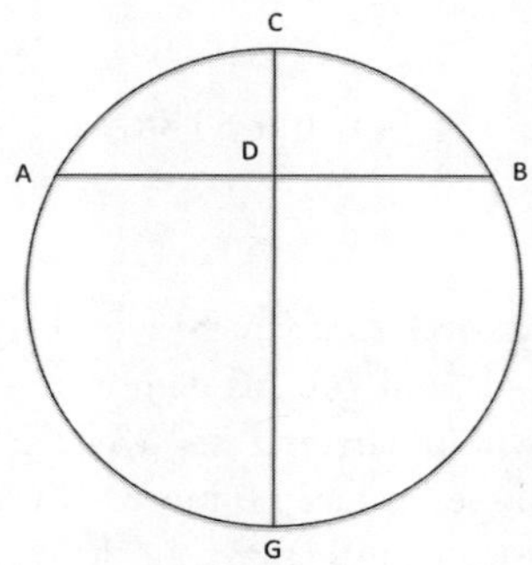

Let (A B) be the chord (C D) the versed sine, which bisects the chord (A B) and is perpendicular to it, consequently if produced it will pass through the centre of the circle.

Produce it to (G) then, having the chord (A B) given the square of half of it will be the rectangle under (A D. D B) but by the 25th. of the (3rd.) the rectangle (A D. D B) = the rectangle (C D. D G). Thus we have the area of the rectangle (C D. D G) and one side (C D) to find the other side (D G). By dividing the area by the side (C D) we get the side (D G); which added to (C D) completes the diameter of (C G) hence the rule.

Divide the square of half the chord by the versed sine, to the quotient add the versed sine, and the product will be the diameter.

J. Duckwarth.[15]

M^r T Hirst | Northgate Offices | Halifax | Yorkshire

RI MS JT 1/HTYP/5
RI MS JT/1/T/512
Portions of letter are LT Transcript Only

1. *Tuesday night*: Louisa Tyndall annotation: 'Tuesday night Feb. 1. (p. mark) 1848'.
2. *a youth*: possibly the J. Duckwarth whose name is on the enclosed piece of mathematical work. Almost certainly a student from Queenwood, the school where Tyndall taught from 1847–8. See Introduction.
3. *Kuppers*: Conrad Kupper.

4. *every time I rise from his book*: Tyndall was an avid admirer of Emerson's works; indeed, he is said to have read all of Emerson's writings. R. A. Haugrud, 'Tyndall's Interest in Emerson', *American Literature*, 41:4 (January 1970), pp. 507–17.
5. *Transcendentalist, and Idealism*: Emerson delivered an address entitled 'The Transcendentalist' that was first presented in 1842. It is possible that Tyndall is referring to this piece, though this would imply that he had access to a pirated copy of 'The Transcendentalist', as it was not yet published legally in Britain. Alternatively, Tyndall may be referring to Emerson's participation in the American Transcendentalist movement. This movement was characterized by a belief in the essential goodness of man and nature. Idealism is a philosophical concept that emphasizes the importance of assessing the ideal in relation to the reality of experience.
6. *two words he uses so emphatically – I ought*: If Tyndall is indeed referring to Emerson's 'The Transcendentalist', this phrase appears therein: 'The popular literary creed seems to be, "I am a sublime genius; I ought not therefore to labour."' R. W. Emerson, 'The Transcendentalist', *Nature: Addresses and Lectures* (Boston: James Monroe and Company, 1849), p. 337.
7. *Dawson's*: George Dawson; see letter 0339, n. 2 of this volume.
8. *Dawson's reference to a principle of duty*: Tyndall's journal entry 10 December 1846: 'Making plans for valuation went to the old assembly room at 8 o'clock to hear Mr Dawson lecture on the life and times of Oliver Cromwell. The lecturer is a young man, rather low crowned, but thoughtful looking – his address was most interesting tho' simple and without declamation. I anticipated something different having heard his name linked with that of Carlyle – They don't cleave the clouds together – he adduced the charges made against Cromwell and those he did not refute he left without a sting – He is evidently of Carlyle's school, "a great law of duty impelling him onward"' (RI MS JT/2/13a/168).
9. *D^r^. Inglis*: not identified.
10. *the Guardian*: possibly *The Guardian*, a British newspaper founded in 1821.
11. *bel*: an abbreviation for 'believe'.
12. *Mr. C.*: probably Richard Carter.
13. *Given the chord*: Louisa Tyndall annotation: 'Enc'. This scrap appears to have been enclosed in Tyndall's letter to Hirst.
14. *find the diameter*: the following illustration is a reconstruction based on a sketch from the Louisa Tyndall transcript.
15. *J. Duckwarth.*: not identified; presumably the student Tyndall mentions at the beginning of the letter.

From Frederic Richard Lees 5 February 1848 0343

Leeds, Feb. 5/48

Dear Sir,

I should have been happy to have made your personal acquaintance; but, in default of that, I shall not fail to communicate with you by means of Transcendentalism[1] and Truth-Seeking.[2]

Should you be able to obtain other subscribers,[3] I shall be glad.

Yours very truly, | Fred^c^ R^d^ Lees.

RI MS JT/1/TYP/11/3746
LT Transcript Only

1. *communicate with you by means of Transcendentalism*: transcendentalism is a philosophical orientation characterized by a belief in the essential goodness of man and nature, most often associated with the writings of Ralph Waldo Emerson. Lees may also have been referencing a British periodical about Transcendentalism, although the specific periodical has not been identified.
2. *Truth-Seeking*: Lees edited a transcendentalist publication entitled *Truth Seeker* from 1844 to 1850 (*ODNB*).
3. *obtain other subscribers*: Lees is probably referring to subscribers for *Truth Seeker.*

From Josiah Singleton [19 or 20] February 1848[1] 0344

My dear son John

full many a day
Has come, rolled on, and passed away
To swell the tides of that dread sea
Which mortals call eternity
(To which not only days but all
Who tread this dark terrestrial ball
Are floating, down life's ceaseless stream
And fading like the hurried dream
That melts before the glowing beam
Of mornings earliest ray)

Yes many a day has come and gone
Rolling its silent moments on
Loaded for some with bitter care
Or hope deferred, or dark despair
Or, chance, amid life's changing scene
Those fleeting moments may have been
Alas how short their stay!

Full freighted with earth's small delights
As when some loving heart unites
With heart of kindred passions
In bonds, how transient, but as bright
As nature ever fashions
Full swiftly by these days have fled
Numbering their thousands with the dead
I hope, friend, you're not tired
For I have not as yet begun
What has and what has not been done
How many have been fired
With schemes as baseless as the air
Or castles that one builds up there
Pretty while building one may swear
But which, confound them will not bear

A moment's retrospection
I've yet said nothing 'true enough'
I hear you say a pack of stuff
Almost without connexion
I'm much afraid if truth be told
My father's[2] brains are growing old
Or something very like it
His harp I know in former days
Resounded to far nobler lays
With bolder hand to strike it
It may be so my son but still
What mortal with the noblest will
To climb the steep Parnassian hill[3]
Such labour could endure –
As to supply a craving pen
With language, ideas and then
Drink only aqua pura!!!
Please to remember you took care
With steaks and chops and such like fare
To feast the inner man
With brandy too on coach's top
The ravages of cold to stop
But here's a different plan
No, brandy, no, nor wine nor beer
Nor porter nor such vulgar cheer
But water pure and bright and clear
Although I must not say I fear
No better ever ran
But to return, full many a day
Has come, rolled on and passed away
Since your epistle came to hand[4]
Posted in Cork 'Ould Ireland'
And many a time and oft I thought
Of paying as I knew I ought
All that to you was due John
But then 'tis rather hard to send
Your letter to a distant friend
This you'll allow is true, John
When all the while you do not know
Where it has pleased that friend to go
I could not send, could you John?
But now your whereabouts is clear
I'll shortly send you never fear
Half dozen for your six
And if you're careful to reply
It seems to me that you and I
Will correspond like bricks.

You've seen a mummy 'praps, my son
At any rate a sketch of one
    But those have all been dead
But if you'll come to my bed side
In Dr Weeding's[5] house in Ryde[6]
You'll have a chance each other day
In spite of what the sceptics say
    To see one there in bed.
For first they strip me to the skin
Then wrap a sheet from foot to chin
    Quite wet and cold around me
And then with blankets five or six
They wrap me round and in that fix
    They leave me where they've bound me
Thus hands and arms and legs all fast
This durance has an hour to last
    And sometimes two or near it
And if your nose should itch and want
A scratch, your hands are fast you cant
    There's nothing but to bear it
Then I've a splash bath very nice
Get well rubbed down and in a trice
    Am dressed and off and walking
Then, breakfast, till eleven I sit
And read or write or chat a bit
    Tho' seldom I've much talking
Then comes the man with cold wet sheets
    To clothe my naked shoulder
At first it brought a sobbing fit
    But now I'm grown much bolder
And then I walk and then I dine
And then I labour at the line
A writing here or writing there
Wishing to write nigh everywhere
Till four o'clock when James[7] again
Comes with the sheet and stops my pen
Then finish letters, then to post
For fear a single day be lost
And then I walk and tea and read
Or write till <u>time to go to</u> bed
And now farewell – 'tis time I'd done

I am | Your friend | J. Singleton.[8]

RI MS JT/2/13a/292–294
LT Transcript Only

1. *[19 or 20] February 1848*: from Tyndall's journal: 'The following reached me on the 21st' (RI MS JT/2/13a/292).
2. *my father*: William Singleton, a schoolmaster at Broom Hall, Sheffield. C. A. Russell, *Edward Frankland: Chemistry, Controversy and Conspiracy in Victorian England* (Cambridge: Cambridge University Press, 1996), pp. 39, 62.
3. *climb the steep Parnassian hill*: Parnassus is a mountain in central Greece which, according to mythological tradition, is a home to music, literary creativity, and poetry (*OED*).
4. *Since your epistle came to hand*: letter missing.
5. *Dr Weeding's*: not identified.
6. *Ryde*: a British seaside town on tie Isle of Wight.
7. *James*: not identified.
8. *J. Singleton*: Josiah Singleton, another teacher at Queenwood. He was also the brother of Anne Singleton Edmondson and brother-in-law of George Edmondson. See Introduction for more on Singleton at Queenwood.

To Thomas Hirst 1 March 1848 0345

Queenwood Mar 1/48.

Dear Tom

We are in want of a young fellow here[1] to assist in the Junior Classes and as I was thinking the matter over between 4 and 5 oC this morning I remembered Booth[2] of the Bird Cage – I should like to push the lad forward if I could – do you think he would be willing to join us, and if so would he be eligible? – Is he a good arithmetician? does he spell and read well and has he any Knowledge of mathematics. What religion is he and what is his general character, do you think he could render himself engaging among lads? Could he procure a testimonial from some respectable person in Halifax. In short I want as full an account of him as possible and I think you are the most likely person to furnish such an account. Should he come he will have opportunities of

<Handwritten pages missing; LT Transcript only henceforth>

advancing himself which would be difficult to attain elsewhere – he will attend our chemical lectures and have access to a valuable philosophical apparatus – What is he able to earn now and what salary do you think he would expect if he came here (x) Now if you can manage to furnish me with this information, without exciting any hopes in Booth that might not be realized, we might be able to open a way for his abilities if he have any. His demand as to salary must be low as Mr. Edmondson has already had an offer of this kind from a party in Lancaster.

Thank Jemmy[3] for me for his punctuality with regard to the Tracings.[4] I wish I could get a diagram of the gradients of the Manchester and Leeds[5] and the West Riding Union.[6]

With good wishes | I am dear Tom | J. Tyndall

(x) he will be boarded in the house.

RI MS JT/1/HTYP/6
RI MS JT/1/T/513
Portions of letter are LT Transcript Only

1. *We are in want of a young fellow here*: i.e., at Queenwood College, the school where Tyndall taught from 1847–8. See Introduction.
2. *Booth:* Francis Booth, a friend of Hirst's in Halifax.
3. *Jemmy*: James Craven.
4. *with regard to the Tracings*: Craven was a surveyor, so this probably refers to some survey work he had undertaken.
5. *the Manchester and Leeds*: The Manchester and Leeds Railway was founded in 1839.
6. West Riding Union: a railway line that became part of the Manchester and Leeds Railway in 1846. Sir F. Pollock, R. Campbell, O. A. Saunders, A. B. Cane, J. G. Pease, W. Bowstead, and E. Potton (eds), *The Revised Reports: Being a Republication of Such Cases in the English Courts of Common Law and Equity* (Boston: Sweet & Maxwell, Limited, 1908), p. 385.

To Thomas Hirst and James Craven 9 March 1848 0346

9th Mar. 1848.

My Dear Sons,

Your last two letters[1] caused me no small exhilaration of soul – I am rejoiced to hear that you are delving into those brains of yours,[2] you will find there a richer product than is yielded by either the mines of Low Moor or the quarries of Southowram[3] – of Tom I never had much fear, but of Jemmy[4] my hopes often sunk to zero – they are now I confess rising and that more on account of his reading than of his republicanism[5] – you did wisely in joining the Mechanics Institute,[6] but let me still keep this as a frontlet before your eyes, that a connexion with a literary Society is valueless if it do not prompt to individual exertion – appropos subject of discussion 'Individual Competition as opposed to Co-operation' the last may be good – the first cant – England however may melt into one wide cooperative league but it will prove a stagnant pool if individual culture be unheeded –

Jemmy cant make his epistle egotistical by letting me know what he is doing. I am much pleased to find that he has attacked Euclid.[7] how far has he got, and what time does he lay out for the completion of the 1st 6 books?[8]

I should like very much to receive a brief abstract of the amount of information he has derived from Joyce[9] –

I know of no method so good as devoting with firm purpose a portion of each day to your studies you are sure to get along even though the time you study may be short. I am sometimes struck by the bulk to which my short Journal entries[10] have swelled, simply because they were regularly made –

Jemmy asks me for a subject for an essay, this I shall not attempt to give – I would advise him to choose one which is embraced by his own feelings, it would be far better if he would look into his own spirit for a subject than if he could commit all Howitt's articles[11] to memory

The subject of capital punishment I dont like, it is so hackneyed and such a load of twaddle has been heaped upon it – the question I conceive can only occur in two lights 1st is it just? 2nd Is it useful the 1st is of course the true standard and if Wm Howitt[12] be able to prove the negative he has more brains than I gave him credit for.

Neither can I see the great utility likely to flow from the discontinuance of hanging –at all events I believe that utility to be utterly disproportionate to the fuss kicked up about the matter. There is great danger of kind hearted generous young men being led away by this tinfoil liberality – the only check I see to this is for them to seek knowledge from themselves and not from others – to take the trouble to form their own opinions which on many subjects must be a work of time and patience. Could not Jemmy take a subject within his scope, for instance the school and the office – the student and the apprentice, and give his hearers a little of his own fresh feeling upon the matter? – And here I set down a parting word for Jemmy – dont let me lead you away – if your opinions be different from mine let your own be your guide and not mine – If you think Howitts articles good pursue the subject by all means –

I respect that feeling on the part of Booth[13] though he certainly has relinquished great advantages –

I care very little for such a movement as the French revolution only so far as it is indicative of the popular current beneath – as an end it would be pitiful – as a means it may be great – the provisional Government are however about to grapple with some terrible questions. they guarantee employment to the workman, if they can solve this question they will deserve eternal honour – I dont like their motto, liberty and equality[14] – as long as society is composed of such heterogeneous particles, we shall now and then want some vigorous heel to trample it into order – but I have spun myself too far

Good bye & believe me | Sincerely yours | J Tyndall

RI MS JT/1/HTYP/7
RI MS JT/1/T/514

---

1. *last two letters*: letters missing.
2. *delving into those brains of yours*: Tyndall was mentoring both Hirst and Craven as they undertook a program of self-education. See Introduction to Volume 2.
3. *mines of Low Moor or the quarries of Southowram*: Low Moor and Southowram are both villages in West Yorkshire, England.
4. *Jemmy*: James ('Jemmy') Craven.
5. *republicanism*: taken in the Irish context, this refers to support for Irish independence from Great Britain (*OED*).
6. *the Mechanics Institute*: Tyndall had taken night-time classes at the Mechanics' Institute in Preston in 1842 (*ODNB*).

7. *Euclid*: Euclid's *Elements,* a Greek treatise on geometry written around 300 BC. Editions of Euclid's treatise were still used to instruct students in geometry in the nineteenth century.
8. *1st 6 books*: probably O. Byrne, *The First Six Books of The Elements of Euclid* (London: William Pickering, 1847).
9. *Joyce*: not identified; possibly J. Joyce, *A System of Practical Arithmetic* (London: 1838), an elementary textbook.
10. *short Journal entries*: Tyndall kept a journal, in which he regularly recorded his daily activities.
11. *Howitt's articles*: William and Mary Howitt founded *Howitt's Journal* in May 1847. This journal focused on radical politics and popular literature (*ODNB*).
12. *Wm Howitt*: William Howitt (1792–1879), a writer from Heanor, Derbyshire (*ODNB*).
13. *Booth*: Francis Booth; see letter 0345, n. 2 of this volume.
14. *liberty and equality*: the motto of the French Revolution was, 'Liberté, Egalité, Fraternité'.

From Frederic Richard Lees 22 April 1848 0347

Edinbro' Apr. 22/48

My dear Sir,

Your letter followed me here,[1] on my way to the Highlands; I am obliged for your kind offer, and will avail myself of it, if needed.

The Truth-Seeker[2] can be sent after you to Germany, for a small postage (½d I think, per copy), if you like. The subscription is 3/6 half year, 7s. per year, which you can remit to Leeds.

As to Liebig,[3] it is his proved theory of Respiration,[4] etc., which (as shewn in the works referred to) complete our proofs, and clears our subject of every difficulty. That theory was all the better for saying nothing of Teetotalism[5] directly. In his 1st Editions[6] he showed that Alcohol could not nourish – that it 'robbed the blood of oxygen', impaired mechanical force as an antiseptic – (which accounts for its lessening appetite, i.e. reducing life by opposing change) etc.

There is nothing in the Note in the last Edition[7] (which I had seen) that alters the case; for I beg to differ from his mere hypothesis or explanation.

It is quite true that Alcohol is partially decomposed in fuel, but thereby it prevents the natural elements of food (as fat, which is both cheaper and better) from being consumed; and 'experiments' show that it actually diminishes the heat of the blood, which experience confirms.

Besides, before it is burned up itself, it excites (i.e. poisons) various organs, and venomizes the blood: and therefore is bad fuel.

The fact of a Master (by the aid of tobacco, alcohol, or opium) saving natural food required by his servants, does not prove the physiological fitness of alcohol, but the reverse, for a good appetite, exercise, rapid waste or change of matter, etc. all indicate health and its enjoyments. But as he admits that Fat is a better and cheaper fuel than alcohol, by 22 per cent, and yet as the use of alcohol by servants leads to their desiring less fat – it follows that the diminished desire arises from diminished vitality. His theory might be very well for the Master, but not for the slave!

True, therefore, alcohol preserves fat; but so does a pantry; and I don't see why we should carry our fat (or pantry) about with us, when we can everywhere get fat when we need it.

The experience of the Arctic whalers, and Ross[8] and Parry's[9] people, show that men can better resist cold by fat than by alcohol. Hence, though I too am thin, I prefer to buy any fat I want for my vital lamp, at the butter or bacon shop, to burdening myself with it in my rambles over hill and dale.

Yours very truly, | F.R. Lees.

RI MS JT/1/TYP/11/3746–3747
LT Transcript Only

---

1. *Your letter followed me here*: letter missing; see letter 0343 of this volume.
2. *The Truth-Seeker*: Lees edited a transcendentalist publication entitled *Truth Seeker* from 1844 to 1850 (*ODNB*).
3. *as to Liebig*: Justus von Liebig (1803–73), a German organic chemist and professor at the University of Giessen. Liebig is most famous for his innovative teaching laboratory at Giessen.
4. *his proved theory of Respiration*: Liebig's theory of respiration was actually somewhat controversial at the time. See W. Brock, *Justus von Liebig: The Chemical Gatekeeper* (Cambridge: Cambridge University Press, 1997), pp. 183–214.
5. *Teetotalism*: complete abstinence from alcoholic beverages (*OED*). Lees was a prominent advocate of teetotalism.
6. *1st Editions*: probably the first edition of J. von Liebig, *Die Organische Chemie in Ihrer Anwendung auf Physiologie und Pathologie* (Braunschweig: Verlag Friedrich Vieweg u. Sohn, 1842). This was translated into English as J. von Liebig, *Animal Chemistry, Or Organic Chemistry in Its Application to Physiology and Pathology,* trans. W. Gregory (Cambridge: J. Owen, 1843).
7. *last Edition*: the most recent edition of either *Die Organische Chemie* or its English translation, *Animal Chemistry.* Specific edition not identified.
8. *Ross*: Sir John Ross (1777–1856), a naval Arctic explorer and whaler who participated in the search for the North West Passage in the ship *Isabella* (*ODNB*).
9. *Parry's*: Lieutenant William Edward Parry (1790–1855), who captained the ship *Alexander* with John Ross in the exploration of the Northwest Passage (*ODNB*).

To Thomas Hirst and James Craven 2 June 1848 0348

Queenwood, 2nd June 1848.

My Dear Boys –

I am sorry to break in rudely upon your anticipations – I did promise myself the pleasure of seeing you at midsummer, and believe me the word 'pleasure' is not used in its conventional sense, but I find that this is now impracticable – I intend to spend 5 weeks among the republicans of Paris[1] – I have improved very much lately in French conversation and to extend my knowledge in this particular will seize upon the present opportunity – with regard to the chemicals I dare say you would like to perform some pretty experiments. Mind however a knowledge of chemistry is not to be got by such experiments

if there be any course indicated by the lectures you have heard which you would like to pursue, I shall be glad to lend you all the assistance in my power – if you would wish merely to make oxygen, hydrogen, explode soap bubbles, cause sulphur to detonate <u>and</u> kick up sundry rows, I will point out what apparatus you will require – *[6/.]* will go a good way upon these matters as the substances we use at first are cheap – I will answer Tom's questions[2] to the best of my ability when I have a little more time. I am at present getting up half a dozen maps for the lads to take home at midsummer and this absorbs all my time – if Jimmy[3] has patience with the 2nd book[4] he will trample its difficulties under foot – if I were near him I would gladly make his track more smooth but it will be better for himself if he can get through it without assistance

Believe me my dear Tom & Jemmy | most sincerely yours | J. Tyndall

Tom's remarks as to the necessity of modesty are quite correct but Hudibras[5] says

all smatterers are more brisk and pert
Than they which understand an art
As little sparkles shine more bright
Than glowing coals which give them light[6]

and I feel satisfied that the man young or old who manfully wrestles with a difficulty will be more truly humble than he who accepts another man's solution – the very fact of his applying himself to the naked case argues a free enquiring spirit, to such a one nature is full of unanswerable problems, and comparing his actual state with his ideal – what he is with what he might be, he is effectually prevented from being puffed up – however the lecture has a special object in view which if Tom was aware of it would clear up the matter to his satisfaction –

RI MS JT/1/HTYP/8
RI MS JT/1/T/515

---

1. *5 weeks among the republicans of Paris*: according to his Journal, Tyndall began his trip to France on 15 June 1848 (RI MS JT/2/13a/338). Edward Frankland accompanied Tyndall on this trip. The February 1848 revolution in France had ended the Orleans monarchy and established the Second Republic. In June 1848, the citizens of Paris rebelled against a conservative turn in the government.
2. *Tom's questions*: letter missing.
3. *Jimmy*: James Craven.
4. *2nd book*: probably the second book of O. Byrne, *The First Six Books of The Elements of Euclid* (London: William Pickering, 1847); see letter 0346 of this volume.
5. *Hudibras*: a satiric poem written by Samuel Butler and published in several parts beginning in 1663. The poem satirizes the Presbyterian Church and the Parliamentarian posi-

tion on the Civil War.

6. *All smatters ... them light*: It appears that Tyndall was imitating the style of *Hudibras* rather than directly quoting from it, as these lines do not appear in Butler's poem.

From Anne Edmondson 6 June 1848 0349

To Edward Frankland and John Tyndall

My much esteemed Friends –

On the occurrences of the last few days,[1] which have distressed us beyond what you could have intended or imagined, I take the liberty of addressing you. I am prompted to do so by my warm attachment to yourselves by the interest I take in the individual who is the presumed cause of all this and by my very natural anxiety for my husband's[2] happiness and what is inseparable from it, his success in his new undertaking.[3] Under the influence of these combined feelings, I am emboldened to ask you to pause and reflect ere you come to a final decision to withdraw your influence beneficial as we deem it from this infant establishment. Are you sure that your motives for making a change will bear calm investigation? Have you endeavoured to remedy the evils of which you complain, and if not are you acting as you would that others should act by you?

If I understand the matter aright you assign as the cause of your leaving us the conduct of a certain party[4] which is such as to risk best interests of the establishment. You refuse to lay before him the grounds of the opinion you have formed of him but expect that your view of the case shall be conclusive: thus you place his employer in very embarrassing circumstances, either he must dismiss the individual unconvicted of error, to struggle for bread as he best may a comparative stranger in his own country, or he must retain him at equal risk (as you believe) since his presence is detrimental to the welfare of the place. Why you refuse to speak to the individual of faults which you do not scruple to lay before his employer I am at a loss to imagine, or why, as I have since heard you determine whether he remains here or not you still must go I cannot understand but I think if you take the trouble to estimate if such be possible the amount of injury which one party or both must sustain should you persist in your determination you will be startled and perhaps think it well to ascertain if things be as you imagine.

Excuse me if you think I am interfering beyond a woman's province in thus writing, the office of peace maker is confined to neither sex and to that only do I aspire. I would if I might diffuse over your ruffled spirits the soothing influence of a clear conscience in this affair, for though you may now think the remark uncalled for believe me there will a time come whether you stay or go when you will be thankful to be able to look back to this period without self condemnation. Allow me then without grieving you to say how I view this matter. I think in speaking of a fellow labourer to his employer in the manner you have done without giving him an opportunity of defending himself you have acted contrary to the spirit of the laws of our country, contrary even to the world's sense of honour and diametrically opposed to the precepts of our holy religion.[5] I think were you in the place of the accused you would agree with me and condemn any decision against you in consequence of the accusation as unjust and partial. I believe could you divest of prejudice towards the individual state openly to him your objections and hear his statements in reply you would find much which you have been commenting upon very harshly would bear a different construction and you would be brought nearer to our view of his character and conduct.

We have never been as you suppose so blinded by partiality as not to see faults in him but whatever we have seen we have unscrupulously pointed out to him and hitherto we have not met with such as we should deem irremediable. If you have reason to think otherwise where is your friendship for us and your interest in our establishment if you do not enable us to point them out to him likewise? I fear my friends I shall both weary and grieve you but my motives must be my excuse, I am unwilling to think that we must necessarily be parted from you. I esteem you too much to think with indifference on the probability of your places at our family board being occupied by others, neither would I willingly lose the co-operation of the accused and his friend;[6] and I believe this storm once blown over this prejudice laid aside there would nothing remain incompatible with your working together as closely as would be required to conduce to the general harmony.

This is what I would wish to see brought about not as a favour to myself (I seek none) but because I believe it to be the right termination of the unpleasant affair and because it would be the only proof you could give of restored confidence in us. In conclusion I must allude to another circumstance which my husband mentioned to me. You say you have by the uncomfortable state of things been induced to resort to a practice to which you were formerly unaccustomed[7] you know to what I refer, I am not sure that you mean by this anything that I might have remedied, but if you do you cannot confer on me a greater kindness than to point out wherein I have failed and what I can still do for you and your successors to make Queenwood[8] a more desirable home. Finally allow me to repeat though I may have grieved you by my plain speaking it is my sincere regard for you that has prompted me to risk your displeasure, my husband is ignorant of my intention of writing therefore the responsibility is entirely my own, pass this into his hands if you think well but feel quite at liberty to do so or otherwise as you please with that exception it is intended for your own perusal only.

I am your sincere friend | and well wisher | Anne Edmondson

RI MS JT/2/13a/331–333
LT Transcript Only

---

1. *the occurrences of the last few days*: see letter 0350 of this volume for Tyndall's account of these events. Furthermore, according to Tyndall's journal, he and Frankland had told George Edmondson of their intent to leave Queenwood on 5 June 1848 (RI MS

JT/2/13a/331).
2. *my husband's*: George Edmondson.
3. *his new undertaking*: i.e., the newly-opened Queenwood College, where Tyndall and Frankland were teaching.
4. *a certain party*: John Yeats, a teacher at Queenwood. According to Tyndall's journals, Yeats had engaged in a campaign to persuade Edmondson to fire another teacher, Josiah Singleton (RI MS JT/2/13a/328–331); see Introduction to Volume 2. See also C. A. Russell, *Edward Frankland: Chemistry, Controversy and Conspiracy in Victorian England* (Cambridge: Cambridge University Press, 1996), p. 57.
5. *our holy religion*: Anne and George Edmondson were both members of the Religious Society of Friends, or Quakers.
6. *his friend*: not identified.
7. *a practice to which you were formerly unaccustomed*: allusion not identified.
8. *Queenwood*: the school where Tyndall and Frankland taught from 1847–8. See Introduction to Volume 2.

To Anne Edmondson 8 June 1848 0350

Respected Madam

The subject of the letter[1] with which you have favoured us is too grave to permit of our wasting much time in prefatory matter, we shall therefore enter at once upon the discussion of this to us truly painful question.

You ask us to pause and reflect, we are ready to comply with your request, nay, we have done so, our determination[2] is by no means an impulsive one: whether our opinions be right or wrong, of this we assure you they have not been hastily formed; our minds have been made up on this subject for months and though recent events may have precipitated matters, our resolution was formed altogether irrespective of these events. We believe that our motives will bear any investigation to which you may think proper to subject them. The very nature of things proves that these motives must have been disinterested, what personal advantage could we expect to reap from any change of school government that Mr E.[3] could possibly make? Certainly none. Our sole motive then is a dictate of duty in obeying which each of us can truly affirm that he acts in a manner directly opposed to his pecuniary interests.

We are at a loss to conceive what is meant by the expression 'you expect your opinion to be conclusive',[4] if it is intended to convey that you imagine us to expect you and Mr Edmondson to subscribe to our opinion we beg to affirm that we have never entertained a hope of the kind. So far from wishing our opinion to be conclusive our intention was to keep it entirely to ourselves and this we certainly should have done had not circumstances called for its utterance. Chimerical as we deem the idea of the gentleman's dismissal[5] we may in passing say that it is a circumstance which we should earnestly deprecate, we believe that he deserves better treatment at your hands. Whatever his motives may have been, whether personal or otherwise he has exerted himself to the utmost of his ability to promote the interest of Queenwood,[6] let the policy, prudence or honesty of the course pursued be questioned as it may, one thing is certain that it has been followed with zeal and activity.

Most heartily do we subscribe to the inference which flows immediately from the following remark 'I would if I might diffuse over your ruffled spirits the soothing influence of a clear conscience in this affair', whatever may be here implied our experience enables us to recognise the philosophy of the sentence as true. The calm that we now enjoy is we believe the direct offspring of a clear conscience, we may possibly labour under a mental hallucination but our peace of mind is undisturbed, if we sin we do so unwittingly, our inward sentinel has become a dumb dog, nay worse, a deceiver, for he stamps with his approval all that we have said and done.

We have thus far endeavoured to follow in proper order the course indicated by your letter, omitting many reflections as to the injustice of condemning a fellow labourer without a hearing. We forbore making any allusion to these reflections inasmuch as we have them all summed up in the paragraph which next presents itself for our consideration.

That paragraph contains as weighty a charge as man could prefer against man, we receive it in the spirit which prompted it, a spirit of kindness and charity, and we thank you for expressing your opinion so frankly and unreservedly. Believing that your justice is equal to your candour we feel assured that a simple recurrence to facts will entirely alter the complexion of the affair and reverse the terrible judgment you have recorded against us. We think it needless to suggest the necessity of calmness and patience in coming to a conclusion such as you have here avowed, your remarks upon this head shew that you are fully alive to the importance of these conditions, we are unwilling therefore to attribute the formation of this your opinion to haste or irritation, but we do think that premises which admit of such a conclusion ought to have been analysed with more than ordinary care, had this been we feel assured that their utter instability would have been proved and we should have been spared the infliction of so unmerited an accusation.

The charge runs as follows, 'I think in speaking of a fellow labourer to his employer in the manner you have done without giving him an opportunity of defending himself you have acted contrary to the spirit of the laws of our country contrary even to the world's sense of honour and diametrically opposed to the precepts of our holy religion'. We will not stop here to ask how a pair of characters such as you have here depicted could possibly be the objects of that esteem which you so kindly profess towards us, we will not ask you to reconsider this charge, as the evident deliberation with which it was penned shews that you consider it has already engaged a sufficient share of your attention, and though we are accused of having at once violated the laws of honour, of man, and of God, we shall do all that in us lies to divest ourselves of all personal connexion with the accusation and shall endeavour to relate facts as we believe a party wholly disinterested would relate them.

On Wednesday May 31st Mr Yeats[7] came to us while we sat at dinner and requested our presence in Mr Edmondson's room at 3 o'clock. Shortly after this Mr E., Mr Yeats, Mr Singleton[8] and those who now have the honour of addressing you found themselves seated together in No 10.[9] Mr Y. commenced by remarking that he had come to clear up an unpleasant affair with Mr Singleton, that he cared little for the opinion of any man, least of all on a matter of business for that of Mr S. In this strain of studied insult he proceeded until we, feeling our position to be anomalous, arose and declined having anything to do in a quarrel so purely personal. Our wish was to leave the room immediately but being entreated by Mr Edmondson we remained. The matter which gave rise to our being called together was laid before us and in compliance with Mr Yeats' own request we delivered our opinion which was condemnatory of his conduct. The true position of Mr E. in this investigation seemed to us to be that of an umpire or mediator, but instead of remaining thus neutral he defended one of the parties as strenuously as if that party had been his client and we found ourselves reduced to the painful necessity of expressing an opinion opposed to that which our respected principal evidently sought to attain.

Our subsequent interview with Mr E. was unsought by us, he came into our room and of his own free will again introduced the subject. Whatever remarks we then made flowed necessarily out of the nature of our conversation. We sought no opportunity to make a charge against the gentleman you consider aggrieved, our opinions were drawn out of us by Mr E. himself and we relied upon his ability to affix a proper value to these opinions. An hour before he had defended Mr Yeats as strenuously as even you Madam could desire, we felt that he was mistaken upon that as well as upon other points and we told him so, the next day proved our correctness in one particular at least. Mr Yeats condemned himself, which condemnation was acquiesced in by Mr E.

But why do we proceed in this strain, who can possibly form so correct an opinion of our conduct as Mr E. himself, our words were spoken in his ears, to him we refer for our justification, nay he has already vindicated our conduct to you in our own presence, how you Madam whose information must necessarily have come through him can found so grave and so opposite a charge upon that information we are totally at a loss to imagine.

We are unconscious of having denied to our fellow labourer an opportunity of defending himself. Mr E. has heard our opinions and is free to make what use he pleases of them, we never made secrecy a condition of our conversation. To the matters of fact mentioned by us we are still prepared to bear testimony and the opinion which we then expressed we still substantially retain. The circumstances which have resulted in these opinions would probably have led you to others totally different. We have therefore not the slightest hope of being able to make a convert of you, nor have we the least desire to do so. Our sincere opinion is that Mr

Yeats has served you most assiduously, that he has at all times shewn an anxiety to anticipate your wishes and to execute your will. One feeling alone can result from such a course of conduct and this feeling we are by no means anxious to disturb. Our opinions shall merely determine our own act, we do not intend to make use of these opinions publicly or privately against Queenwood, our desire is to depart in peace, sincerely trusting – nay believing – that our successors will co-operate more harmoniously for the interest of the institution than we have hitherto been able to do. We are ready to admit the abundant flexibility of Mr Yeats' character, we have never seen in him anything approaching to obstinacy but always precisely the reverse. We are perfectly satisfied that any particular error pointed out by you he would at once endeavour to eradicate, and we further believe that recent circumstances will cause such modifications of the state of matters at Queenwood as to render a repetition of our act in the case of our successors unnecessary.

Our friendship we believe to have been consistent, we have gone as far as we dared in expressing our disapproval of existing things, one method of disapproval has been noticed by yourself and Mr E. We have even gone further than this silent expression of dissent but received no encouragement to proceed, that any remonstrance of ours could have altered matters we were far from imagining. We had no reason to expect that our interference would have been better received than that of your own brother, his was evidently unpalatable, there was no pointing out faults, the alleged misconduct was defended and Mr Singleton exposed to insult for the utterance of a conviction which a night's reflection proved to be strictly correct.

There is we believe but one point more in your letter to which we have not alluded, it is thus expressed. 'I believe could you divest yourselves of prejudice towards the individual &c.' x x x x 'I believe that this storm blown over, this prejudice laid aside &c.'.

On reading your letter with that attention which everything of the kind coming from the respected writer demands, we meet the legible fact that it has emanated from a mind whose opinion is already formed. What more than this can be said of us? We beseech you Madam to weigh matters carefully and determine whether it is not as reasonable to expect prejudice on your side of the question as on ours. Nay, we should not be surprised if after giving the matter due consideration you should conclude with us that your position in the present instance is more exposed to this influence than our position. Mr Yeats, as in duty bound, has always paid you the most implicit attention, he has never, openly at least, attempted to injure us. We are therefore free from personal hostility to him while we are not sure that you are equally free as regards personal attachment. You will admit that the latter is as fruitful a source of prejudice as the former. Would there be no reluctance on the part of a lady whose conclusions are in general marked by coolness and sound judgment to acknowledge even to herself that she had leaned too long upon a hastily formed opinion? We feel satisfied that if every possible agency, secret and open be taken into account, that the probability of prejudice will not be found on that side which truth and justice have compelled us to espouse.

Thus far through life have we striven to enact the dictates of our consciences which we have no reason to believe have ever deceived us. We rely upon their fidelity and have little fear of subsequent self-condemnation if at the present juncture we act according to the light afforded us. We voluntarily throw ourselves upon the world to struggle for bread as we best may, steadfastly believing that our present act will in after life receive the sanction of our judgment.

In conclusion Madam we beg to say that though your letter is so severely condemnatory of our conduct, though the charges therein contained would, if valid, prove our presence not only at Queenwood but in society a pest, still we thank you for its frankness and for the numerous professions of kindly feeling which it contains. Our object in this reply has simply been to justify ourselves in your eyes. We have not at the present moment the remotest wish to change your convictions as regards others, for we believe that the interest of Queenwood demand that these convictions should remain substantially as they are. We entreat you to pardon our deficiencies, and if anything rude or undesirable have escaped our pen attribute it kindly to the novelty of our position for in thought, word or deed we would not willingly offend you.

RI MS JT/2/13a/333–336
LT Transcript Only

---

1. *the letter*: see letter 0349 of this volume.
2. *our determination*: i.e., the decision to leave Queenwood and pursue doctoral degrees in Germany.
3. *Mr. E.*: George Edmondson.
4. *'you expect your opinion to be conclusive'*: the quotes in this letter all come from letter 0349 of this volume.
5. *The gentleman's dismissal*: the 'gentleman' was Josiah Singleton, an instructor at Queenwood who had recently left under contentious circumstances; see letter 0349, n. 4 of this volume. See also C. A. Russell, *Edward Frankland: Chemistry, Controversy and Conspiracy in Victorian England* (Cambridge: Cambridge University Press, 1996), p. 56.
6. *Queenwood*: the school where Tyndall taught from 1847–8. See Introduction to Volume 2.
7. *Mr Yeats*: John Yeats, a teacher at Queenwood; see letter 0349, n. 4 of this volume.
8. *Mr Singleton*: Josiah Singleton; see n. 5.
9. *No 10*: not identified; probably a reference to a building or room at Queenwood.

From Elisa de Lamartine 16 July 1848 0351

Mr John Tyndall or Mr Frankland.
Sir,

I am extremely sorry to say that I have not any means of obtaining tickets for the Chamber of Representatives.[1]

M. de Lamartine having retired from government,[2] I have abstained from making any application of any sort; but I should imagine that if you were to ask Mr Bastide,[3] Minister of Foreign Affairs, he would have much pleasure in sending you tickets for the Diplomatic Tribune.[4]

I regret not being able to assist you at this moment being resident in the country.[5]

With best compliments, | E. de Lamartine.[6] | 16th July.

RI MS JT/2/14/34
LT Transcript Only

---

1. *Chamber of Representatives*: the publicly-elected lower body of the French Parliament.
2. *M. de Lamartine having retired from government*: Alphonse de Lamartine (1790–1869), poet and politician, led the provisional government of the Second Republic in Paris following the February Revolution of 1848. In June 1848, he was thrown out of office after the revolution was crushed.
3. *Mr Bastide*: Jules Bastide (1800–79), then the Minister of Foreign Affairs in France.
4. *the Diplomatic Tribune*: a seating area where ambassadors and other foreign persons of importance could observe the Chamber of Representatives.
5. *resident in the country*: i.e., Madame de Lamartine was away from Paris.
6. *E. de Lamartine*: painter and sculptor Elisa de Lamartine, née Mary Ann Elisa Birch, the English-born wife of Alphonse de Lamartine (see n. 2).

From William Ginty 30 July 1848 0352

30th July 1848

........[1] Thomas[2] is still R.E.[3] at Wigan[4] and Milton[5] in Lincolnshire[6] – Roberts'[7] conduct has got him into a scrape at last. Mr Wright[8] fined him a sovereign – but with no effect, he again absented himself in a drunken spree for 3 days – he then dismissed him but has since taken him back – fining him £8.8 and compelling to write a letter of apology acknowledging its justice. Some have been dismissed lately[9] – more likely to follow – Topling[10] is back again – the lime burning[11] not suiting it appears. You did not say whether you found your friend[12] safe or not on your return to Paris – I wish I had been with you. I shall certainly go there as soon as I can (by the way that ear of wheat was not enclosed) (send it) Peb[13] is jolly well and at her tricks again.

Why don't you write oftener?

very sincerely | W.G. Ginty.

RI MS JT/1/TYP/11/3638
LT Transcript Only

---

1. ....: Louisa Tyndall annotation: 'first part missing'.
2. *Thomas*: not identified; probably not Thomas Hirst, who was still working for Richard

Carter.

3. *R.E.*: likely 'Railway Engineer' or 'Royal Engineer'.
4. *Wigan*: a town in Lancashire, located on the important West Coast railway line.
5. *Milton*: an estate in Lincolnshire, near Peterborough.
6. *Lincolnshire*: a county in east England.
7. *Roberts'*: not identified; possibly a relative of Margaret Roberts Ginty.
8. *Mr Wright*: first name unknown, not to be confused with William Wright.
9. *Some have been dismissed lately*: not identified.
10. *Topling*: not identified.
11. *lime burning*: probably a reference to the process of producing calcium oxide (quicklime) in a lime kiln.
12. *your friend*: probably Edward Frankland, who had accompanied Tyndall on his trip to Paris.
13. *Peb*: a nickname for Margaret Roberts Ginty.

To Edmond Sayers [July] 1848 0353

My dear Bob[1]

I am sincerely glad to find that my letter[2] sponged your anxiety away and afforded amusement to your friends – had I known it was to come under the cognizance of so many bright eyes I should have polished my periods, sprinkled my pages with the rosewater of poetry and said many pretty things which now, alas, must lie like unseen nosegays[3] under the dusty canopy of my sensorium[4] –

I am much obliged for your kind invitation[5] and should like much to accept it – You have with a few trowel strokes shapen a possibility which perhaps for 5 minutes in the year occupies my attention only to be dismissed again as impracticable. I am going to Germany[6] in a month, there I shall turn my purse inside out and come back a pauper. I resign the substantial certainty of my present situation[7] merely guided by an idea which is leading me I know not whither. Who would join her fortunes to such a Will O the Wisp?[8] I don't expect to meet one, and if I should I would rather think of putting her in a glass case like a beautiful Egyptian Ibis[9] to be worshipped in due season than of causing her fate to oscillate with mine between the extremes which I imagine await me.

I like methodism[10] as well as any other <u>ism</u>. An honest man will be so under any <u>ism</u> while no formula whether Wesleyan[11] or Calvanistic[12] can transmute the dross of selfishness and insincerity into nobleness and truth. The Great Spirit which from time to time expresses himself audibly among the sons of men dwells far below the scum of sects – into this shall methodism, churchism and many other isms one day sink and a purer lovelier and more practical faith – a faith which Jesus taught and John[13] understood shall bend with benignant influence over our altered world.

J. Tyndall.

RI MS JT/1/TYP/12/3973
LT Transcript Only

1. *My dear Bob*: Louisa Tyndall annotation: 'from JT. | for R Sayers | Summer 1848'.
2. *my letter*: letter missing.
3. *nosegays*: bundles of flowers or herbs (*OED*).
4. *sensorium*: the human brain, understood as the centre of thought and nervous energy (*OED*).
5. *your kind invitation*: letter missing.
6. *I am going to Germany*: in October 1848 Tyndall began studying for a doctoral degree at the University of Marburg.
7. *my present situation:* i.e., Tyndall's teaching position at Queenwood College, where he taught from 1847–8. See Introduction to Volume 2.
8. *Will O the Wisp*: a 'will-o'-the-wisp' is a mysterious object that appears suddenly and leads those who view it astray (*OED*).
9. *Egyptian Ibis*: a long-legged wading bird frequently depicted on Egyptian artifacts.
10. *methodism*: in spite of the lower case initial, it appears that this is a reference to Methodism, a Protestant branch of Christianity.
11. *Wesleyan*: John Wesley's religious teachings. Wesley (1703–91) was a founder of Methodism (*ODNB*). In 1848, there was controversy within the church between Wesleyans and Calvinists expressed in a series of anonymous pamphlets.
12. *Calvanistic*: a reference to Calvinism, a branch of Protestantism that arose in the Reformation and was based on the teachings of John Calvin (1509–64).
13. *John*: one of the Twelve Apostles of Jesus; also possibly a reference to the author of the Gospel of John.

To Thomas Foy — 7 August 1848 — 0354

Sir

Last night returning from Southampton[1] I came to the resolution of writing to you. At that time I intended to refer to the opinion which I once entertained of you, but second thought convinces me of the impolicy, if not the impertinence of such a course. The experience of the last three years enables me to guess the figure at which you estimate the opinion of those who would wish to see you what they once believed you to be – an honest man.

However this apart; there is a matter of business between you and me. In the month of March 1845 when I was badly able to afford the cash you borrowed from me the sum of three pounds nineteen shillings. When I wanted the money I wrote to you[2] from Halifax, you replied[3] that you were embarrassed and I believed you, resolving at the time never to add to your embarrassment. I should be glad to believe even now that the part you have acted was not the result of want of principle, if I thought so I would, if I know myself, change the present occupation of my right hand and stretch it forward to assist you; but all I know, and all I have heard concerning you renders such a conviction on my part impossible. I believe that if I never reaped a shilling advantage from the sum, if immediately on the receipt of it I enclosed it to the secretary of some Mendicity Institution[4] that the money would be better employed than if left in your hands. I shall therefore make an effort to regain it. This then is to inform you that in less than eight weeks I shall be leaving this country,[5] that I am willing to take the debt in weekly instalments of 10/ which you can hand over to Archy McLachlan, that if this offer be declined by you I shall without hesitation put the matter into the hands of a Solicitor.

J.T.

RI MS JT/2/13b/373–374
LT Transcript Only

---

1. *Southampton*: a city in Hampshire, England.
2. *I wrote to you*: letter missing.
3. *you replied*: letter missing.
4. *Mendicity Institution*: a charitable institution in Dublin that provided meals to the poor. 'Mendicity: About Us & History', *Mendicity*, at http://www.mendicity.org/about.htm [accessed 16 April 2014].
5. *leaving this country*: in 1848, Tyndall moved to Germany to pursue a doctoral degree at the University of Marburg.

To Thomas Foy 11 August 1848 0355

There is perhaps one and but one class of individuals entitled under any circumstances to treat others with contempt, namely those who are above the reach of it themselves. Are you among the latter? Your 'silence' I have already experienced, your 'contempt' is an airy nothing for which no true man will care a jot, in fact you can't afford the thing. I will await your convenience for a 'day or two' and shall peruse with interest your annotations on the phrase '<u>your injured friend</u>'.[1]

J.T. [2]

RI MS JT/2/13b/374
JT Journal Transcript Only

---

1. *'<u>your injured friend</u>'*: the quotations in this letter presumably came from a missing letter by Foy.
2. *To Thomas Foy*: This fragment was transcribed in Tyndall's journal entry for 7 August 1848. Tyndall wrote: 'Foy wrote lest I should consider that he had treated my letter with silent contempt, and subscribed himself "your injured friend". Wrote to him as follows' RI MS JT/2/13b/374). For Tyndall's previous letter to Foy, see letter 0354 of this volume.

To Thomas Hirst 11 August 1848 0356

Queenwood Friday[1] –

My Dear Tom,

Until now I have not had time to look at the question you have sent me,[2] much less to *[unlap]* the roll of memory and lay before you the occurrences of the last 6 weeks – Even now you must be content with a brief outline and console yourself with the prospect of hearing the details some wild wintry night not yet unravelled from the web of time, when you and I shall meet – I with with the frost of years upon my 'pow' and you with a time scar upon your now beardless cheek – I started from Southampton on the 15th of June accompanied by our chemist[3] – a young man to whom I would like to introduce you – if he perseveres I have no doubt of his name being familiar to you yet. After a qualmish[4] passage of 9 hours reached Havre[5] where we put up for the night – we had some difficulty in getting through the claws of the Gens d'armes[6] at the Custom House as we had come without passports – on the following day we started for Rouen[7] where my companion had a friend[8] we remained at Rouen two days, during which we saw all the curiosity including the Cathedral from the base of which Wm of Normandy[9] set sail to invade England – the square where Joan of Arc[10] was burnt in which stands the heroic woman's statue – a voyage down the Seine the scenery of which is delightful – on Monday the 19th we entered Paris and put up in the third story of a high house taking each a bedroom at 25 francs a month and no sitting room – we breakfasted at our lodgings walked all day and dined at a Restaurant – a capital dinner, including Soup, flesh, fish & desert & half a bottle of wine for 11d! – On the morning of the day on which the insurrection[11] commenced I left Paris for London stopped that night at Amiens[12] where the intelligence of the battle reached me – the next morning the *[*rappel[13] *]* was beaten and a host of national

<*Handwritten pages missing; LT Transcript only henceforth*>

guards volunteered to go and assist their brethren at Paris. I saw them start – a fine body of men – each furnished with a substantial loaf which was transfixed by his bayonet! Arrived at London the Times[14] contained the news of the horrible carnage and moreover mentioned the very street where I had left my friend as being the scene of a bloody encounter. I was in a sad state of anguish, and directly after having finished in London I set out again to seek him. All was tranquil when I arrived, groups of military were encamped in the streets. Dragoons, horses, hay, and straw thronged the public squares, the paviors[15] were at work levelling the barricade and the floating multitude seemed as gay as usual. Of course the scenes of the hottest encounter engaged my attention first. Paris will long bear the mark of the affray. The Rue St. Jacques; Rue St. Antoine, Faubourgh, St Antoine, Clos St. Lazare,[16] &c will long be memorable. Having remained about 3 weeks, during which we visited the palaces of St. Cloud and Versailles, we bade good bye to the metropolis of fashion. Lamartine[17] we did not see, but we received an exceedingly kind note from Madame Lamartine[18] to whom we had written for tickets[19] for the National Assembly. Having left Paris we journeyed northwards, through Douay, Valenciennes[20] &c to Brussels; here we remained 3 days, one of which was spent upon the plains of Waterloo.

We then went to Antwerp, where are deposited the great works of Reubens[21] and the blacksmith artist. Afterwards to Ghent. I can't afford space to describe the magnificent churches of Belgium. Statues, pillars, cedar pulpits set amid trees of marble – the leaves branches, and fruit most exquisitely chiselled – I will defer the remained until the time of meeting aforesaid

Goodbye | Tyndall

Now for your question

$$x + y = \frac{4xy}{x + y} + 6$$

xy = 108

Clear the first equation of fraction by multiplying by x + y

$(x + y)^2 = 4xy + 6(x + y)$

$(x + y)^2 - 6(x + y) = 4xy = 4 \times 108 = 432.$

add to each side the square of half the coefficient of (x + y)

$(x + y)^2 - 6(x + y) + 9 = 432 + 9 = 441$

a glance will prove to you that the left side of the equation is a complete square extract the square root of both sides and we have

$x + y - 3 = 21 \qquad x + y = 24$

$\qquad\qquad\qquad\qquad xy = 108$

which reduces it to a simple equation. In some equations a good deal of tact is required in the grouping of the quantities this by practice alone can be obtained had I multiplied (x + y) actually by itself I should perhaps have got myself into difficulty.

I hope Jemmy[22] will enjoy his ramble – I often wish to be near you both once more, as I believe I could render myself more useful to you than heretofore. As an educator at Queenwood[23] I have myself been educated far less by a contact with my colleagues than by a contact with the lads.

RI MS JT/1/HTYP/9-10
RI MS JT/1/T/979
Portions of letter are LT Transcript Only

---

1. *Queenwood, Friday*: Louisa Tyndall annotation: 'Fri. Aug 11/48 | See Entry in T's Journal "solved Tom's equation."' This quotation can be found in Tyndall's journal entry for 11 August 1848 (RI MS JT/2/13b/374).
2. *the question you have sent me*: letter missing.
3. *our chemist*: Edward Frankland.
4. *qualmish*: nausea-inducing (*OED*).
5. *Havre*: Le Havre, a major port on the river Seine in northwestern France.
6. *Gens d'armes*: armed men (French), i.e., policemen.
7. *Rouen*: a city in Normandy, France.
8. *my companion had a friend*: possibly the 'Mr Scott' with whom Tyndall and Frankland visited the Cathedral at Rouen on 17 June 1848, according to Tyndall's journal (RI JT 13a, f. 340–341).
9. *W^m^ of Normandy*: King William I of England (1028–1087), also called William the Conqueror. In 1066 William led an invasion force from Normandy to England to claim the English crown after the death of his cousin, King Edward the Confessor.
10. *Joan of Arc*: also called the Maid of Orléans, Joan of Arc (*c.* 1412–31) was a military leader who believed that God had called her to support King Charles VII against the English

during the Hundred Years' War. She achieved several important military victories before being captured by the English and burned at the stake.

11. *the insurrection*: this probably refers to the 23 June 1848 uprising of Parisians against the French Second Republic.
12. *Amiens*: a city in northern France.
13. *rappel*: a beat of drums used to summon soldiers to arms (*OED*).
14. *the Times*: this presumably refers to *The Times*, a London newspaper.
15. *pavoirs*: people who lay paving (*OED*).
16. *Rue St. Jacques; Rue St. Antoine, Faubourgh, St Antoine, Clos St. Lazare*: all of these are street names or areas in central Paris that were occupied by revolutionaries during the Revolution of 1848. Some three thousand individuals were killed.
17. *Lamartine*: Alphonse de Lamartine (1790–1869); see letter 0351, n. 2 of this volume.
18. *kind note from Madame Lamartine*: Elisa de Lamartine; see letter 0351, n. 6 of this volume.
19. *written for tickets*: Tyndall and Frankland wrote to Madame de Lamartine requesting tickets to the Chamber of Representatives; see letter 0351 of this volume.
20. *Douay, Valenciennes*: Douay (Douai) and Valenciennes are both communes in northern France.
21. *Reubens*: probably Sir Peter Paul Rubens (1577–1640), a Flemish Baroque painter.
22. *Jemmy*: James Craven.
23. *Queenwood*: the school at which Tyndall taught from 1847–8. See Introduction to Volume 2.

To Thomas Foy 16 August 1848 0357

Sir

My engagements here prevent me from devoting the amount of time I could wish to a reply to yours of Sunday.[1] If your feelings were wounded it is a proof that one object aimed at in my letter[2] has been attained. It was penned with this deliberate intention, though certainly with very little hope of such a result, as I had begun to imagine that you had no feelings to wound.

I don't think I am made of very combustible material. If you can prove me so constituted you may, if you think fit, set yourself down as the discoverer of the fact. I entirely disclaim all knowledge of that 'inflammation' which you fancy your absence on Sunday week created. My opinion of your conduct was formed previous to my going to Southampton and your presence or absence was a matter of exceedingly small moment to me. I wrote to you actuated by no sudden gust of passion, but with the settled conviction that it was my duty to do so, that I was writing to a man whose honourable feeling had vanished, the very broad cloth on whose back was a shining falsehood inasmuch as it deceived the public into the belief that its wearer was a gentleman.

This was the sad and sober feeling which influenced me when I wrote to you. I am indebted to no 'cowardly rascal' for the materials of my opinion. Years ago I heard how you could amuse yourself after my departure from Manchester[3] and yet could not afford to pay me. You were employed at Swansea, this I learned from Hubert Couran[4] and I wrote to you there. You may deny having received my letter but what guarantee have I for the truth of this denial. Some months afterwards you met Tidmarsh in London and asked him for my address but you never wrote to me. I asked him how you looked and his reply was – without knowing anything of the transaction between us – that you were 'splendidly dressed'. You have mentioned the name of Mr McLachlan, I believe him to be an honest man and that if you imagine the contrary you do him wrong. This is my opinion of him and this being the case I would accept him with brotherly affection if he had not 6d in his pocket sooner than such a character as I deemed yours to be, however elaborate might be the manmillinery[5] of its exterior.

One sentence in your letter amuses me; you seem to have woven the principle therein contained into your actions and referred to it as their justification 'I may with you or anyone else be affected by circumstances' a convenient creed no doubt for those who wish to make use of it, but a creed to which no man of principle will fly for refuge. Take the advice of one who notwithstanding all his 'abuse' and all his 'revenge' would still wish to see you rescued from the effeminacy and folly, to gratify which you have unfortunately made shipwreck of whatever nobleness or manhood you ever possessed, put this creed of circumstances under your feet and endeavour to be honest in spite of circumstances.

I have given your integrity a long trial and I would give it another if I could. Had I a wish to gratify my 'vengeance' nothing could be easier; but I would like to throw you once more upon your principle; my letters are intended to cut deep. I speak my mind fully without exaggeration or diminution. If what I have said shews you the peril in which you have placed your reputation, and if a regard for its safety prompts you to adopt the only course of action by which that safety can be ensured, I shall be infinitely better pleased than if in reply to my first letter you had at once responded by prompt payment.

Your friend | John Tyndall

RI MS JT/2/13b/375–376
LT Transcript Only

---

1. *yours of Sunday*: letter missing.
2. *my letter*: Tyndall had written to Foy requesting repayment of a debt; see letter 0354 of this volume.
3. *after my departure from Manchester*: Tyndall worked as a railway engineer in Manchester beginning in 1843. See Introduction to Volume 2.
4. *Hubert Couran*: not identified.
5. *manmillinery*: excessive male preoccupation with clothing (*OED*).

To Thomas Foy 29 August 1848 0358

Sir

I am content that for the present you shall mistake my character. I have written to you[1] in a manner which naturally gives truth to the opinion of my 'warm friendship' and 'bitter enmity' but in all my writing I had a higher object in view than the mere recovery of what you owe me; for this I care as little as perhaps any man of your acquaintance. I had reason to believe you dishonest, I now believe you to be unfortunate; did I still entertain the former opinion I should most assuredly act as I threatened to do in my first letter.[2] I accept your own account of yourself[3] without asking for extraneous proof, inasmuch as your letters carry a kind of internal evidence of general correctness and truth. With regard to the saving of money, it is an art in which I never hope to excel, nor is it an art to which I attach much importance. My present situation[4] is worth £200 a year and my employer is most anxious to retain my services at this salary; surely if a love of money actuated me I should not think of going to Germany[5] where I shall pay for my living and instruction and in a pecuniary point of view return a pauper. This I do purely because I consider the line of duty lies that way and that I can render myself more useful to humanity by extending my knowledge and struggling to be a wiser and a better man.

I confess that the dislike which I have for some time felt for the philosophy of human character propounded by St. Simon, Fourier, and Owen[6] has by your case been heightened into disgust. Theoretically I have long been aware of the impossibility of extracting a single noble impulse from such a creed, but when I find its practical results set forth bodily in you, when I find it sapping the foundations of morality and furnishing a ready excuse by which folly stifles the promptings of duty, I am constrained to cry 'alas for the society where such principles are permitted to prevail'.

It would be unwise on my part to enter into any discussion with you – it would be productive of no good result. We should set out with the determination to differ and the more lengthy our argument the wider would be our divergence. I have waded through your present opinions myself though I never allowed them to interfere with the discharge of my engagements; these opinions constitute a phase through which almost every thinking mind in our present English society has had to pass. If you love truth and give yourself the trouble of seeking it up, I have not a shadow of doubt as to your abandonment of your present creed – in 9 cases out of 10 it finds laziness to be its best supporter.

It would be extremely easy to start a dozen questions which are incapable of solution according to your principles. It would be easy to confound you, but I don't think that in doing so I should convince you just now. You consider me rather in the light of an antagonist than of a wellwisher, and this opinion would be sufficient to nullify any attempt that I could make. All I have to say is, that the doctrine you have propounded is rejected by the most philosophic minds in the world, that the great men of Germany have long since scouted it from their frontiers, that the profoundest thinkers in France and England are arrayed against it and that if you and I look within ourselves instead of looking at the 'broad face of society' we shall find ample data for its refutation without citing any external authority whatever.

I have no wish to prolong my remarks on this subject but if you think it worth your while to look into the matter again I shall be happy to recommend a book or two to your notice. One important item in the educational course of your parents you seem to slur over. They instilled into you, you say, a love of clean linen and for this you offer God thanks! On their behalf I would suggest that they taught you to love a clean conscience also, and that cleanliness in this respect is of infinitely greater importance than in the former. That you were so taught, although you appear to be culpably forgetful of the fact, my first connexion with you compels me to believe; and I would further hope that along with the anise and cumin of clean linen you will one day acknowledge the necessity of attending to the weightier matter.

Your remark concerning the Irish character is perhaps too true, though coming from your pen it sounds strange, but I cannot see the meaning of it; surely you don't mean to insinuate that a clean shirt is a strange phenomenon to Tidmarsh. My personal knowledge of him would render such an insinuation foolish in the extreme to my ear, knowing as I do the refinement with which his family is able to surround him. This however it would be well to remember, that there is an infinite distance between a love of cleanliness and a cultivation of expensive habits at other peoples' cost.

I have written this last sentence with reluctance as my wish is to avoid everything approaching to severity in this letter. With regard to paying me I wish it to be your own free act and shall therefore relinquish all intention of putting my threat[7] into execution. My time here will be up on the 29 of September, but I shall probably remain a week longer. It is useless for me to tell you that my views of humanity have not been altogether formed in a college, that I have had as free and as extensive intercourse with the world as most young men, and that I have not looked upon it with an unobservant eye, all this you know already, but the amount of true knowledge to be drawn from such sources is at the utmost exceedingly meagre and worthless unless it be referred to a proper source. What you and I want is a knowledge of ourselves and not a knowledge of society. Let us look inward and see whether we cannot find underlying our own natures a great spiritual agency of which the world and its transactions are merely the scum.

Your friend | J. Tyndall

RI MS JT/2/13b/377–379
LT Transcript Only

---

1. *I have written to you*: Tyndall had written to Foy requesting repayment of a debt; see letters 0354, 0355 and 0357 of this volume.
2. *act as I threatened to do in my first letter*: Tyndall had threatened to turn Foy's debt over to a solicitor; see letter 0354 of this volume.
3. *your own account of yourself*: Foy's letters to Tyndall are missing.
4. *My present situation*: Tyndall was working as a teacher at Queenwood College in Hampshire.
5. *going to Germany*: in October 1848 Tyndall began studying for a doctorate at the University of Marburg.
6. *St. Simon, Fourier, and Owen*: utopian socialists Claude Henri de Rouvroy, comte de Saint-

Simon (1760–1825), Charles Fourier (1772–1837), and Robert Owen (1771–1858).
7. *my threat*: see n. 2.

To James Craven [September] 1848 0359

With regard to reading, I would advise you to confine yourself rigorously to a certain course for a certain period.[1] I have wasted a great deal of time in allowing my thoughts to be diffused over too many subjects at once. We get very little profit by such a method of study – it completely annuls the power of concentration. Above all things, remember this – that the number of books you read is no index to the amount of benefit you derive from reading. I know one or two most extensive readers who could talk for half a century about various systems of Philosophy, and can tell you the opinions of this and that great man upon such and such subjects; and yet the intellectual power of these readers is truly contemptible. They are merely so many conduit pipes through which information from some other spring finds a passage – throw them into circumstances which demand the exercise of original power, and they get instantly tangled and helpless. This comes from their having contented themselves with driving a retail trade in the opinions of other men, without enquiring into the reasons of these opinions.

The best way of reading is, I imagine, the following: – Don't spend too much time in unravelling the difficulties of your book during the first perusal – give yourself a general acquaintance with it first. Then commence afresh, and go over it chapter by chapter, looking at the question in all possible lights, and noting down your observations in writing – thus proceed, till you complete your treaties; you will sometimes perhaps find that towards the end you understand the subject better than the author. This may seem tedious, but if the book be worth reading at all it is worth understanding, and the means I have here described, I imagine, are the best to this end. Once master of this subject, it does not lie like a crudity upon your intellect – you go forth with fresh strength to meet the next difficulty.

There is a temptation to which many young men of brilliant parts are exposed – the temptation to appear learned in society. When asked 'Have you read so-and-so?' you don't like to say no, and thus, when you find books talked of, you are tempted to read them merely for the purpose of saying you have read them. If you allow yourself to be overcome in this way, there is no end to your task – and what is worse, there is no profit – you defraud yourself of your own manhood, and convert yourself into a kind of hamper basket for stowing away the products of braver minds. Take time with your reading – let your patience be inexhaustible, and instead of being led away by the goose-cackle of society, think that your mission is to lead society, not to follow it. Now, are you tired of me? I don't think you are, for if I did I should not have written so much. I have a kind of faith that what I say to you you will not reject. I reckon upon your diligence, for I feel a personal interest in your progress. We shall probably meet in after years. Let me see you a man.

RI MS JT/2/Hirst Journal/86–87
Hirst Journal Transcript Only

1. *With regard to reading ... a certain period*: from Hirst's journal, 4 September 1848: 'Jimmy received a letter from Tyndall, which among the rest contains some advice as to a course of reading which I must quote, as such advice, tested by his experience, must be worthy of attention' (RI JT/2/Hirst Journal/86). This letter is the extract Hirst copied into his journal.

From Thomas H. Carroll 24 September 1848 0360

Carlow Sentinel Office. | Sep. 24, 1848.

My dear Sir,

Having been absent during the last fortnight, at Kingstown,[1] I had no opportunity of replying to your note of the 14th inst.[2] and I feel assured you will do me credit when I state that the omission was not occasioned by neglect, or from any indifference arising from some imaginary notion that you changed your principles.

But supposing your principles had undergone a change, arising from matured reflection, or experience, I do not see why even that change should cause a disagreement between old friends. I shall feel most happy to hear from you as frequently as possible, and to see you here for a few days, should you visit your native country.

It is needless to detail the sad incidents connected with our history during the last two years,[3] but the inference may be legibly read in the reckless countenance of the Irish peasant. We had just escaped the consequences of famine and pestilence through Divine favour, when a Rebellion[4] (if the late outbreak could be so termed) burst on the country, and between the agitation of the public mind – the malignant and widespread disaffection of the Peasantry, the tide of emigration flowing rapidly – and a general combination against the payment of rents in the South, you may picture to yourself the hopeless condition of this unfortunate country. In this and the neighbouring counties the Peasantry are well affected – and industrious; but the failure of the potato crop a third time has been attended with considerable disappointment, if not dismay. But I sincerely hope after this struggle is over – and the minds of the Peasantry turned to the growth of other articles of food – and the Irish Gentry become more industrious, humble and provident, better days will dawn on this country.

It is unnecessary for me to state what I have before repeated, how glad I shall be to receive any communication from you suited to the columns of a Journal,[5] literary, political, Scientific or otherwise. I am during a considerable period less of a politician myself, having devoted the greater portion of my leisure time to the study of Geology and Agricultural Chemistry – but to say that I have attained ordinary proficiency in either science would be giving too flattering an account of myself.

Whether it be from close study, from the application of your mind exclusively to dry scientific subjects, or otherwise, I am not aware, but I fear you have become too morbidly sensitive in supposing that I pay no attention to your communications. Now let me inform you that your letters from Paris were most interesting, and were perused with gratification, and that they one and all appeared in the Sentinel,[6] copies of which were forwarded as directed. I should like to know whether they reached you or not. I hope you will not in future cast any suspicion on my motives for being certainly at best but an irregular correspondent – but I shall never plead guilty to the neglect of a friend for whom I always entertained a high personal respect and in whose prospects and pursuits I felt more than ordinary interest. Again I repeat the sooner and more frequently I hear from you, the more gratified I will feel.

Yours faithfully, | Thos. H. Carroll.[7]

RI MS JT/1/TYP/11/3492
LT Transcript Only

---

1. *Kingstown*: a town in County Dublin, Ireland. The town's name was changed from Dún Laoghaire (alternate spelling: Dunleary) to Kingstown in 1821. In 1920, the town's inhabitants changed its name back to Dún Laoghaire.
2. *note of the 14th inst.*: letter missing.
3. *our history during the last two years*: Carroll is alluding to the Great Famine (also called the Irish Potato Famine), which began in 1845 and would last until 1852. The famine resulted in mass starvation and a spike in emigration from Ireland.
4. *a Rebellion*: probably the Young Irelander Rebellion, which took place in July of 1848 in the town of Ballingarry, County Tipperary. A small group of Irish nationalists attempted to begin a revolt and engaged in an armed conflict with local police constables; the constables famously retreated to the home of Margaret McCormack, where they took the McCormack family hostage. The conflict dispersed when police reinforcements arrived.
5. *suited to the columns of a Journal*: i.e., the columns of the *Carlow Sentinel,* of which Carroll was editor and proprietor.
6. *the Sentinel*: the *Carlow Sentinel*; see n. 5.
7. *Thos. H. Carroll*: Thomas Carroll (*d.* 1861), editor of the *Carlow Sentinel* for more than 30 years. Following the death of Henry Malcolmson in 1843, he also became the paper's proprietor.

To Thomas Hirst and James Craven 5 October 1848 0361

Ovr Darwen[1] Thursday[2]

My Dear Boys

I'm at last in the North[3] – Some day next week I shall peep into Halifax possibly I may accept Tom's[4] kind invitation cannot say just yet. I shall remain here until monday or tuesday, then proceed to Manchester, thence to Halifax, thence to Leeds – thence to London – thence to Rotterdam thence up the Rhine to Frankfort,[5] thence to Marbourg[6] in Hesse Cassel[7] where I purpose becoming stationary – Just drop me a line by return directed to the Post Office Preston –

as usual yours | J. Tyndall.

RI MS JT/1/HTYP/11
RI MS JT/1/T/971

---

1. *Ovr Darwen*: Over Darwen, a town in Lancashire.
2. *Thursday*: Louisa Tyndall annotation: 'postmark on back is Oct. 5/48'.
3. *in the North*: according to Tyndall's journal entries from 5 and 6 October 1848, he was visiting Lancashire (RI MS JT/2/13b/388).
4. *Tom's*: Thomas Archer Hirst.
5. *Frankfort*: a common nineteenth-century spelling of Frankfurt.
6. *Marbourg*: in October 1848 Tyndall began studying for a doctoral degree at the University of Marburg. See Introduction.
7. *Hesse Cassel*: also spelled Hesse-Kassel. A German electorate that bordered on the Netherlands.

From Charles Mackay 14 October 1848 0362

7, Torriano Avenue, | Camden Road Villas,[1] | October 14th, 1848

My dear sir,

There needed no apology for your kind letter,[2] nor the expression of a fear, that, even if your effusion had been thrice the length, I should have lacked patience to peruse it. To those who write from their convictions, and because they cannot choose but write, sympathy and appreciation are doubly pleasant; and I can assure you very sincerely, that I have derived great gratification from the knowledge that my writings[3] have met so cordial a reception from a mind like yours.

I trust your residence in Germany[4] will be productive of even more than the benefit you anticipate; and shall be most happy, whenever you feel in the mood to write to me, to know how you are getting on; I shall by no means think your letters an '<u>infliction</u>'.

With best wishes, I remain, | Yours very truly, | Cha. Mackay. [5]

John Tyndall Esq[re] | [11 Rockingham St | Leeds][6]

RI MS JT/5/15c
LT Transcript Only

---

1. *Camden Road Villas*: an address in London, England.
2. *your kind letter*: letter missing.
3. *my writings*: Mackay was best known as the author of *Songs and Poems* (London: Cochrane and McCrone, 1834) and *Memoirs of Extraordinary Popular Delusions and the Madness of Crowds* (London: Richard Bentley, 1841).
4. *Residence in Germany*: Tyndall was about to leave Britain to pursue a doctoral degree at the University of Marburg.
5. *Cha. Mackay*: Charles Mackay (1812/14–89), Scottish poet, journalist, and novelist; see also n. 3.
6. *John Tyndall Esq[re] ... Leeds*: address given by Louisa Tyndall.

To E. D. Steuart 17 October 1848 0363

London 17th. Oct. 1848

Madam,

This day week I wrote to you[1] sending my address, No 9 Don's Street West Lambeth London expecting on my arrival in town to meet a letter containing the commission which you wished me to execute[2] at Frankfort.[3] In this expectation I am disappointed – possibly a letter may arrive by this night's post if not I fear it will be too late as I start tomorrow morning from Blackwall[4] for Rotterdam. In my letter of this day week I enclosed the half of a £20 bank of England note No 76508, the other half I left in the hands of Mr Hebdon, Old Bank, Preston, Lancashire who was to forward it on being requested by you to do so. I was compelled to make this arrangement because I was unable to procure a letter of credit.[5]

My address in Germany will be

John Tyndall | care of Professor Bunsen | Marburg | Hesse Cassel.[6]

This will not be far from Frankfort and should your letter not reach me tonight I will endeavour to have your commission executed personally or through the medium of a friend.

believe me Madam | very sincerely yours | John Tyndall.

RI MS JT/1/TYP/10/3326
LT Transcript Only

---

1. *wrote to you*: letter missing.
2. *commission which you wished me to execute*: unknown reference.
3. *Frankfort*: a common nineteenth-century spelling of Frankfurt.
4. *Blackwall*: a district of London.
5. *a letter of credit*: a letter from a bank verifying that they had granted Tyndall a line of credit and would pay debts Tyndall might incur.
6. *Hesse Cassel*: also spelled Hesse-Kassel. A German electorate that bordered on the Netherlands.

To Thomas Hirst 25 October 1848 0364

via Frankreich | [Postmark] 25 Oct 48

Mess$^{rs.}$ Hirst & Craven | Horton Street | Halifax | England | *[*Franco*]*

For Tom

'These things have I spoken unto you that you should not be offended' He knew well that this way would not be always on the delectable hills and therefore he developed a toughness within them able to face the evil when it came – The reason I write is that in your aftn[1] reading you will find men of profound attainments and devout hearts differring from Fichte[2] and if you be guided too much by the beliefs of others you will get perhaps into a whirl which will take you off your own legs, which be they strong or be they shimbleshanked[3] are far better to you than any other mans can possibly be – Look to the notice preceding the advertisements to the catholic series[4] – 'It cannot be hoped that each volume will be essentially catholic in its nature &c–' You will find Fichte [sneering] mere Knowing – you will find Emerson saying that he is satisfied if he can know a little and treating action in his turn with a half [sneer] – the secret of the matter is to see the correctness of both views not to cling with a party spirit to either, nay Emerson sometimes seems to contradict himself, but there is no real contradiction in the case, it is due to the imperfection of language which must shew its subject in details – a revolving light at sea will shew successively – soda,[5] [strontium], & borax[6] – yellow red green and black, but a full revolution is necessary to see the whole – you may see the green another man may see the red and say so very earnestly, but you must swear by your own eyes, or look round the matter and find that the man was strictly right from his point of observation. J.T.

(x)One of the best practical helps to this that I have found is to keep your pen in your hand when reading and set down in the spot any corroboratory or analogous thought which flashes thro' your own mind – The author thus becomes only an awakener he makes us possess ourselves without possessing us

I have intentionally forgotten the poetry!

RI MS JT/1/T/522

1. *aftn*: afternoon.
2. *Fichte*: based on the content of the letter, Hirst was probably reading Fichte's *Grunddage der gesamten Wissenschaften* (English title: *Foundations of the Entire Science of Knowledge*), first published in 1794.
3. *shimbleshanked*: slang for weak or misshapen.
4. *catholic series*: the Catholic Series was published by Chapman Brothers in London from 1845–8. The series included volumes by J. G. Fichte and R. W. Emerson.
5. *soda*: a compound of sodium, carbon, and oxygen.
6. *borax*: a compound of sodium and boron.

To Thomas Hirst and James Craven 11 November 1848 0365

Sat. Nov. 11. 1848

My dear Boys,

However far the conclusion of this epistle may dip into the future, and it is impossible to define the precise extent of it in this respect, a beginning has been made on the afternoon of Saturday November 11th 1848. I saw Tom[1] at Leeds and Jemmy[2] doubtless received my note[3] from London. Since then I have been in the dark, be mine the task to throw a little torchlight on my proceedings since.

We[4] had a rough passage from London to Rotterdam, the usual time is I believe about 24 hours, we however contrived to spend two days and two nights on the deep, during the first night we rode quietly at anchor the horizon looming as black as pitch, and every thing boding a tempest. next morning we found ourselves surrounded by vessels like our own seeking shelter. large masses of floating wreck hinted that the tempest had not been perfectly harmless, however it was only of a nights duration and next day we stood out to sea, all that day we ploughed along and towards 10 oC at night the steward recommended me to take a little brandy and to go to bed. I took his advice, swallowed half a glass of brandy and stood up, no further movement was required on my part, the attraction of gravitation did the rest. the vessel gave a lurch and I was pitched forward into the sleeping cabin, where I devoutly fell upon my knees and delivered up the brandy with immense interest. I went to bed but such another night I hope I shall never spend, sleep was out of the question, my head, hips and shoulders were hammered mercilessly against the mahogany sides of my berth, had I not been made of tough materials I should doubtless have been shaken to pieces.

On the morning of Friday the 20th Oct. we reached Rotterdam took the train to the Hague and stopped there a few hours, returned to Rotterdam and got lost amid the bewilderments of canals, drawbridges, wharfs and shipping – thanks to a young Dutchman who could speak French we got to our Lodging – we had even forgotten the name of the street or what is more likely found it unpronounceable –

On Saturday morning at 6 oC we entered the river steamer a fine vessel with an elegantly furnished saloon. It is usual I believe for tourists to travel by Railway from Rotterdam to Bonn and thus save themselves a journey of about 200 miles along the river where the banks are flat and the scenery insipid – there is something noble however in a fine broad river irrespective of the fringe of land it which binds it, the sun shone by day and the stars by night while the waters rolled calmly along, with fishingboats creeping up and down its edges and stretches of pebbled beach contrasted with the bright green verdure of the fields on either side – The dinner on board would turn John the Baptist into an epicure – it was too good, – irresistible in fact – Boiled beef, sausages, fish, cauliflower, fried potatoes, roast pork, apple sauce, pudding, roast veal, roast goose chesnuts &c &c – these dishes duly classified came in, course after course, tempting the palate long after the stomach was satisfied.

At 5 oC on Sunday we reached Dusselldorf in the neighbourhood of which is Rheydt[5] the dwelling place of Kuppers.[6] we had two or three hours to spare so had the pleasure of hearing a gorgeous mass in the cathedral – a forest of candles flamed about the altar, the light from which was reflected by numberless figures of saints and winged angels in polished brass, amid this scene the priest stood like a necromancer half obscured by the fume of incense, he raised the host, a little bell tinkled and the whole congregation fell down and worshipped – On monday morning at 4 oC reached Cologne and having two hours to spare sallied forth to see the cathedral. after some searching we found it, the gloom of its immense aisles was rendered 'visible' by the light of a few small candles stuck here and there at intervals, early as it was an old priest with snow white hair and deep sepulchral voice was saying mass, while a scanty congregation of superior saints or superior sinners devoutly responded to his orisons – verily this thing of religion however in its outward manifestations it may be twisted and tortured by circumstances finds a root in human nature which is deeper than all sensuous experience and lies below our modern Science of logic – true it is that circumstances may instigate it, may call it in a particular manner outwardly forth, but it has more to do with the creation of circumstances than circumstances have to do with its creation – in good truth they have nothing at all to do with it.

As I know Tom is fond of this subject let me say a word or two more on the matter. To argue logically with an irreligious man in favour of religion is idle in the extreme – I do not use the term irreligious in the sense immoral, but merely as applicable to those who reject all knowledge, all life, of a higher character than that derived through the medium of the senses – you could never by force of logic convince McArthur,[7] nay, should you make the attempt it would indicate on your part a state of mind which would infallibly make you the convert instead of him – If a man possess only four senses you cannot possibly bring the fifth within the domain of his reasoning – my figure is defective for in the nonreligious man the fifth sense is latent not unpossessed – Such a sense as far as mere human agency is concerned can only be awakened by contact with a religious man – you may chop logic till doomsday – a single fervent word fresh from the heart of a religious man will find a response where bushels of syllogisms are unavailing – let your words be living words not dead chaff and the chances are that they will bear fruit unto a higher life.

But to return – At Bohn[8] the Rhine begins to assume a character which it has not heretofore possessed. you suddenly dive into a country of strongly marked character, stiff precipices and towering basaltic hills rise around. within a few miles of Bohn and on the left hand side as you go up stands the 'castled crag of Drachenfels'[9] mentioned by Byron in Childe Harold – the country around it is indescribably lovely – Byron wrote truly –

> The river nobly foams and flows –
> The charm of this enchanted ground,
> And all its thousand turns disclose
> Some fresher beauty varying round
> The haughtiest breast its wish might bound
> Through life to dwell delighted here
> Nor could a spot on earth be found
> To nature and to me so dear
> Could thy dear eyes in following mine
> Still sweeten more these banks of Rhine[10]

Though not in absolute need of his Lordships complementary blessedness still I could appreciate his feelings.

There the castle stands however like a sky pillar perched upon a wild stormy looking pedestal – on the opposite side of the river is 'a noble arch in proud decay'[11] while the foliage and vineyards clothing the hills to their very summits were of the richest autumnal colours.– Thro scenery such as this we travelled two days – the river bank all along presenting evidences of tremendous volcanic agency – I often wished to take a stick in my hand and wander as free as a red deer through the mountains.

On the evening of Tuesday the 24th we reached Mayence[12] – thence the same night by train to Frankfort[13] where I remained that night – next day by an omnibus to Marburg where I arrived at 8 oC P.M. it being precisely 8 days since my departure from London.

I am now a student[14] – I work tolerably hard – up at 5 oC every morning – work at German until 7½, breakfast, from 8 to 9 a lecture on Physics – 9 to 10 Professor Bunsen on Chemistry – 4 to 5 a private lesson from the professor of mathematics[15] – I could not possibly be better suited than I am – the day I spend in the laboratory have tea at 6 oC and study mathematics until 10 at which hour I punctually go to bed – I'm not likely to get fat on this work for so far I have stood it very well. before I seal this I must endeavour to say a few words on chemistry, as by this time you will have got Fresenius.[16] I now halt.

I find myself so busy that I must defer the chemistry until I next write – could you manage to let old Mrs Piercy and my young friends[17] see this letter or tell them something about it, and when you write let me know how they are. really they are as kind to me as if I was a son or a brother.

You and Tid,[18] John R.[19] & Ginty might club together, you can each send a *[little]* almost as cheaply as one

Good bye now my sons and may the good Gods be your protectors

ever yours affectionately | J Tyndall

I promised John Richardson that you would send him this to read, this will save me the repetition of the scroll which would be no joke

RI MS JT/1/HTYP/12–14
RI MS JT/1/T/516

---

1. *Tom*: Thomas Archer Hirst.
2. *Jemmy*: James Craven.
3. *my note*: letter missing.
4. *We*: Tyndall was travelling with Edward Frankland.
5. *Rheydt*: a town in western Prussia.
6. *Kuppers*: Conrad Kupper.
7. *McArthur*: not identified.
8. *Bohn*: more commonly spelled Bonn; a city on the Rhine river.

9. *'the castled crag of Drachenfelds'*: Lord Byron, *Childe Harold's Pilgrimage*, III.lv.10.
10. *The river nobly foams...these banks of Rhine*: Lord Byron, *Childe Harold's Pilgrimage*, III.lv.40–49.
11. *'a noble arch in proud decay'*: Lord Byron, *Childe Harold's Pilgrimage*, III.lv.26.
12. *Mayence*: also spelled Mainz; a city on the Rhine river.
13. *Frankfort*: a common nineteenth-century spelling of Frankfurt.
14. *I am now a student*: in October 1848 Tyndall began studying for a doctoral degree at the University of Marburg. See Introduction to Volume 2.
15. *the professor of mathematics*: probably Christian Ludwig Gerling (1788–1864), the professor at the University of Marburg who supervised Tyndall's doctoral work in mathematics. Gerling had obtained his own doctorate in 1812 under the supervision of Carl Friedrich Gauss (1777–1855). He became a professor of mathematics at Marburg in 1817 and spent his career at that university.
16. *you will have got Fresenius*: probably C. Fresenius, *Elementary Instruction in Qualitative Analysis,* trans. J. L. Bullock (London: John Churchill, 1843).
17. *Mrs Piercy and my young friends*: not identified. Mrs Piercy may be a relative of Ada Piercy.
18. *Tid*: John Tidmarsh.
19. *John R.*: John Richardson.

To Thomas Hirst [21] November 1848 0366

[*[Leighlin]* Nov. 23/48 | *[Halifax]* Nov. 25/48][1]

My Dear Tom –

My religion has not yet been unable the make the organs of speech its proper instrument, it has not yet unravelled itself into fit utterance, it has no dialect, but while reading your letter[2] I experienced many feelings which the ancient dialects embody forth in the phrases 'praise the lord O my Soul[3] &c my feelings were certainly those of joy and gratitude, the source of which completely baffles my powers of analysis. 'We mark' says Emerson 'with light in the memory the few interviews we have had in the dreary days of routine and sin with those souls who made our souls wiser, that spoke what we thought, that told us what we knew, that gave us leave to be what we inly were'[4] such a light-streak has been your interview with Caryle[5] – for my own part I owe to him and Emerson more than to any other men living – you feel another being since you read this little book, can you not connect this change in your feelings with the doctrine of regeneration, can you not even by means of your present light see a beauty and a truth in that discourse with Nickodemus[6] where the phrase 'born again'[7] is introduced which escaped your attention heretofore – you have already drawn a circle round all sectaries, you have a key to all their mysteries, you will be able to appreciate the truth of popery and of protestantism, of Mahomed[8] and John Wesley[9] – your religion is a general formula of which theirs is but particular cases, at bottom they are all true, and the differences, hostilities, wars, and burnings with which each sect has propagated its own views are simply the result of elevating a part and worshiping it as a whole – Thus we solve the silly problem of the freethinker who demands how all can be true since they are so opposite. It has long been fashionable to extol the private judgment of mankind, it is the phrase of protestantism, which was driven to it as an apology for its separation from Rome, but you see the folly of submitting this question to argumentation, the impossibility of making it a mere intellectual topic. Jesus refuses throughout do do so, his words are 'If any man shall do his will he shall know the doctrine whether it be of God or whether I speak of myself'[10] So says in effect every great thinker, Emerson says 'Every mans words who speaks from that life must sound vain to him who does not dwell in the same thought on his own part'[11] Fichte says 'There is no logical means by which this insight can be forced upon man.'[12] Carlyle says 'Get first into the sphere of thought by which it is so much as possible to judge of Luther or any man like Luther otherwise than distractedly, and then we may begin arguing with you'[13] The views of Jesus on this point are best collected from the writings of John;[14] Paul[15] is only half a christian, he is a disputant to the back bone; and he it was who introduced, by his choosing argument as the test of truth, confusion and war into the early christian church; hence arose the antagonist principle the infallibility of the pope; men saw the hopelessness of getting on with that chartist[16] christianity where every screechowl bedlamite[17] proclaimed his right to judge for himself and to establish his views by argument, any thing was preferable to this riotous anarchy, and hence the gigantic imposture before alluded to.

The religious life is progressive, if you are true you will see deeper and clearer this day twelvemonth than you do now, I make this remark because there is a tendency in the age to call every thing nonsense which it does not comprehend; forgetting that the power of comprehension depends upon previous culture, hence the wisdom of patience, let us reserve our judgment on such men as Emerson and Caryle, and on the particular difficulties of their writings, and attribute these difficulties to a defect in ourselves rather than in the writers; do not be discontented if you cant see to the bottom of every thing just now – 'have patience and these oracles will tell thee all'[18]

If religion then be not a question for the intellect how is it to be propogated, how are men to be convinced, that men can be so convinced the history of all religions prove, no founder of religion has leaned upon his power of reasoning, there would seem to be a magnetic sympathy between soul and soul, so that when one utters its earnest thought the other is as it were polarised by the 'utterance' and brought into the same magnetic state – a sincere word was never lost to a sincere mind, a sincere speaker is always a blessing to a sincere hearer; Religion is not a <u>persuasion</u>, it is a <u>life</u> and as Fichte remarks 'it is the first characteristic of true life to create other life from itself'[19]

the leading idea of Past and Present[20] I conceive to be the application of the principles handled in Heroes & Hero Worship[21] to politics, it aims rather at elevating the whole system of politics than at obtaining any single measure or series of measures, we have no Morrison's pill[22] for social maladies;

Inflammation of the lungs is a most dangerous complaint, it was once near cutting me off I got cold neglected it, had to go to bed, the doctor was called and pronounced me in danger, he bled & blistered me, in a week I was able to go out; I went to fish, got wet, fell ill again and remained in bed for 6 weeks afterwards, The effect of the attack is still discernable at times and will doubtless cling to me always, I sincerely trust that by this time your mother[23] is restored to you in the full enjoyment of her usual strength, and now I will close –

bel- me[24] dear Tom <yours affect>ionately Tyndall

I don't know the address of the office – so will send this to Grove Cottage.[25]

RI MS JT/1/HTYP/15–16
RI MS JT/1/T/1016

---

1. *[Leighlin] Nov. 23/48 | [Halifax] Nov. 25/48*: postmark given by Louisa Tyndall. The letter was probably written a day or two earlier, as Tyndall would have been writing it from Marburg.
2. *your letter*: letter missing.
3. *praise the Lord O my Soul*: Psalm 103:2.
4. '*We mark...inly were*': Quote from an address to the Senior Class in Divinity College, Cambridge, Sunday Evening, July 15, 1838. R. W. Emerson, *Nature; Addresses, and Lectures* (Boston: James Munroe and Company, 1849), p. 142.
5. *Caryle*: Thomas Carlyle.
6. *Nickodemus*: Saint Nicodemus is a figure who was noted in John 3:1–21 for his faithfulness to Jesus.
7. *'born again'*: John 3:3 and 3:7.
8. *Mahomed*: an alternate spelling of Muhammed; the Muslim prophet and founder of Islam.
9. *John Wesley*: Wesley (1703–91) was an Anglican cleric and founder of the Methodist movement (*ODNB*).
10. *'If any man ... of myself'*: John 7:17.
11. *'Every mans words...his own part'*: R. W. Emerson, 'The Over-Soul', in A. Hobson Quinn (ed.), *Emerson's Essays* (New York: Charles Scribner's Sons, 1920), p. 148.
12. *'There is no ... forced upon man'*: J. G. Fichte, *The Characteristics of the Present Age*, trans. by W. Smith (London: John Chapman, 1847), p. 258.
13. *'Get first ... arguing with you'*: T. Carlyle, 'Lecture IV: The Hero as Priest, Luther; Reformation: Knox; Puritanism', in T. Carlyle, *On Heroes, Hero-worship, & the Heroic in History: Six Lectures* (London: James Fraser, 1841), p. 213.
14. *the writings of John*: i.e., the Biblical Gospel of John.
15. *Paul*: the Apostle Paul, to whom fourteen of the twenty-seven books of the New Testament are attributed.
16. *Chartist*: the People's Charter of 1838, whose proponents became known as Chartists, demanded democratic reform of the political system. Tyndall was criticizing a democratic ideal of Christianity that allowed men to interpret religion for themselves.
17. *bedlamite*: an insane person (*OED*); a reference to a famous asylum, the Bethlehem Hos-

pital London, known as Bedlam.

18. *'have patience and these oracles will tell thee all'*: the precise quotation has not been identified. Probably an allusion to the tragedies of Sophocles.
19. *'it is the first characteristic of true life to create other life from itself'*: J. G. Fichte, *The Popular Works of Johann Gottlieb Fichte,* trans. W. Smith (London: John Chapman, 1848), p. 261.
20. *Past and Present*: a reference to T. Carlyle, *Past and Present* (London: James Fraser, 1837).
21. *Heroes & Hero Worship*: T. Carlyle, *On Heroes, Hero-worship, & the Heroic in History: Six Lectures* (London: James Fraser, 1841), a collection of six lectures given in 1840.
22. *Morrison's pill*: James Morrison (1770–1840), a merchant, sold popular 'vegetable universal medicines' commonly called 'Morrison's pills'. P. Branca, *Silent Sisterhood: Middle-class Women in the Victorian Home* (Abingdon, Oxon: Routledge, 2013), p. 67.
23. *your mother*: Hannah Oates Hirst.
24. *bel- me*: believe me.
25. *Grove Cottage*: postscript given by Louisa Tyndall.

To William Ginty [10 December 1848][1] 0367

...[2] prove to him that the echoes of ancient melodies still warble in my heart and that amid the tag rag and bobtail of worldly interests and occupations the feelings of my youth still bloom with undimmed radiance. Full of the thought I seized my stick and sallied out amid the pine forests which clothe our rugged hills. Here I wandered alone until night fell.

'And the broad moon rose circling into sight'.[3] During the ramble I must say that I thought less of you than of myself, the mysterious workings of my own spirit, my relations to the scenes around me, to the torrent, the crag, the mountain's head – all was mystery; to one truth however did my spirit cling – that goodness love and blessedness is the sustaining principle of all ...

I never witnessed a more enchanting scene than I enjoyed last night, the sun had gone down as I stood upon the summit of an extinct volcano, with the walls of a ruined fortress round me, on the right were the crimson traces of sunset empurpling the mountains, on the left shone the full moon burnishing the surface of a river which stretched amid the meadows far away.

(I heard) a melancholy account of the sufferings of the poor people in Leighlin – arranged that twelve of them should spend Christmas day comfortably.

RI MS JT/5/13a
LT Transcript Only

---

1. *[10 December 1848]*: Louisa Tyndall annotation: 'Dec 11/48'. However, 11 December was a Monday, so this letter has tentatively been dated to Sunday, 10 December 1848.
2. ...: this extremely fragmentary letter appears in Louisa Tyndall's unpublished biography of her husband. Louisa Tyndall introduced the letter as follows: '1848. | One Sunday Tyndall determined to write to an old friend [Ginty] to' (LT, 'Biography', vol. 1, p. 18).

3. *'And the broad moon rose circling into sight'*: Lord Byron, *Don Juan*, II.clxxxv.4.

To Thomas Hirst and James Craven [Nov. or Dec.] 1848 0368

My dear Boys

I believe Fresenius[1] does not contain a table of the elements and without them he would be unintelligible. I have copied a list with the symbols used to express them. Always the initial letter or some two letters which form the chief part of the <u>sound</u> in pronouncing the name – you will find these symbols in combination in Fresenius – for instance Soda is composed of one atom of Natron or Sodium (the names for the same substance) and one of oxygen and it is written Na O. Alumina is composed of two of aluminum and 3 of oxygen and is therefore written Al O and so of the rest. The elements are divided into two grand classes the metals and the metalloids – those marked with the X are metalloids the rest are metals. They are distinguished by the latter possessing that lustre which we call metallic, while the former possess it not, but more particularity by the fact that the metals are conductors of electricity and the metalloids not. A metalloid combining with oxygen usually forms what is called an acid from its taste – for instance sulphuric acid is expressed by SO one of sulphur and 3 of oxygen – nitric acid by NO one of nitrogen to 5 of oxygen, and so on. There may be and are other acids formed with nitrogen and oxygen, and with sulphur and oxygen than nitric acid and sulphuric acid but the termination -ie always indicates that acid which has the greatest amount of oxygen <u>in its composition</u>. Do not let the number of these elements frighten you many of them are very rare and will never trouble you – you must read quickly on until you come to the groups – you will meet the term 1st. group Potass, Soda, Ammonia, the latter with the symbol NH or NH O I forget which, both are right in a certain sense. Here you must halt and must verify every statement he makes with regard to every substance in the six groups – for instance he says the salts of soda[2] are soluble in water, take a little of the carbonate and prove the truth of this. You must have a little bottle with a thick neck in which you must put a cork and through the cork insert a little candlewick fill the bottle with naptha and this is the spirit lamp you must have. You will also want a blowpipe which any druggist will tell you where to procure. I use the term salts which I must explain – The metals unite with oxygen thus forming oxides or '<u>bases</u>' – thus iron rust is simply formed by the oxygen of the atmosphere uniting with the outer particles of the iron and is written FO. In the same sodium unites with oxygen to form soda NO when an acid unites chemically with one of these bases the product is called a <u>salt</u>. Thus carbonate of soda is composed of carbonic acid CO and soda NO and is written NO, CO. And now I shall bid you goodbye for the present.

Ever yours affectionate | Tyndall

Aluminum	Al	Didynamin		(Mercury)	
(Antimony)		Fluorine X	F	(Hydrargyrum)	H
(Stibium)	S	Glucinum	Gl	Molybrenum	M
Arsenic X	H	(Gold)		Nickel	N
Barium	B	(Aurum)	A	Nitrogen	N
Bismuth	B	Hydrogen X	F	Osmium	O
Boron X	B	Iodine X	I	Oxygen X	O
Bromine X	B	Iridium	I	Palladium	P
Cadmium	C	(Iron)		Phosphorus X	PH
Calcium	C	(Ferrum)	F	Platinum	P
Carbon X	C	Lanthanum	L	(Potassum)	
Cerium	C	Lead		(Kalium)	K
Chlorine X	C	Plumbean	P	Rhodium	R
Cobalt	CO	Lithium	L	Ruthenium	R
Chromium	C	Magnesium	M	Selenium X	S
(Copper)		Manganese	M	Silicium X	S
(Cuprum)	C				
(Silver)		Tellurium X	T		
(Argentum)	A	Thorium	Th		
(Sodium)		(Tin Stannum)	S		
(Natrium)	N	Titanium	T		
Strontium	S	Wolfram	W		
Sulphur X	S	Uranium	U		
Tantalum	T			Vanadium	V
				Yttrium	Y
				Zinc	Zn
				Zirconium	Z[3]

RI MS JT/1/HTYP/19
LT Transcript Only

1. *Fresenius*: probably C. Fresenius, *Elementary Instruction in Qualitative Analysis,* trans. J. L. Bullock (London: John Churchill, 1843). Also possibly the original German version, C. Fresenius, *Anleitung zur Qualitativen Chemischen Analyse* (Bonn: Henry and Cohen, 1841).
2. *the salts of soda*: i.e., sodium salts.
3. *Aluminum A ... Zirconium Z*: this table has been transcribed accurately and matches the version given in Louisa Tyndall's typed transcript, including the four offset elements (Vanadium-Zirconium) at the end of the table. Some of the unfamiliar chemical terms may be awkward Anglicisations of German terminology; it is also possible that Louisa Tyndall mistranscribed some of them when she typed the letter.

1849

From Thomas Hirst [4 or 5 February][1] 0369

We spent 8d on naptha[2] to distill water and then found that the great chemist outside[3] carries on the same process in a wholesale way, and sold his article very cheap. What a magnificent chemist He is. If anyone wishes to appreciate him in that capacity let him go into his cellar kitchen, always taking care that mere names – sciences – do not hide from him the wonder, admiration and awe which belong to him. 'We call that thunder of the black cloud Electricity and grind the like of it out of glass and silk, but what is it? What made it?? Whence comes it? Whither goes it? At bottom we do not yet know; we can never know at all'.[4] I fancy I see you smile as you read this and learn that Tom has waded through a work of Carlyle's at last. Aye, that I have, and I'll soon swim through 'Past and Present'.[5] You will not consider it vanity or self-praise when I tell you that Carlyle's Hero Worship[6] has done me a power of good. I feel a different being to what I was. I have got a glimpse of that religion that exists apart from logic, that can receive no assistance from logic; in fact that mysterious 'heart business' you tried to drive into me. Before, I was trying to see the truth there was in different creeds and could only see a huge jumble of useless forms with no life, no meaning in them. I could not see – blind as I was – that there was a religion apart from them; in short I was beginning to think religion a nonentity, parsons licensed jugglers, and creeds things that we ought 'to believe or to believe that we believe'[7] because they were time honoured things and what our fathers are said to have believed.

Why could I not see that they were but symbols of this Religion, that can only be approached by such forms, that seems almost to defy all symbols? 'As long as men did believe in these symbols they were right'.[8] Yes, I have come to the conclusion that I am better as I am, that there is no good in my trying to believe that I believe, that at any rate I will not be guilty of idolatry.

T.H.

RI MS JT/2/13b/417–418
LT Transcript Only

1. *[4 or 5 February] 1849*: exact date of letter's composition unknown. Tyndall received this letter on 6 February 1849 and transcribed it in his journal.
2. *naptha*: liquid petroleum (*OED*).
3. *great chemist outside*: i.e., God.
4. *'We call that ... know at all'*: T. Carlyle, 'Lecture I: The Hero as Divinity', *On Heroes, Hero-Worship, and the Heroic in History* (London: James Fraser, 1841), p. 12: 'We call that fire of the black thunder-cloud 'electricity,' and lecture learnedly about it, and grind the like of it out of glass and silk: but *what* is it? What made it? Whence comes it? Whither goes it?'
5. *'Past and Present'*: T. Carlyle, *Past and Present* (London: Chapman and Hall, 1843).
6. *Hero Worship*: Carlyle, *On Heroes, Hero-Worship, and the Heroic in History.*
7. *'to believe or to believe that we believe'*: T. Carlyle, 'Lecture IV: The Hero as Priest', in *On Heroes, Hero-Worship, and the Heroic in History*, p. 202: 'A true man believes with his whole judgment, with all the illumination and discernment that is in him, and has always so believed. A false man, only struggling to "believe that he believes", will naturally manage it in some other way.'
8. *'As long as men did believe in these symbols they were right'*: not identified; possibly a reference to Carlyle, 'Lecture IV: The Hero as Prophet', in *On Heroes, Hero-Worship, and the Heroic in History*, p. 102: 'If the wild idolatrous men did believe this, and with their fiery hearts lay hold of it to do it, in what form soever it came to them, I say it was well worthy of being believed.'

To Thomas Hirst and James Craven 9 February 1849 0370

Marburg Feb 9th

My Dear Jemmy[1] and Tommy

<u>My sons</u>

Your laboratory has been an expensive affair, £2 is a high figure, I did not intend you to get Fresenius Quantitative[2] for some time to come, but as you have it it is safe stock; now the best way to make the laboratory cheap will be to extract the greatest possible amount of good out of it, the philosophy of this remark is so evident that you may feel it with your finger ends.

For the present never mind the experiment with potass[3] and Carbonic acid – I have not done it myself; Carbonic acid <u>is</u> a gas, you have heard of persons in descending a coal shaft drop dead suddenly, it was through the inspiration of this noxious gas, it is expired from the lungs in respiration and thus renders air once breathed unfit for <u>further</u> use it is composed of two atoms of Oxygen united to one of Carbon CO_2[4] and is thus prepared

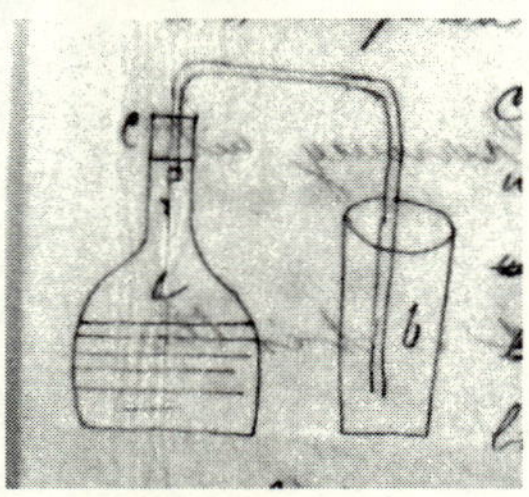

chalk is a carbonate of lime, composed of carbonic acid and lime, by immersing chalk in a stronger acid than CO_2 the latter is driven away and can be collected place some small lumps of chalk or pounded marble in the bottle a (you see the way the glass tube passes through the cork, by the way you can bend your glass tube over the flame of a little oil lamp) pour water in sufficient to cover it and add a little Hydrochloric acid (muriatic acid) or nitric acid, almost any acid the gas will be immediately given off and can be collected in a vessel b, it is so heavy it can be collected as shewn in the figure, it fills the glass like an invisible water, it will extinguish a lighted taper, and if a bit of candle lighted be stuck in the bottom of another glass it will be extinguished by simply pouring the CO_2 upon it like water, the agent of course is invisible. CO_2 decomposes Lime water and gives a white precipitate of Carbonate of lime, take a glass of lime water and breathe into it through a glass tube for some time, it will become milky – just try this – see fig. –

A sand bath is merely hot sand, it is good for evaporation, a water bath is simply hot or boiling water upon which a second vessel containing the substance is laid, thus you get the heat of the water without its moisture. You are perfectly right as regards the litmus paper, there is no better way of testing an acid or alkaline solution. acid of course turns it red alkali merely deepen the blue color. 'Is the menstrum worth keeping'?[5] No! –

The substance KO is called a base, T[6] is an acid, bases and acids combine to form salts and in the case in question they combine in the following proportions (why they do so nobody knows you might as well ask why Jimmy's[7] nose is so like a button) $KO\overline{T} + HO\overline{T}$ that is two atoms of acid to one of base, hence the name bitartrate of potass,[8] just as you are a member of the class bipoda simply because you have two legs. I will briefly point out the reactions in the various groups which are necessary for qualitative analysis, the rest will come forward in Quantitative which is still some distance in prospect – you may for the present omit the antimoniate of potass.[9]

Ammoniac[10] = Make the experiment by mixing with lime and rubbing, make the experiment by mixing with potass heating and testing for white fumes with glass rod moistened by Hydrochloric acid. NB. even though the solution should have no smell, these reagents will develop the smell, muriate of ammonia[11] for instance has no smell.

Barytes,[12] Test with ammonia, carbonate of ammonia or of potass and sulphuric acid – the last most important
Strontian[13] Test as in the case of Barytes, but do not omit the burning with Alcohol – this is very much used to detect Strontian –
Lime Test as in the former cases, but dont omit oxalic acid, it is the Chief –
Magnesia[14] you can try ammonia and the carbonates, but especially phosphate of soda
Alumina[15] Test with ammonia and potass
Chromium Test D°[16] D°
With Tin you will have no difficulty, you are instructed to make that sweet scented gas sulpheretted hydrogen,[17] this must bubble through the solution of Tin, common tin foil can be dissolved in a little hydrochloric acid and used
Mangan[18] no difficulty have you got a blow pipe?
Nickel[19] no difficulty ———————— You can make hydrosulphuret of ammonia[20]

By this time you will be expert at the use of the reagents and do not need my assistance further. In my next I will say something as to the testing for substances in a strange solution, for instance take the pipe water of Halifax and test it with a solution of nitrate of silver the chance is that it will shew the presence of salt by turning milky[21]

I am stronger now a good deal than when I came to Marburg first.[22] Change of place and regimen gave me a heavy headache for the first fortnight but that has vanished and at present I enjoy robustious vigour – I can adhere to my rule without danger, surely 7 hours are sufficient to spend in bed, when you see me again you will look upon a fine blooming young fellow, there is a seam in my forehead, if I could smooth it off with an enamel of putty it would add much to my appearance. Ellershaw[23] is like Noah's dove, where will he rest finally? – I hope he is a good boy and endeavours to apply to his own special case the wholesome homilies he used to deliver to me long ago. Remember me fondly to Tom Perkinton,[24] is Rawdon[25] yet finished – I verily think I shall have to finish Pye Nest[26] myself. what a joke it would be.

I have not got an English edition of Fresenius,[27] a german one[28] lies <at> my elbow, in page 196 of which and Section 117 are to be found instructions as to the qualitative analysis of substances soluble or insoluble in Water but soluble in Hydrochloric and nitric acid. Detection of the Base – I know it is the same section in the English edition but the page doubtful, let one of you mix a few substances (known) together in water or acid and let the other test – the latter will not know what is in the mixture, and he will find the discovery a pleasure, use easy substances such as those 1st mentioned and a few at a time <at> first. NB. you must make the experiment with soda with a blowpipe, <*1 word missing*> pipe shank will do temporarily, this is the best method for the detection of soda. You will find the instructions arranged like Euclid[29] almost, I think you will be able to understand all.

Bridges algebra[30] is a capital book for self instruction. I devote about 4 or 5 hours a day to mathematics, and have scampered over a tolerably wide range during the last three months. You have treated the idea of my portrait so profanely that I dont know whether to send it to you or not. does Jimmy continue to grow? has he reached my mark yet? and now I must bow myself out of the present page and make my exit with a <u>fehle mich</u>[31] *

*a solution at parting – I recommend myself.

One of the students[32] sketched the enclosed,[33] it must satisfy you until I go to Frankfort,[34] he has made me as demure as a Saint –

Will Tom have the goodness to put stamps on the enclosed I will settle with him some day

RI MS JT/1/HTYP/17–18
RI MS JT/1/T/517

---

1. *Jemmy*: James Craven.
2. *Fresenius Quantitative*: C. Fresenius, *Anleitung zur Quantitativen Chemischen Analyse* (Braunschweig: F. Vieweg & Sohn, 1845). Possibly also the English translation, C. Fresenius, *Instruction in Chemical Analysis (Quantitative),* trans. J. L. Bullock (London: John Churchill, 1846).
3. *potass*: a chemical compound containing potassium.
4. *two atoms of oxygen united to one of carbon* CO^2: carbon monoxide, a highly poisonous gas.
5. *'Is the menstrum worth keeping'?*: probably a quote from a previous letter. A menstrum is a chemical solvent.
6. *T*: a chemical abbreviation for tartar.
7. *Jimmy's*: James Craven.
8. *bitartrate of potass*: a compound of potassium and tartar, now called potassium bitartrate.
9. *antimoniate of potass*: a compound of antimony, potassium, and oxygen.
10. *<u>Ammoniac</u>*: sal ammoniac is a hard white opaque crystalline salt (*OED*).
11. *muriate of ammonia*: a salt of ammonium and chloride.
12. *<u>Barytes</u>*: baryte (barite) is a mineral.
13. *<u>Strontian</u>*: strontium carbonate, a mineral (*OED*).
14. *<u>Magnesia</u>*: a term used to refer to both magnesium oxide and magnesium hydroxide (*OED*).
15. *<u>Alumina</u>*: aluminium oxide (*OED*).
16. *D*^{o}: ditto, i.e., repeat the above.
17. *sulpheretted hydrogen*: now called hydrogen sulphide, a poisonous gas with a characteristic rotten-egg smell.
18. *<u>Mangan</u>*: a term for minerals that include the element manganese (*OED*).
19. *<u>Ammoniac</u> ... <u>Nickel</u>*: in this list Tyndall offers a series of tests by which the identity of each substance could be determined.
20. *hydrosulphuret of ammonia*: now called ammonium sulphate, the compound produced by the reaction of sulphuric acid and ammonia.
21. *pipe water of Halifax ... turning milky*: the natural water in the Halifax UK region contains high levels of peaty acids from the nearby moors.
22. *when I came to Marburg first*: in October 1848 Tyndall began studying for a doctoral degree at the University of Marburg. See Introduction.
23. *Ellershaw*: not identified.
24. *Tom Perkinton*: see letter 0340, n. 8 of this volume.

25. *Rawdon*: not identified.
26. *Pye Nest*: a location in Halifax where Hirst and Craven lived.
27. *English edition of Fresenius*: probably C. Fresenius, *Elementary Instruction in Qualitative Analysis,* trans. J. L. Bullock (London: John Churchill, 1843).
28. *a german one*: C. Fresenius, *Anleitung zur Qualitativen Chemischen Analyse* (Bonn: Henry and Cohen, 1841). Henry and Cohen published only the first edition. Later editions were published in Braunschweig by F. Vieweg & Sohn.
29. *Euclid*: Euclid's *Elements,* a Greek treatise on geometry written around 300 BC. Editions of the treatise were still used to instruct students in geometry in the nineteenth century. Hirst and Craven were probably using O. Byrne, *The First Six Books of The Elements of Euclid* (London: William Pickering, 1847); see letters 0346 and 0348 of this volume.
30. *Bridges algebra*: probably B. Bridge, *A Treatise on the Elements of Algebra* (London: R. Watts, 1821).
31. *fehle mich*: missing you (German).
32. *the student's*: not identified.
33. *the enclosed*: enclosure missing.
34. *Frankfort*: a common nineteenth-century spelling of Frankfurt.

From Jean Lancaster 13 February 1849[1] 0371

[Feb 13. 49.][2]

Why did I e'er behold Thee

Why did I e'er behold thee
Too lovely as thou art
Why did I e'er allow thee
To steal away my heart
Time was I knew thee not love
O graceful time for me
Before my eyes met thine love
For then my heart was free
O that I could forget thee
And give my heart its rest
Which palpitates for thee love
Within this ardent breast
But ah, I try in vain love
From all thy charms to fly
Thou art my life and soul love
And without thee I shall die.[3]

RI MS JT/1/TYP/11/3738
LT Transcript Only

1. *Jean Lancaster*: a friend of Tyndall's from the Ordnance Survey.
2. *Feb. 13. 49*: date given by Louisa Tyndall. It is unclear whether this is the date of receipt or the date of writing.
3. *Why did I e'er ... without thee I shall die*: this poem was published in *The Times* and attributed to 'T.F.'. 'To Harriet', *The Times*, 2 December 1837, p. 66.

To Thomas Hirst and James Craven 30 March 1849 0372

[(Mar 30 1849)][1]

My Dear Tommy & Jemmy[2] –

I wish I could write phonography[3] or some other condensation of penmanship that so I might throw all I have to say to you into half a dozen lines – you will find the following sketch of some use in your researches. 1st you must procure some sulphuret of iron, or rather you must make it yourselves, the process is described in Fresenius[4] §26 page 30 German edition, from this sulphuretted hydrogen is easily procured a most valuable reagent; you must also make some hydrosulphuret of ammonia, the process is in Fresenius §27. These two reagents are indispensable – well you take the fluid you wish to seek and make it weakly acid with hydrochloric (muriatic) acid, add, sulphuretted hydrogen – it gives no precipitate (I omit all until I reach the 4th group) add ammonia to the same fluid until it is converted from acid into alkaline and then add hydrosulphuret of ammonia, the following substances are precipitated –

<table>
<tr><td>Nickel
Cobalt } black
Iron</td><td>Manganese – flesh color – Aluminum

Zinc – white – oxide of chromium</td><td>}</td><td>you dissolve the whole precipitate in hydrochloric acid and the following scheme will shew you how to manage the solution –</td></tr>
</table>

1st add solution of potash[5]

<table>
<tr><td colspan="3">In solution
Zink,[6] Alumina, chromium
boils</td><td colspan="3">Precipitated
Nickel Cobalt Manganese and Iron; dissolve this precipitate again in hydrochloric acid and add Sala-mmoniac[7] and ammonia</td></tr>
<tr><td rowspan="3">precipitated by boiling chromium</td><td colspan="2">remains dissolved
Zink & Alumina</td><td colspan="2" rowspan="3">In Solution
Cobalt, Nickel, Manganese
add acetic acid until the solution is weakly sour, heat, and then lead a stream of sulphuretted hyd. gas thro the solution N.B. it must continue for about 10 minutes</td><td rowspan="3">Precipitated
Iron
a reddish brown</td></tr>
<tr><td colspan="2">carry sul. hyd. gas thro solution</td></tr>
<tr><td>precip.
Lime</td><td>dissolves
Alumina</td></tr>
<tr><td colspan="3"></td><td>Precipitated
Nickel, Cobalt
For the present I will say no more of Nickel and cobalt –</td><td>In Solution
Manganese
you can prove whether Manganese is there or not by adding hyd. sulp. of ammonia, it gives a precipitate</td><td></td></tr>
</table>

Supposing hyd. sul. of ammonia gave no precipitate, proceed as follows, add carbonate of ammonia, if 2nd group be there they are precipitated with the exception of magnesia to test for the latter you <must> take a little of the filtrate and add phosphate of <Soda> if there it will be <precipitated> it must be <separated of> course by filtration and the remaining fluid searched for the 1st group. Suppose carbonate of amm. gave a precipitate – Dissolve in hyd. chlo ac. and divide solution into 3 parts – 1st evaporate to dryness, add alcohol, set on fire, if *[strontion]* be present the edges of the flame are red. 2nd add solution of sulphate of lime. Barytes precipitated. 3rd add sulphuric acid, which seperates the Barytes & *[Strontion]* and leaves almost the whole of the lime behind – to detect the latter add oxalic acid, or oxalate of potash – lime precipitated –

With regard to the substances of the fourth group you must buy peroxide of manganese, for iron & zinc you can take a bit of the metal & dissolve it in the proper acid, dilute nitric acid for instance, not too dilute, you know a salt is a compound of an acid and a base. well as salt which contains an atom of base for every atom of acid and vice versa is a <u>neutral</u> <u>salt</u> – if the acid predominate it is an acid salt, if the base it is a basic salt. for example $NaO\ CO_2$ is a neutral salt, – but $\overline{T}2NH_4HO$ is an acid salt there being 2 atoms of acid to one of base (vide Fres. at the end of am^{a}. 1st group, in German edition it is wrong T being only <u>once</u> repeated) for examples of Basic Salts – see <u>Magnesia</u>. – carbonate of potass & phosphate of soda *[throw]* such down – in each case you will find the <u>number</u> of <u>atoms</u> of the base predominant – this division is altogether independent of the effect of the salts upon litmus – a neutral salt will sometimes redden it – See Zinc 4th group –

I wish I could galvanize time as people do india rubber to improve its elasticity – if it could be so stretched I would write to you at greater length, thus far however you have no cause to complain. Tom's discovery of the young chemists[8] in Halifax was I suppose attended with feelings similar to what mine would be should I discover myself some fine day unexpectedly surrounded by a troop of 31st cousins[9] here in Germany – The spring is beginning to peep in upon us, and nature is preparing for new life – man on the contrary sharpens his implements of death[10] – troops of Bavarian dragoons[11] pass daily on their way to Schleswig Holstein,[12] Hesse Cassel[13] sends 6.000 in a week or two they will be slashing away amid blood fire and vapour of smoke – little children in the streets are shouting at the top of their voices

> Schleswig Holstein Stammverwandt
> Bleibe treu mein vaterland[14]

You can guess the meaning. Tom has covered the tracing paper very judiciously <the> change of ink was no improvement – as from front and rear the strength is nearly alike. We have now our vacation, the students are all away and I devote the whole of my time in the laboratory to Electro Chemistry – it is deeply interesting, I intend to follow it up as far as it has been developed, we had a magnificent course on the subject during the winter, the whole matter was beautifully explained & illustrated by splendid experiments. here ends the neutral sheet.

believe me my Dear boys | affectionately yours | J Tyndall

A letter <reached me>[15] yesterday from Tidmarsh – he is on the way to Sidney![16] poor fellow I have known him long, we <were as brothers.>[17]

RI MS JT/1/HTYP/20–21
RI MS JT/1/T/1018

---

1. *(Mar 30 1849)*: date given by Louisa Tyndall.
2. *Jemmy*: James Craven.
3. *write phonography*: i.e., in shorthand.
4. *Fresenius*: probably C. Fresenius, *Elementary Instruction in Qualitative Analysis,* trans. J. L. Bullock (London: John Churchill, 1843).
5. *potash*: a chemical compound containing potassium.
6. *Zink*: zinc.
7. *Salammoniac*: ammonium chloride.
8. *young chemists*: not identified.
9. *a troop of 31st cousins*: i.e., if Tyndall discovered distant relatives living in Germany.
10. *sharpens his implements of death*: probably a reference to preparations for the First Schleswig War (1848–51). From 1848–51, Danish troops and German nationalists fought the first Schleswig War, a conflict over whether Schleswig-Holstein ought to be part of Denmark or part of the German Confederation. The war ended in 1851 with a Danish victory; Schleswig and Holstein remained part of Denmark until 1864.
11. *dragoons*: mounted infantries (*OED*).
12. *Schleswig Holstein*: two Duchies which, in 1849, were part of Denmark.
13. *Hesse Cassel*: also spelled Hesse-Kassel; a German electorate that bordered on the Netherlands.
14. *Schleswig Holstein Stammverwandt | Bleibe treu mein vaterland*: excerpt from the German song 'Wanke nicht mein Vaterland!', 2.5–6. Translation: 'Schleswig Holstein steady stand / Stay true to my fatherland!'
15. *letter reached me*: letter missing.
16. *Sidney*: Tidmarsh moved to Sydney, Australia in 1849.
17. *we were as brothers*: Tidmarsh worked on the fifth division in the C District of the Ordnance Survey with Tyndall between 1840 and 1842.

To Thomas Hirst 16 April 1849 0373

[Ap. 16/49][1]

My Dear Tom,

The old christians and the serious folk who lived in the old time before them had the habit of blessing God for every thing, whether a gift or a deprivation it was all the same. Job exclaims 'The Lord gave and the Lord hath taken away &c'[2] Paul[3] glories in tribulations,[4] another cries it is well that I am afflicted,[5] and they one and all have a hymn of praise for the bounties of heaven, for the bright sunshine and the fruitful dew. These old people seem to have thrown discredit on the creed of circumstances, they seem to have possessed the power of bending all circumstances to their profit and out of seeming evil to extract good. A stout wholesome feeling this, and much needed in our own times. It refreshes with its presence wherever it oozes out, and when I hear my friend and fellow thinker Tom muttering 'I am rather glad than otherwise'[6] I fancy that an echo reaches me from those healthy primeval ages when men rested more upon their own inherent energies than upon external props and pillows – That your mother[7] should call you young and inexperienced is the most natural thing in the world – It is moreover quite true; but experience is a thing of secondary character, what is it at bottom but the solidification of a few energetic thoughts, who are the creators of experience? men – For my part I accept your resolution 'to live honestly by your own labour'[8] as an indication of a healthy, manly, fearless view of the case, and I would lay a small wager that notwithstanding the nullity of his progenitors good intentions, Tom Hirst will compel the world to afford him bread & butter.

Emerson is strongly tinged sometimes with the philosophy of Kant,[9] and without an acquaintance with the critical philosophy of the latter it is almost impossible to unravel some of his passages. Kant sets a thing called Reason down as a kind of master principle in human nature, this reason is in a state of perpetual longing after unity, to gratify its longing it compels Intellect into its service whose office it is to discover relations and connect facts by laws – when you look at the balance wheel of a watch you are not satisfied. Reason prompts intellect to enquire further, to ask after the cause of its motion, and to harmonize its seeming incongruities; intellect sets to work and not until the dependence of each particular part and its relation to the whole are understood can Reason find repose. This thing called Reason then, lying as it does behind intellect is altogether beyond the inspection of the latter and must be accepted as a fact and not explained as an effect. Men may strive to push their hidden feelings forward, as it were, so to enable the understanding to look at them, but in most cases they will find the attempt futile, The man who pretends to know the why for all these things is a shallow man; and you must not think meanly of yourself if you find many of your emotions beyond the plumbline[10] of your intellect – Resignation and humility with regard to such emotions is the best philosophy – A man shews his wisdom sometimes by quietly & humbly yielding himself up to the spirit within him without any attempt to determine its mathematical relations – The passage you have selected is difficult, it is grounded on the Kantean philosophy (though I am not going to say that without a knowledge of the latter the substance of the thought is unattainable) Kant proves, that Space and Time, which are the forms of all ideas, flow from the mind – Space is the form of all sensible appearances, and knowledge of Space and Time is not a knowledge of objects from without but only a knowledge of our mental organization. 'Hence Emersons expression '"know thyself" and "study nature" become at last one maxim'[11] I have once gone through the philosophy of Kant but only superficially

– it will require my entire undivided thought to get to the bottom of it, and on that account I have referred the sifting of it to a future day, not having time at present. Perhaps we <shall> read it together –

I have not seen yeast, but the extract is exceedingly good.[12] The prophet[13] is doubtless the embodiment of some spiritual force possessed by Launcelot[14] himself – when a man turns his thoughts inward he will find many such offering themselves, they are best & most agreeably illustrated under the forms of an angel, a prophet, a demon, a bunch of fairies &c.

Wishing you health & courage in your progress my Dear Tom, & with kind wishes to your mother, believe me

You affectionate friend | J Tyndall

RI MS JT/1/HTYP/22
RI MS JT/1/T/518

1. *Ap. 16/49*: date given by Louisa Tyndall.
2. *'The Lord gave and the Lord hath taken away &c.'*: Job 1:21.
3. *Paul*: the Apostle Paul, to whom fourteen of the twenty-seven books of the New Testament are attributed.
4. *Paul glories in tribulations*: probably a reference to Romans 5:3: 'And not only so, but we glory in tribulations also: knowing that tribulation worketh patience'.
5. *another cries it is well that I am afflicted*: possibly a reference to Psalm 119:71: 'It is good for me that I have been afflicted; that I might learn thy statutes'.
6. *'I am rather glad than otherwise'*: a reference to a missing letter from Hirst. Tyndall quotes this letter again in letter 0376 of this volume.
7. *your mother*: Hannah Oates Hirst.
8. *'to live honestly by your own labour'*: a reference to a missing letter from Hirst.
9. *Kant*: Immanuel Kant (1724–1804), German philosopher best known for *Kritik der praktischen Vernuft* (1788) (English translation: *Critique of Pure Reason*).
10. *plumbline*: a standard by which one can test or judge (*OED*).
11. *'Hence Emerson's expression "know thyself" and "study nature" become at last one maxim'*: R. W. Emerson, *The American Scholar: An Oration Delivered Before the Phi Beta Kappa Society at Cambridge, August 31, 1837* (Boston: James Monroe and Company, 1837), p. 8.
12. *I have not seen yeast, but the extract is exceedingly good*: a reference to C. Kingsley, 'Yeast', *Fraser's Magazine for Town and Country*, 38 (July–December 1848), pp. 689–711; see also letter 0374 of this volume. Kingsley published *Yeast: a Problem* in book form in 1851.
13. *The prophet*: a character in 'Yeast'; see n. 12.
14. *Launcelot*: a character in 'Yeast'; see n. 12.

From James Craven 17 May 1849 0374

Halifax. May 17th, 1849.

My dear Tyndall,

Tom in his letter[1] to you has given you an account of our proceedings both with respect to our Geological enquiries in this neighbourhood and the progress and difficulties we have met with in our Chemical studies. I have little – very little news – to communicate, however I intend to give you a tolerably large dose (as my paper seems to indicate) of the subjects which I introduced in my last.[2] You say they afforded you 'the pleasure of a most healthful cachination'.[3] I am glad of it. I doubt not that ere you have got to the end of this you will do the same – and some benefit at all events will have arisen from this subject – for I guess that it is not often that you indulge in a hearty laugh. You proceed to say I am greatly entangled with Combe's opinion.[4] I am certainly in favour of his views and consequently see things in a very different light from many around me – instead of the perpetual interferences of God which Protestants of all denominations conjure up at will, to account for all the evils which are continually arising, I see or rather in most instances can trace them to the <*1 word excised*> of laws established by a great wise Being which in themselves are wise and just. You laugh and say 'it is incomplete and utterly useless as an agent of Religious Instruction'.[5] But my notions are that if Combe's hypothesis be right the Bible is virtually set aside – for where is the need of Prayer? I grant you that it is not entirely dispensed with – but the command so often reiterated is in a great measure unneeded. The deceitfulness of sin – the wiles of the Devil and all the wiles to which he resorts are all humbug – for man has always good and evil placed before his eyes and happiness or unhappiness must necessarily follow as the result. In short it seems to me if not directly to set aside the Bible, it does so, indirectly, inasmuch as most of the precepts and commands there laid down are nullified. On this argument I do not lay much stress – but on the next one I must say I find a great deal which I cannot in the least grasp. I cannot get it between my finger and thumb and hold it – in fact I cannot reduce it to such a state that I can reason upon it and draw my conclusion. I often think – well, the Bible certainly as a book is divine – the writers seem elevated above the common level of men – and their writing is perfectly in accordance. But then when I begin to look into it and there see that such a fact must be most truly and faithfully believed in – when I see that this is the sole ground on which I can hope to gain happiness hereafter – when I consider the improbabilities which are upon the face of the fact – the mystery which surrounds it – both as to its actual occurrence and numerous other particulars – do you not think it right – nay, can you expect that I should give an unequivocal assent to the whole? Does it not seem to you in discordance with a Great Being's justice and mercy? For Him to say that unless we believe a most curious fact for which no true motive can be assigned? I must confess this is a curious question enough – especially when you know that God has given me powers intellectually for no other purpose than to examine – to compare and to decide upon the various phenomena brought before my notice. Is it not right that upon this all-important question I should bring these powers into play? Surely they should form the principal basis on which I ground my faith. Why then are the facts etc. so mysteriously connected that this mainstay is or cannot be brought into play!

I have gone on this rambling cross-questioning long enough. You will understand no doubt what I mean – and let me tell you I would not have so freely told them any one, but knowing you to be one who has been in a similar position, and knowing that you are always ready to assist, I at once tell you my doubts, believing that in matters of so grave an import we must not make ourselves to believe that we do believe, but genuine and sterling faith must be the basis. Perhaps too I have gone a step further than the spot where my opinions rest; but at all events it is a stage which is but little removed and in the same direction. I have read 'Yeast'[6] – I have a little paragraph (the last in the novel) inserted in my journal, I will give it you as it is about the only one which I know that agrees the nearest with my present views.

The character Launcelot[7] is one which is a facsimile of yourself nearly, at least you are the nearest one that I can remember just now. The writer gives his opinions thus:– 'For Launcelot alone, who has cast off all outward forms, customs, respectabilities, in the hope that he that loseth this life in the old system may find it in the new;[8] – in the hope of reaching some ground of truth and righteousness, some everlasting rock stratum whereon to build, utterly careless what the building may be, how contrary so-ever to prejudice and precedent, provided God and nature help him in its construction, for him alone I see land – while all who set sail with him are tossed back at starting by the breakers or swept into dull whirlpools, there to revolve and rot; for him, far across the deep Atlantic islands bloom. But alas! between are vast, raging waters, foul mudbanks, thick with dragons and syrens; and many a bitter day and blinding night, in cold and hunger, spiritual, perhaps physical, await him. For it was a true vision that John Bunyan[9] saw, and one which as the visions of wise men are wont to do, meant far more than the seer thought, when he beheld in his dream that there was indeed a land of Beulah[10] and Arcadian shepherd paradises,[11] on whose mountain tops the everlasting sunshine lay; but that the way to them, as this year 1848 is preaching to us, lies past the mouth of Hell, and through the Valley of the Shadow of Death'.[12]

This is really beautiful and when one looks round at all the hypocrisy that exists on this one subject Religion one wonders that there is so much prejudice – for outwardly they confess that God is just, merciful and gracious, but when weather or any little matters do not come about in a desirable manner they at once complain, which has been the case with everybody lately on the first occurrence, it having been fine and rain being wanted to bring things round. Thus I contend that by finding fault they prove that they themselves are neither meek nor lowly nor possess any of the Christian virtues – this same thing is asserted in nearly every action of their lives. Yet tell them they are not what they pretend and you offend them. Surely it is a strange world.

But I must pull in or else some dire calamity will overtake you. You see you do not often laugh and I am afraid that something serious might happen were I to continue this strain longer, as the violent convulsions which you have suffered for the last few minutes are such as to betoken that the dose already administered is a pretty strong one. So now for the sugar, first however taking this pill. 'Read the Book carefully, seriously, reverently, and compare it with the unfolding of your own life and you will by this means penetrate its meaning – <u>the moment you cast about for exterior support</u> that moment you leave the right track'.[13] Do you mean I am not to examine it with reference to external objects – that I am not to question the motives, and the opinions therein contained or what – there's the rub! Am I or am I not to reason upon it? or am I blindly, because the style and the writing is such as to touch a chord in my heart, to follow it without further enquiry – Such I think is not your meaning, for it does not seem exactly sensible.

As I said at the beginning I have very little sugar and what I had the mice have run away with it, so you must be content with what you can get. Plainly speaking there is not a particle of news – i.e. I have not seen or heard of anything likely to interest you. My brother in town[14] is coming on gloriously. He is wanting me very much indeed to go to him when my time for recruiting comes, and as my intentions of taking a continental jaunt have changed both in consequence of the unsettled state of the continent and the opportunity of having good company and someone to show me round London as well as of seeing John I have determined to go there in the middle of July. By the bye Continental affairs are but in a very dilapidated state[15] – poor France a spark would set in a blaze, so inveterate is one party against the other. All indeed seem agitated except England where a happy contrast at present exists. Our mills, our workpeople, are all busy – trade and everything is coming round, and everything seems to bespeak a good harvest. Probably it was this which caused the House of Peers to give their assent to the Navigation Laws[16] which at the 2nd Reading was carried by a majority of 10. You will be aware that had this not have passed Stanley[17] would in all probability have been Prime Minister and formed his own Cabinet since Lord John[18] would not carry on.

But I must now draw to a close. I trust the remarks herein contained will not lead you to think any worse of me for if I thought they would I would not have written them.

My brother in his letter wishes to be remembered to you as he always does, but owing to my neglectfulness all are forgotten. The same with my father[19] and others among the rest myself. Trusting then that you are in the enjoyment of good health and that you will continue so,

Believe me dear Tyndall | Yours very very sincerely, | James Craven.

P.S. If this tracing paper is not good to read on let us know and we will adopt our former kind.

RI MS JT/1/TYP/11/3530–3533
LT Transcript Only

---

1. *Tom in his letter*: letter missing.
2. *my last*: letter missing.
3. *'the pleasure of a most healthful cachination'*: the letter in which this quote appears is missing. Cachination (cachinnation) means loud laughter (*OED*).

4. *Combe's opinion*: George Combe [Comb] (1788–1858), the Scottish phrenologist (*ODNB*). Craven was probably reading Combe's recent pamphlet *On the Relation Between Religion and Science* (Edinburgh: MacLachlan, Stewart & Co., 1847), in which Combe argued that scientific knowledge rendered belief in a supernatural Deity unnecessary to explain the natural world, and that religion was not necessary for a system of morality.
5. *'it is incomplete and utterly useless as an agent of Religious Instruction'*: the letter in which this quote appears is missing.
6. *'Yeast'*: 'Yeast', *Fraser's Magazine for Town and Country*, 38 (July–December 1848), pp. 689–711; see letter 0373, n. 12 of this volume.
7. *Launcelot*: 'Lancelot Smith, Gentleman' is a character that appears in 'Yeast' and other stories from the July–December 1848 edition of *Fraser's Magazine*; see n. 6.
8. *The writer ... may find it in the new*: a quote from 'Yeast', p. 711; see n. 6.
9. *John Bunyan*: John Bunyan (*bap.*1628– *d.*1688) was a Christian English writer and author of the highly successful Christian allegory *The Pilgrim's Progress* (1678) (*ODNB*).
10. *land of Beulah*: in Bunyan's *The Pilgrim's Progress,* the land of Beulah is an idyllic land that the pilgrims reach just before the River of Death.
11. *Arcadian shepherd paradises*: Arcadia, a secluded region of Greece, is often described in classical literature as a pastoral paradise or utopia.
12. *'For Launcelot alone ... through the Valley of the Shadow of Death'*: a quote from 'Yeast', p. 711; see n. 6.
13. *'Read the Book ... leave the right track'*: the letter in which this quote appears is missing.
14. *My brother in town*: John Craven, biographical details not identified. The reference in this letter implies that John Craven also lived in Halifax.
15. *Continental affairs are but in a very dilapidated state*: Europe was still recovering from the revolutions of 1848, which had begun in France.
16. *House of Peers to give their assent to the Navigation Laws*: in May 1849, the House of Peers (the British House of Lords) approved the Navigation Bill after hearing it for a second time. This bill repealed the Navigation Acts, protectionist limitations on trade that had been in place since the seventeenth century. See, for example, 'NEWS OF THE WEEK', *The Spectator*, 12 May 1849, p. 1: 'ONE subject has overridden all others in Parliament this week – the Navigation Bill; which stood for the second reading in the House of Lords ... Ministers have induced the Lords to adopt the principle of the bill, on the second reading; ostensibly on the broad grounds, of desirable concessions to foreign allies and customers'.
17. *passed Stanley*: Edward John Stanley (1802–69), second Baron of Alderley and a prominent Whig MP (*ODNB*).
18. *Lord John*: Lord John Russell (1792–1878), Prime Minister from 1846–52 and 1865–6 (*ODNB*).
19. *my father*: not identified.

From Thomas Hirst [27 May] 1849[1] 0375

Thus you see what a little world of science we move in. When I look back upon my past life, and recollect how miserably slow time dragged on, how often morning after morning would dawn upon me, unlit by one single ray of hope, or of purpose to be achieved; and then compare it with the present, when every moment is, comparatively, occupied and the day seems far too short for the work to be done; – I feel thankful for the change. Distant as you are too, every such retrospect adds another link to the chain of friendship between us, for honesty and my own heart give you the praise. I know John Tyndall far too well, to be afraid that he will think this flattery, or consider it hypocrisy when I say further that I spend hours in looking forward to the day, when we shall renew our old intimacy, and when I shall be better able to appreciate its advantages. Still; uneasy mortal that I am; doubt and fear creep in yet. I have in my mind's eye many men, able men too, that have devoted their energies to study and science, have pursued the same track in which I am following,[2] yet their state is far from enviable, at least to my view. They are so many calculating, classifying machines; they always suggest to me some such reasoning as the following – 'To what end is all this study of phenomena? Granted that I shall be able to tell the composition of that rock and assign its proper place in that great hill – that I <u>shall</u> be able to analyse the soil I tread on – Is it for this I was sent into this world? If not <u>what</u> is it for? If its effect will be to reduce me to the pitiful machines that assuredly do exist – if it makes me like an Augustus Comte,[3] who, unless hearsay lies, believes neither God, Devil, nor Religion: I'll have none of it'. Such clouds however, Thank God are only transient. A little more examination shews me, that the evil lies in the Man not in the Study, in his spectacles, not in the things he looks at. Yet I shudder sometimes for fear that I too, may be looking too much on the surface of things: My old want! I wish somebody to tell me how far the visible may be followed and not lose sight of the invisible, the spiritual. At such moments I take up Carlyle or Emerson and the medicine is generally efficacious. They tell me that one grand end that nature subserves to man is 'Language', that 'Nature is the symbol of spirit',[4] and a short reading of Nature's book with such a dictionary, restores me at once to confidence. But it is time I closed this strain – I have just been studying Carlyle's Chartism:[5] a work that just 'clenched' (to use a Yorkshire term) some vague opinions I possessed: No doubt you recollect the time when I was a strenuous advocate of Universal Suffrage and Ballot Boxes. It was a plaster that I thought would heal the sore, a little further experience shook my confidence in the remedy and a vague suspicion arose that the disease was more internal. Carlyle came and told me what I suspected. The only man that I ever <u>read</u>, who has looked any deeper than the surface – the *[Pikes and Monster]* Petitions of the malady. Your 'sketch for the Preston Chronicle'[6] appeared the week following, it was very interesting and as a wholesale manner of letting your friends know your whereabouts, and the company you kept, it was a good idea. This letter has been often interrupted and will consequently be fragmentary. I must leave it again now, so will add the conclusion just before we post it. Meanwhile I remain – my dear Tyndall

Yours very affectionately | T.A. Hirst

Jimmy[7] having filled his folio shoot[8] and pronounced it ready for the post I must just conclude this. We intend to go to the last of Richardson's course of lectures this evening on Pneumatics. I met Mrs Wright today I forgot to tell her I was writing, though she enquired very kindly about you. Poor woman, she is far from being the sprightly body she was when I first knew her. She always refers to the old time when she had you, Tidmarsh and Hall as lodgers. 'Three such lodgers' says she 'as I shall never find again'. I often see little Ada Piercie,[9] she grows a fine girl but looks very delicate, her sisters I never see.

RI MS JT/1/HTYP/27
LT Transcript Only

---

1. *[27 May] 1849*: Louisa Tyndall annotation: 'Fragment of Hirst's letter enclosed to Carlyle; probably written May 27'. The letter to Carlyle is letter 0376 of this volume.
2. *track in which I am following*: Hirst, at Tyndall's urging, was working to improve his education. He had enrolled in courses at the Halifax Mechanics' Institute and was reading a number of books on literature and the sciences.
3. *Augustus Comte*: Comte (1798–1857) was a French philosopher and the founder of the modern positivist movement. Comtean positivism argues that only scientific inquiry, not theology or metaphysics, can produce certain knowledge.
4. *'Language', that 'Nature is the symbol of spirit'*: 'Language' is a chapter from Emerson's 1836 essay 'Nature', and 'Nature is the symbol of spirit' is a quote from the first page of that chapter. R. W. Emerson, *Nature*, New Edition (Boston: James Munroe & Company, 1849), p. 23.
5. *Carlyle's Chartism*: T. Carlyle, *Chartism* (London: James Fraser, 1840).
6. *'sketch for the Preston Chronicle'*: Tyndall occasionally sent his hometown paper news or stories about his life in Germany. This specific item in the *Preston Chronicle* has not been identified.
7. *Jimmy*: James Craven.
8. *folio shoot*: not identified.
9. *Ada Piercie*: Ada Peircy.

To Thomas Carlyle 6 June 1849 0376

Marburg University Hesse Cassel[1] Germany | June 6th 1849

Dear Sir

Many times during the last 5 years has the thought of writing to you suggested itself to me; I have often thought that a closer connexion than that which your books afford would be of great value to me, but I have been repelled, partly by the fear that you would consider me an intruder and partly by the conviction that I ought to be able to fight my way alone; these influences would probably continue to produce the same effect were myself alone concerned; but my object now in writing to you is to bring under your notice, altogether without his knowledge, a worthy and an able young man.

For the last four years in conversation and correspondence I have been a kind of mouthpiece, uttering to him in my own dialect what I had previously received from you. Four years ago he became an articled pupil in an engineers office in the North of England[2] where the writer worked as an assistant; we remained together about two years and since our seperation[3] have corresponded frequently; his letters disclose the unfolding of a vigourous healthy mind; about 6 months ago, through some legal trickery, he lost his patrimony[4] – he writes to me 'I am rather glad than otherwise; I will work for my sustenance':[5] this, and what he says in the accompanying fragment of a letter[6] received a few days ago, will throw a stronger light upon his character than any description of mine could possibly do.

Thomas Hirst is approaching 20 years of age – his physique is one of the finest I have ever seen – he is upwards of 6 feet in height and stout in proportion – his intellectual power is very great – his moral stage is best unfolded by what he says himself; it would be to me a sweet occupation to watch over his progress but this proceeds in so quick a ratio compared with mine, that a point must soon be reached where my 7 or 8 years longer experience will, as the Germans say, be vernichtet.[7] In anticipation of this I would place him in your hands, a word of counsel and encouragement, even at long intervals, will be sufficient for him – I know how he will prize it.

I have lately visited Eisenach[8] and Wittenberg,[9] indeed the foregoing was sketched in pencil upon a hill summit with the Wartburg[10] in sight – I should never have made the journey had I not read your Heroes & Hero Worship,[11] for I should have been readily drawn into the current of those who, unable to estimate men of Luther's[12] calibre, mention them with a compassionating shrug. My young friend has also read the book and in a former letter mentions it as follows – he and another young charge of mine[13] not yet ripe enough for your notice are studying chemistry together – 'We spent ᵈ8 in naptha to distil water and then found that the great chemist outside carried on the same process in a wholesale way, and sold his article very cheap, What a magnificent chemist He is. If any one wishes to appreciate Him in that capacity let him go into His cellar kitchen, always taking care that mere names do not hide from him the wonder, admiration and awe which belong to Him. "We call that thunder of the black cloud electricity and grind the like of it out of glass and silk, but what is it? What made it? Whence comes it? Whither goes it? At bottom we do not yet know; we can never know at all." I fancy I see you smile as you read this and learn that Tom has waded through a work of Carlyle's at last – aye that I have, and I'll soon swim through "Past and Present" – You will not consider it vanity when I tell you that Carlyles Hero Worship has done me a power of good. I feel a different being from what I was, I have got a glimpse of that religion that exists apart from logic, that can receive no assistance from logic, in fact that mysterious "heart business" you tried to drive into me. Formerly I was trying to see the truth there was in different creeds and could only see a huge jumble of useless forms, with no life, no meaning in them. I could not see, blind as I was that there was a religion apart from them; in short I was beginning to think religion a nonentity and parsons licensed jugglers'.[14]

He also mentions 'Chartism'[15] – the effect he ascribes to it is I suppose similar to that experienced by all young men who take time to read it. It is pleasant to find ones own experiences thus flashed back and to find Mr Emersons assertion illustrated that 'there is one mind common to all individual men'[16]

One reflection alone has emboldened me to draw so heavily upon your patience – for myself I ask nothing, though no one could prize your acquaintance more – for my young friend a word of counsel when you have time to give it – I have strong faith that he will one day prove himself worthy of your pains, that in his own circle he will be a valuable man – his address is

Thomas Hirst | Horton Street | Halifax | Yorkshire–

and my name is <u>John Tyndall</u>

Thomas Carlyle Esq[re] | <u>Chelsea,</u> | London.

RI MS JT/1/HTYP/25–26
RI MS JT/1/T/147

---

1. *Hesse Cassel*: also spelled Hesse-Kassel. A German electorate that bordered on the Netherlands.
2. *an engineers office in the North of England*: Richard Carter's firm in Halifax. See Introduction to Volume 2.
3. *together about two years and since our seperation*: Tyndall met Hirst in 1845 when Hirst moved to Halifax, working as surveyor for the West Yorkshire Railway. Tyndall and Hirst were separated in 1847 when Tyndall left to work at Queenwood College. See Introduction to Volume 2.
4. *he lost his patrimony*: the exact circumstances are unclear, but Hirst felt that he had been cheated out of his inheritance following his father's death in 1842.
5. *'I am rather glad ... my sustenance'*: quotation from a missing letter. Tyndall also quoted this passage in letter 0373 of this volume.
6. *the accompanying fragment of a letter*: see letter 0375 of this volume, which was enclosed in this letter to Carlyle.
7. *<u>vernichtet</u>*: destroyed (German).
8. *Eisenach*: a town in the German state of Thuringia, famous as the location where Martin Luther translated the Bible into German.
9. *Wittenberg*: a city in the German state of Saxony-Anhalt. Best known as the city where Martin Luther nailed his 95 Theses to a church door, one of the instigating events of the Protestant Reformation.
10. *the Wartburg*: Wartburg is a castle in Eisenach.
11. *Heroes & Hero Worship*: T. Carlyle, *On Heroes, Hero-Worship, and the Heroic in History* (London: James Fraser, 1841).
12. *Luther's*: Martin Luther (1483–1546), German monk and instigator of the Protestant Reformation. Luther accused the Roman Catholic Church of straying from the Bible by teaching that mankind could earn salvation; Luther argued that humans would never deserve salvation but could only be granted it as a gift from a merciful God. The spread of Luther's teachings led to a split in European Christianity.
13. *young charge of mine*: James Craven.
14. *'We spent 8d in naptha ... parsons licensed jugglers'*: this is an extended quote from letter 0369 of this volume.
15. *'Chartism'*: T. Carlyle, *Chartism* (London: James Fraser, 1840).
16. *'there is one mind common to all individual men'*: R. W. Emerson, 'Essay I, History', *Essays: First Series* (London: James Fraser, 1841), p. 1.

To Thomas Hirst [10] June 1849[1] 0377

My Dear Tom,

I have just returned from a pilgrimage to the scenes of Luthers[2] life and find a mountain of work before me – thus am I necessitated to put you off with a short letter in reply to your last[3] long and valuable communication – 'have patience with me and I will pay thee all'[4] –

Last Monday I was seated in the room of the Wartburg[5] in Eisenach[6] where Martin flung the inkbottle at the devil[7] – the spot where it struck the wall is still shewn, there is no ink the plaster is merely struck away. thence to Leipzig, Kuppers[8] is there, but being attired as a thorough-bred traveller I felt reluctant to call upon him, so passed on to Dresden merely dropping him a note letting him know my proximity – Dresden is something like what Paris was at the end of last June[9] – shattered and burnt in many places[10] – From Dresden I travelled to Wittenberg[11] and saw Luthurs grave his house and old furniture – upon the spot where he burnt the popes bull[12] is planted an oak tree protected by railing and surrounded by flowers

as I rose on friday morning last a 'long rule of levelled light'[13] streamed through a gap in the building and smote his statue which stands in the market place, it was early, all was still, and the scene was suggestive of many thoughts –

I see no flattery in you Tom but I see a feeling which readily leads you to over estimate the effect of our intercourse – had you never seen me you would have been excited by some other agency perhaps a better – I shall not attempt to assign a value to your letter by comparing it with any material good – the quantities are mathematically incommensurable – but this I will say, that what you have written will be a sweet remembrance and precious encouragement to me – I have lately written to Carlyle[14] and mentioned you to him, he may write to you or he may not, if the former good – if the latter we can fight our way without him.

I was once obliged to leave Halifax on account of ill health if you remember – Jemmy[15] will never forget it – I took 'Chartism'[16] along with me and read it in the green fields between Preston and Goosnarg[17] – It is a most able and eloquent work, and deals, as you say, with the root of the matter – I wish heartily that we had a good translation of it for German students, they are brave young fellows but mad with the phantasm of democracy – a republic they imagine is the panacea for all earthly ills – this is a natural reaction against the irrational King-government which has so long pressed them down, but like other reactions it swings just as far beyond the true centre –

Will you mention in your next letter the nature of the experiments made by Mr Richardson? – I have thought it possible that I might some day deliver a course on electro chemistry in England, but if forestalled I of course would spend no time in arranging materials – an account of the experiments will at once inform me whether I have any thing new to offer.

Poor Mrs Wright I am truly sorry to hear that she is no longer cheerful – She was a brave struggling little woman – a little fiery, but that was necessary to keep her courage up.

Doubtless I shall soon be obliged to release Ada[18] from her bond, she will be fixing her choice upon somebody else, the people here say that I am either married or engaged, I am so inattentive & seem to care so little for lady's society – this however is only apparent

I send a sketch of the ramble to the Chronicle[19] and will request the editor to send you a number

bel- me[20] dear Tom | Your affectionate friend | J. Tyndall

next month I will answer your questions on Chemistry –

RI MS JT/1/HTYP/24
RI MS JT/1/T/519-1

1. *[10] June 1849*: Postmark: 11 June 1849, Preston. The letter was probably written in Germany a day or two earlier.
2. *Luthers*: Martin Luther; see letter 0376, n. 12 of this volume.
3. *your last*: possibly letter 0375 of this volume.
4. *'Have patience with me and I will pay thee all'*: Matthew 18:26.
5. *the Wartburg*: a castle in Eisenach where Martin Luther stayed following his excommunication from the Catholic Church.
6. *Eisenach*: see letter 0376, n. 8 of this volume.
7. *Martin flung the inkbottle at the devil*: according to a popular story, the Devil tried to interrupt Martin Luther as he was translating the Bible into German, but was frightened away when Luther threw a bottle of ink at him.
8. *Kuppers*: Conrad Kupper.
9. *what Paris was at the end of last June*: probably a reference to the effects of the 23 June 1848 Paris insurrection.
10. *shattered and burned in many places*: probably a reference to the effects of the May Uprising in Dresden. This uprising was one of the last events in the German Revolutions of 1849.
11. *Wittenberg*: see letter 0376, n. 9 of this volume.
12. *the spot where he burnt the Pope's bull*: on 10 December 1520, Martin Luther publically burned a papal bull (an official declaration from the Pope) which had commanded Luther to recant his teachings.
13. *'long rule of levelled light'*: possibly a misquote of J. Milton, *Comus*, l. cccxl: 'Thy long levelled rule of streaming light'.
14. *lately written to Carlyle*: see letter 0376 of this volume.
15. *Jimmy*: James Craven.
16. *'Chartism'*: T. Carlyle, *Chartism* (London: James Fraser, 1840).
17. *Goosnarg*: also spelled Goosnargh. A village in Preston.
18. *Ada*: Ada Piercy.
19. *Chronicle*: probably the *Preston Chronicle.*
20. *bel- me*: believe me.

From Thomas Hirst 21 June 1849 0378

Halifax – | June 21st 1849

My dear Tyndall –

You will be surprised on receiving this to learn that I am about to pay you a visit. I have long promised myself the treat but hesitated mentioning it until all was certain, indeed I was contemplating walking into your lodgings some fine morning without more notice & hailing you with a <Yorkshire> 'How goes it, Old Boy', but, however on second thoughts & by Mothers[1] advice (who thinks the distance rather too far for the joke, should you happen to be absent) I write to inform you – I have been rather unwell lately & they seem desirous at home that I should make the trip more especially as our house will be shut up in August for they intend spending a short time with my Brother[2] in Bristol. Carter is quite agreeable too so early in August you may expect me – I shall wait for a letter from you in answer to this wherein you will perhaps give me a little information with respect to the best route, passports &c &c anything indeed that will be necessary and useful in my journey. I am scraping up a little of my German, so as to avoid being in such a situation as poor Kuppers[3] when he could not make Mr Wright understand that he wanted something to eat. –

I read with great interest your ramble to the scenes of Brave Old Martin,[4] & was sorry that I could not have been with you. The Editor of the Chronicle[5] sent me a paper & I have since received from him two other papers in which your letters have appeared, & which you ought to have informed us of. We are working away now at the 5th Group,[6] & by the time I shall see you, we shall have done the 6th too. There are many difficulties which I hope to see solved when in Germany – As to Richardson's Lectures[7] you need not fear their interfering with your project they were confined the Elementary principles of Electricity to the exclusion of Chemistry altogether. I will bring you a programme or syllabus of his lectures which will shew you at once

I have just received a letter from a Mr Phillips at Ilkeley[8] to the Following Effect –

'My dear Sir, A letter[9] has been sent to me from Thomas Carlyle which bears the superscription of John Tyndall, in which letter you are honourably mentioned. And it will give me much pleasure to make your acquaintance & to exchange good offices of help & fellowship with you. I am at present *[recruiting]* an exhausted body amongst these beautiful hills & woods at Ilkley & should be glad to see you here for a day or two (say Sunday & Monday next) or Saturday, if your engagements will permit. My life occupation is that of Secretary to the Huddersfield[10] Mechanics Institution[11] and I may be known otherwise to you as (Here he mentions seemingly an assumed name[12] under which he has appeared before the public in print but the writing is such horrid stuff as if written with a stump in half a minutes time that I cannot decipher the name – but he goes on to say) 'If you cannot come & see me here, you will find my house & me in West Parade Huddersfield any time in July: indeed I believe my house stands there now in spite of my absence – at all events write a line here & believe me with much interest

Very truly yours | Geo. S. Phillips

There is an air of originality & honesty about the fellow that interests me much and if I cannot go over to Ilkley I shall certainly go to Huddersfield & make an acquaintance <with> him & his <house. I> only fear you may <have> led them to <expect> a different fellow from plain Tom <Hirst.> But however more about this new companion when I see you –

In looking over one of your old letters I see you say the route is Hull,[13] Antwerp, Cologne <(by> rail) <Frankfort> & Marburgh[14] – Hotels I <should think> are the <best> places to stop at when <passing through.> My <passport> can be got in Hull I <should> think I must never shew it unless asked & they will give me information as to getting it re-signed. Must I get some of my English money changed into German at Antwerp or where? – So waiting your answer to know whether you will have me. I remain my dear Tyndall

yours very affectionately | TA Hirst

Mr John Tyndall, | care of Professor Bünsen, | <Marb>ourg, | <HES>SE CASEL | GERMANY.

RI MS JT/1/HTYP/28
RI MS JT/1/H/137

---

1. *Mothers*: Hannah Oates Hirst.
2. *my Brother*: Hirst had three living older brothers. He was probably referring to John Henry Hirst (1826–82), who worked as an architect in Bristol. 'Obituary', *The Furniture Gazette* (15 July 1882), p. 45.
3. *Kuppers*: Conrad Kupper.
4. *Brave Old Martin*: probably a reference to Tyndall's visit to Martin Luther's house in Wittenberg; see letter 0377 of this volume.
5. *The Editor of the Chronicle*: this probably refers to the *Preston Chronicle* (see letter 0386 of this volume), but the editor of this paper at this time has not been identified.
6. *5th Group*: i.e., the 5th group of chemical elements, discussed in C. Fresenius, *Elementary Instruction in Qualitative Analysis,* trans. J. L. Bullock (London: John Churchill, 1843); see letter 0368 of this volume.
7. *Richardson's Lectures*: lectures by William Richardson on electricity, galvanism, and pneumatics at Gainsborough, mentioned in the *Lincoln, Rutland, and Stamford Mercury,* 19 May 1848, p. 2. There is also a reference to a lecturer named William Richardson in H. Thomson, *A History of Ackworth School* (London: Samuel Harris and Co., 1879), p. 249.
8. *Ilkeley*: a town in West Yorkshire, England.
9. *A letter*: probably letter 0376 of this volume.
10. *Huddersfield*: a town in West Yorkshire, England.

11. *Huddersfield Mechanics Institution*: this Institution was founded in 1843 as the successor to the Young Men's Improvement Society. 'A short history of the University', *University of Huddersfield* (7 May 2013), at www.hud.ac.uk/archives/briefhistory/ [accessed 21 May 2014].
12. *an assumed name*: George Searle Phillips often published under the pen name 'January Searle' (*ODNB*).
13. *Hull*: Kingston upon Hull, a city in Yorkshire, England.
14. *<Frankfort> & Marburgh*: common nineteenth-century spellings of Frankfurt and Marburg.

To Thomas Hirst 2 July 1849 0379

[July 2/49][1]

My Dear Tom,

The intelligence has indeed fallen upon me unexpected, like the first shower of coriander manna upon the Israelites in the desert of the red sea.[2] One word however before you set out, dont allow imagination to inflate herself to an undue extent dont expect too much, nay expect very little for I should not like to send you back disappointed with me and Marburg. our town is nichts besonders,[3] nay it is positively old and ugly and were it not lifted up into heavens sunshine by the hill on which it stands would be a semiplace of perdition – our University is not grand, it is broken into parts and presents no imposing front – our laboratory[4] presents rather a scoundrel like appearance but dont conclude hastily against it – it holds a man whose superior as a chemist is not to be found within a radius of 8000 miles from the Piece Hall of Halifax[5] – there however right over against me on the summit of a hill with the sun shining upon its white walls, and its tower piercing the air is a fine building an astronomical observatory and physical institute, its interior furnished with costly apparatus and notwithstanding all this I will not say much for it, on the other hand I can lead you into a little room with hacked ricketty benches perhaps the whole not worth 5 and sixpence where a man of genius makes his hearers forget the poorness of his furniture as he crushes the crust of a mathematical Calculus between his fingers, but come however and judge for yourself, if you are a lover of nature you will be pleased Marburg liegt schön[6] – our neighbourhood is smashed and torn by volcanic agency and here & there the basalt juts from the hill summits –

Your best route I believe will be from Hull[7] to Antwerp. Packets sail twice a week – you must ascertain the days and the fare – From Antwerp you come to Colne[8] for a few shillings, you travel always 3rd class of course. from Colne along the Rhine to Mainz[9] – from Mainz to Frankfort[10] by railway, and thence by a rascally old postomnibus[11] to Marburg take the 1st cabin at Coln (Cologne), as I believe you sleep a night on board, If you were over in London your best way would be by Dover to Ostend[12] and thence by Railway to Coln. The fare from Antwerp to Marburg will be about 28p* and it will cost you perhaps 10p for

Nahrungs Mittel[13] &c. vieleicht nicht so viel[14] – I would write to London if I were you and ascertain the time and fare by that route also then you can compare both – you may come as early in August as you please but you must so manage that you can remain till about the 10th of September, we have then our vacation and I will accompany you part of the way back – if you like we will take a run into Switzerland together, that is if it dont cost too much – Chambers[15] has written a narrative of a tour up the Rhine. you can get it for a few pence, buy it – I am wrong, I have just made enquiry, the title of the book is a tour in Holland, the Countries of the Rhine and Belgium[16] – but instead of a few pence it cost 1/6 perhaps it is cheaper now. you will get maps of the river cheap along the way but dont buy a large one, they are expensive –

* this at the highest, it may be less.

I believe Mr Phillips is the January Searle of the truth seeker[17] – in the last number that I have seen was a discourse on the use of books <signed> G.S.P.[18] probably by him – the style was his – he spoke rather freely of both Caryle & Emerson, acknowledged them to be the greatest English thinkers but insinuated that the latter was <an> imitator and the former unhealthy. I believe however the imitation of Emerson had <reference> only to his style not to his matter – & that the unhealthiness of Carlyle consists in his rejecting the agency of worthless men in the effecting of any good end whatever – In attempting to describe Caryle however G.S.P. <resembles> a man swimming amid billows and talking of the expanse of ocean round him, he is swamped in the wave his pen describes, he is a disciple of Caryle to the back bone, a rugged chip of the same rugged block – dont think that I undervalue G.S.P. his utterances though <tinged> by the peculiarity of the man are brave and honest there is an unknown mass of talent in him – When you come you will be able to tell me whether I am right in my surmise –

I'm not sure of my spelling[19]

Poor McArthur![20] His death and the method of it are the most natural things in the world. superior intellectual cultivation allied to an inveterate love of gin, both eternally pulling in contrary directions – he could not bear it – it is in fact unbearable – Had he been true to his bottle and thought no further he would have lived; but he had reached a point from which he could see the <degradation> of his habit, this sight coupled with the <absence> of all hope of successful battle against it led naturally to self slaughter. pious people will call the act wicked and send him to hell by special train without thinking of the tortures and agonies, the vain efforts, and bitter self accusations which took place beforehand. there is a deeper meaning in the fate of such a man than pious people take notice of.

English money – gold – is passable everywhere – I want you <to> bring me a little with you – Write to Mr Edmondso<n> Queenwood College[21] Stockbridge letting him know the day you start from Halifax. I will write to him to send you 10 or 15 or perhaps 20£ which you will like a good boy bring over in sovereigns with you – take care of being robbed – If you come here without my money I will put you in the hand of the State officers.

affectionately yours | Tyndall

Put postage stamps on the enclosed. I will pay you when you come

If you can get a cheap edition of Launcelot[22] buy it for me

You will <perhaps> besides the money from Mr E. receive a pound or so from another quarter –

RI MS JT/1/HTYP/29–30
RI MS JT/1/T/520

---

1. *July 2/49*: date given by Louisa Tyndall.
2. *coriander mana ...red sea*: a Biblical reference (Exodus 16) to God's daily provision of manna, a consumable, to the Israelites during their travels through the desert (*OED*). It is said to have been white in colour, like coriander seeds.
3. *nichts besonders*: nothing special (German).
4. *our University...Our laboratory*: in October 1848 Tyndall began studying for a doctoral degree at the University of Marburg. See Introduction.
5. *Piece Hall of Halifax*: Piece Hall is a Georgian building that stands in the centre of the town of Halifax in West Yorkshire, England.
6. *liegt schön*: 'is beautiful' or 'is in a beautiful location' (German).
7. *Hull*: Kingston upon Hull, commonly called 'Hull', a port city in Yorkshire.
8. *Colne*: Cologne, Germany.
9. *Mainz*: a city in the German duchy of Hesse that lies alongside the Rhine.
10. *Frankfort*: a common nineteenth-century spelling of Frankfurt.
11. *postomnibus*: typically a horse-drawn public vehicle that transported passengers by road (*OED*).
12. *Dover to Ostend*: Dover is a ferry port town in Kent, England; Ostend is a small city in Belgium.
13. *Nahrungs Mittel*: food, consumables (German).
14. *vieleicht nicht so viel*: maybe not so much (German).
15. *Chambers*: William Chambers (1800–83), a Scottish publisher known for his work on *Chambers's Edinburgh Journal*, a popular and inexpensive weekly publication that first appeared in 1832 (*ODNB*).
16. *tour in Holland, the Countries of the Rhine and Belgium*: W. Chambers, *A Tour in Holland, the Countries on the Rhine, and Belgium* (Edinburgh: William and Robert Chambers, 1839).
17. *Mr Phillips is the January Searle of the truth seeker*: George Searle Phillips often published under the pen name 'January Searle'. He also edited the transcendentalist publication the *Truth Seeker* along with F. R. Lees (*ODNB*); see letter 0378.
18. *discourse on the use of books signed G.S.P.*: G. S. P., 'The Choice of Books', *The Truth-Seeker and Present Age: A Catholic Review of Literature, Philosophy, and Religion* (1849), pp. 410–19.
19. *my spelling*: the specific word to which Tyndall is referring is unclear.
20. *Mc Arthur*: not identified.
21. *Queenwood College*: the school at which Tyndall taught from 1847–8. See Introduction to Volume 2.

22. *cheap edition of [Launcelot]*: possibly a reference to 'Yeast', *Fraser's Magazine for Town and Country,* 38 (July–December 1848), pp. 689–711; Launcelot is a character in this story. See letter 0373, n. 12 and letter 0374 of this volume.

To Thomas Hirst 30 July 1849 0380

[Marburg July 30/49][1]

My Dear Tom,

I'm sending a packet to England & seize the occasion to request you to bring me as follows.

" 5 shillings worth of postage stamps.

" A Booksellers catalogue of Scientific books, chiefly astronomical; I want it for a professor here. Lord[2] will assist you to obtain it & it will oblige me much –

Is there not a Second series of essays by Emerson?[3] I think so. I have one containing History, Selfreliance, Compensation, Spiritual laws, Love, Friendship, Prudence Heroism, The oversoul, Circles, Intellect, Art, Nature[4] subdivided into 8 parts Commodity, Beauty &c, Lectures on the times .3. and orations – If there is a series besides this pray bring it to me; it costs I believ 3/6 – I think there is one 'with a notice by Thomas Carlyle' –

Bring a few punches[5] with you I has not seen one since I came to Germany;

I am working hard at present preparing for an examination which I intend to pass in a month so when you come I shant be able to play Jackstones or shoot marbles[6] with you; we shall have doubtless many pleasant rambles together but I warn you to buckle on your harness for a little work; and further to buckle up your heart against the seductions of our Marburg beauties –

I have got the reputation of being cold and zurückhaltend[7] among them –

If you get the money from Southampton or from Mr Edmondson pray enquire of Dr Lees Pontefract road, Leeds, how much I owe him and remit the Sum, I take his Truthseeker[8] – –

I believe I owe nothing more in England;

The state of trade

The state of education

The state of Benevolent institutions generally as in another half year I shall probably be forced to apply for aid –

Does Mr Baines[10] still persevere in the attempt to smooth education

Mr Baines ideal of education is by no means exalted –

post the accompanying[11] for me –

Wishing you a fair passage and no sea sickness

believe me Dear Tom | Your affc[12] friend | J Tyndall

Is Jimmy[13] still in Halifax – if so kiss him for me – bring me a kiss from Ada[14] – & remember me kindly to your mother[15] –

I wish Ada was living in Marburg – we should have such long walks in the woods together

RI MS JT/1/HTYP/31
RI MS JT/1/T/1019

---

1. *Marburg July 30/49*: postmark given by Louisa Tyndall.
2. *Lord*: not identified.
3. *a Second series of essays by Emerson?*: R. W. Emerson, *Essays: Second Series* (London: J. Chapman, 1844).
4. *History, Selfreliance ... Nature*: a list of essays from R. W. Emerson, *Essays* (London: James Fraser, 1841). Also mentioned in letter 0380 of this volume. Later editions of this book were republished under the title *Essays: First Series.*
5. *punches*: *Punch, or the London Charivari*, was a weekly satirical English periodical.
6. *play Jackstones or shoot marbles*: Jackstones is a game in which participants attempt to catch pebbles that have been tossed into the air in a basket. In the traditional game of marbles there is a central ring in which marbles are placed; players take turns trying to flick their own marble at the central marbles to displace them from the ring (*OED*).
7. *zurückhaltend*: restrained (German).
8. *Truthseeker*: Lees edited a transcendentalist publication entitled *Truth Seeker* from 1844 to 1850 (*ODNB*).
9. *sensorium*: the human brain, understood as the centre of thought and nervous energy (*OED*).
10. *Mr Baines*: probably Sir Edward Baines (1800–90), an English journalist, politician, and supporter of widespread education for people of all classes and all ages (*ODNB*).
11. *the accompanying*: not identified.
12. *affc*: affectionate.
13. *Jimmy*: James Craven.
14. *Ada*: Ada Piercy.
15. *your mother*: Hannah Oates Hirst.

To Thomas Hirst 10 September 1849 0381

Marburg 10th Sept. 1849.

Dear Tom –

By the time this reaches you the first gush of feeling will have found vent and you will begin to look at the realities of your position – I cannot multiply words at present – a year hence you will see something beautiful in your present calamity[1] – you will find a profit in sorrow – on one point I would do what I could to set you at rest – you told me some time ago that you were cheated out of your patrimony[2] – I dont know the precise nature of your worldly position but whatever it is Tom dont fear it. I have no estate to share with you, but as long as I have a hand and brain and nine years extra experience, these good and chattels and whatever marketable value they can bring is yours as well as mine – – my words may be altogether superfluous they are suggested by passages from my own early experience, and are merely thrown out for the purpose of stifling this puzzling question of board and lodging and allowing your mind to turn to more profitable thoughts – God bless you my boy – remember the Jewish proverb 'Woe be unto you if ye quit the world without tribulations'[3] – our best men have been taught in this school – sorrow rightly used is purifying and holy – this your own experience will verify far better than I can express.– a calamitous stroke often taps the eternal springs of human nature and covers the heart with herbage –

again God bless you Tom – | your affectione[4] friend | Tyndall

Frederick[5] sends his kind remembrances

Thomas Hirst | Horton Street | Halifax | England

RI MS JT/1/HTYP/32
RI MS JT/1/T/521

---

1. *your present calamity*: i.e., the death of Hirst's mother. Louisa Tyndall annotation: 'Hirst's Mother died Sunday night; he quitted Marburg for Halifax Sept 8'.
2. *you were cheated out of your patrimony*: the exact circumstances are unclear, but Hirst felt that he had been cheated out of his inheritance following his father's death in 1842.
3. *'Woe be unto you if ye quit the world without tribulations'*: quotation or proverb not identified.
4. *affectione*: affectionate.
5. *Frederick*: possibly Frederic Richard Lees.

From Thomas Hirst 18 September 1849 0382

Halifax Sep. 18th/49

My dear Tyndall –

I have escaped for an hour from the confusion & uproar attending the breaking up of a home;[1] none can feel the sweet significance of that term so keenly until they find themselves in situations like mine, at the very moment too when we feel a longing to sit down & muse on events transpired it is distracting indeed. It was amidst such confusion & in such a disposition that your letter[2] found me, and may God bless you Tyndall for its contents. I shall ever cherish the tokens of affection it contains & never forget its cheering effect upon me in that time of my affliction. 'A friend in need is a friend indeed'[3] – I can always meet affliction best in solitude & for that reason I wished to leave you so soon[4] although I knew at last its inutility. It was agony to me at that time to receive sympathy even from yourself although I knew well its sincerity. In such moments as those we would look sorrow calmly in the face. I scarcely remember a single incident of that nights journey. I sat huddled up in my corner for 10 hours & knew nothing, thought nothing, was sensible of nothing but a despairing stupor, until arriving at Frankfort[5] & exposing my features for the first time I met the astonished curious gaze of my companions of the journey – So much does nature take the impression of the mind that the lovely scenery of the Rhine seemed tame to me: & thus I ended my miserable journey. Yet I found no peace even then. it is true I was welcomed by kind faces but their very kindness pained me, & was soon followed by incessant talk of goods & chattels, money & worldly matters, more distracting still. I longed to be away again to escape such desecration, yet I knew not where to go – I wished for some support apart from myself to bear my burden, & found at last that I must be my own physician. At last conviction came upon me that such grief as mine was pure weakness – useless – & after much wrestling resignation crept over me by degrees – this is comfort to the misery I was in & I look forward with hope & without fear to time when after all this bustle I can sit me down in peace & calmly grapple with my position

And now to other matters. I fully expected that now I should be thrown upon the world entirely on my own resources that I must look alone to my hands & brain for my own subsistence. I had fully reconciled myself to this & longed to be up & doing & was consequently almost dissapointed when I learnt that my poor Mother[6] had kept us much in the dark in regard to our circumstances & no doubt expected it would be a little surprise to us to know at her death that we were possessed of a little fortune still. Poor Good Soul she acted wisely for by Gods help the determinations I had formed while under such an impression I will follow still. I had made up my mind to do without & it shall not weaken my resolve – All fear of board & lodging for me is taken away Tyndall. I am master of what I consider a considerable independence & such as will enable me far sooner than I expected to follow the bent of my own inclination at Marburg[7] but more of this another time – Only this much I will say more on the subject just now, and in mentioning it I would have you not to consider it an impulse of the moment called forth by your own generous offers, for it is no such thing the idea I have had since you first went to Marburg except at the time when by my own fancied losses I thought it was impossible. but now my position being clear I wish you to weigh it calmly & not to be influenced by any motives of delicacy but to take it as the act of one brother towards another, for I can only say that your acceptance of my offer will gratify me as much as your refusal would cause me sorrow – My Mothers trustees have considered that on account of their opinion of my ability to manage my own affairs & from a general opinion of my prudence &c &c &c that contrary to the usual custom with minors I am to have the entire control of my own income, it amounts to some £150 per annum: Half that amount will serve my purposes or nearly (& I mention the sum in order that you may fully judge for yourself & not consider it enthusiasm or any other boyish influence) Now I know well your desire to remain longer in Germany Berlin or France & that this affair of money is your sole hindrance. Will you accept of part of this for the purpose? If you do not I shall have to deposit it in some other persons hands to keep it for me. You can see from my statement of the circumstances that what will serve your purpose will be no inconvenience to me & I will trust to that same brain of yours to repay me whenever it may suit you. I wont offer it as a gift for then there would be small chance of your acceptance. Consider this Tyndall my boy, at your leisure, & be as candid with me as I am with you – I shall long remember the kindness shewn me by Mr Knoll his kind, honest & manly conduct will often recur to me. remember me to him kindly & tell him if I ever meet him in Germany I shall grasp his hand with the feeling of a brother, & that a time will come when I shall be able to thank him without the use of a Dictionary – Frankland too & all my other friends I do not forget for I met with nothing but kindness from them during my visit. In the hurry of my departure I forgot many things that I now think of, the settlement of which we must defer to our next meeting. At present then adieu

from your affectionate friend | T.A. Hirst

Sep 25th / 49

Since writing the above I have taken lodgings with your old friend Mrs Wright who I have no doubt will make me as comfortable as possible. The enclosed is a letter parcel[8] for you which I forgot to deliver when I was over. I don't know exactly who it is from but it will probably explain itself – TAH.

RI MS JT/1/HTYP/33-4
RI MS JT/1/H/138

1. *breaking up of a home*: Hirst's mother, Hannah Oates Hirst, had just died; see letter 0381 of this volume.
2. *your letter*: letter 0381 of this volume.
3. *'A friend in need is a friend indeed'*: a common English proverb praising those who aid friends under difficult circumstances.
4. *leave you so soon*: Hirst had been visiting Tyndall in Marburg when news of his mother's death reached him.
5. *Frankfort*: a common nineteenth-century spelling of Frankfurt.
6. *my poor Mother*: Hannah Oates Hirst.
7. *my own inclination at Marburg*: Hirst hoped to study for a doctoral degree at the University of Marburg.
8. *The enclosed is a letter parcel*: according to letter 0384 of this volume, the enclosed consisted of a 'sugar work' (homemade sugar candy in a fanciful shape) from Ada Piercy.

To Edward Frankland [6 or 7] October 1849[1] 0383

An | Den Herrn Dr Frankland[2] | Gasthaus Zum Ritter[3] | Cassel[4] | 1849.[5]

My dear Frankland

I arrived and you were gone – gone in a double sense – I opened your letter[6] and your card appeared

Edward Frankland | Sophie Fich[7] | verlobt[8]

my heart jumped for an instant against my ribs; Noll[9] was beside me, I thrust the document into my pocket and felt somewhat similar to a fellow coming home and finding that his brother had died two hours ago. 'O that I could have had but one word with him before departing',[10] when the probability is that had he even arrived in time he would have had nothing to say – Well the die is cast and I will not indulge in vapoury congratulations; but if a wish as solid as a heart like mine can utter for the lasting happiness of both you and your Sophie be of any value to you you have it – had I an intelligible dialect of my own I would use it, but as this is not the case I must help myself to the accredited form of expression and say from the bottom of my soul God bless you –

Your affectionate friend | Tyndall | E. Frankland

RI MS JT/1/TYP/12/3967
LT Transcript Only

---

1. *[6 or 7] October 1849*: Tyndall's journal records that he returned to Marburg after a trip and found Frankland's engagement announcement on 6 October 1849 (RI MS JT/2/13b/465).
2. *An | Den Herrn Dr Frankland*: To The Honoured Dr. Frankland (German).
3. *Gasthaus Zum Ritter*: the oldest extant house in Heidelberg, Germany.
4. *Cassel*: a city in the German electorate of Hesse-Kassel. The city now spells its name 'Kassel'.
5. *1849*: year probably added by Louisa Tyndall based on her knowledge of when Frankland announced his engagement.
6. *your letter*: letter missing.
7. *Sophie Fich*: Sophie Fick (1821–74), later Sophie Frankland. Frankland had fallen in love with her during his first trip to Marburg in 1847. The couple married in 1851.
8. *verlobt*: betrothed (German).
9. *Noll*: Mr. Knoll.
10. *'O that I could have had but one word with him before departing'*: quotation not identified.

To Thomas Hirst 10 October 1849 0384

Marburg [üf nonio] nüft [Munn] 1849

My Dear Tom,

I hereby certify that you deserve some credit for originality, for, by the Law, the offer you have made me[1] is not a thing common in the world. It argues a degree of courage which I hardly contemplated as belonging to your pulpy texture! – What you dog! would you 'set your life upon a cast and stand the hazard of the dye'?[2] – and to have contemplated the thing beforehand – this tops all, for it takes away every excuse of rashness from the act and reduces it to a case of 'malice prepense'![3] – Somebody mentions a class of people who feel themselves wronged in being born wealthy, inasmuch as their independence has shut them out from all healthful struggle in the world – it was with a feeling somewhat akin to this that I read your letter; I should have felt a grim satisfaction and the strength of ten men in taking the world by the throat, did I know that the fate of my friend depended upon the energy of my gripe![4] but your letter pared me down to my former dimensions and just left me sufficient courage to struggle for myself. yet do I not bless thee Nature? – sweet mother! who, teachest my hands to war and my fingers to fight, and who when difficulty arises, pats thy son cheeringly upon the back and criest 'Go it my boy – the devil himself is gifted with claws only by thy sufferance'.

'Whatever is is right'[5]

So saith Alexander Pope[6] wheth he understood it or not. It would be a holy and wholesome struggle, Tom, yours with the world had you been compelled to buckle on your armour and face it – it is just as holy and wholesome however to let it alone and to fight for something else */besdes/* a bellyfull. – Were I a catholic how I would pray for the everlasting peace of that good mother who thus watched over your welfare[7] – What is it Tom at bottom but a transfusion of life – no death at all – I saw two signboards the other day one bearing the date 1791 the other 1847 – In that little gap thought I fitted the lifetime of my father,[8] and here am I to carry out the same, ready also to disappear when my time comes; shall I mourn because a leaf falls from the everlasting tree? Is his tongue silent? can I not now truly affirm that 'he being dead yet speaketh'[9] – is he not engrafted into me, 'bone of my bone and flesh of my flesh?'[10] Come not even now the brave words as fresh as ever from his grave – 'be just and fear not'[11] – by the Lord I would not give his memory for that of a King, nor my own blood for the Prince Royal's.

I stand here upon a free equipoise,[12] ready to accept your manly offer to some extent and just as ready to let it alone – have you calculated chances? If God gives me health and strength I shall pay all debts and have capital remaining, but I may die, and the cholera is at hand to increase perils; what then Tom? draw bills upon the world to come for payment or exclaim in the spirit of Marcus Antonius 'I have lost all except what I have given away!'[13] –

I dont know what to say; 3 or 4 months in France would be of great advantage to me and you would not press me for payment for a couple of years – I dont know what to say – I should have been vastly content had you been a beggar, and now equally so when this is not the case – I will decide nothing now; for the present I will look to affairs as if you were at the antipodes[14] and if I find you can render me a service I will accept it, without troubling you with many thanks. I thank you not, because the act if rightly performed will be its own reward – 'There are 20 pounds' said I to a man[15] once who wanted it but did not expect it – 'take them' – he did so and owes it to me yet, but the expansion of my pericardium[16] at the time was recompense sufficient to tempt me to a similar act tomorrow had I the means. If I borrow any thing from you it must be done in proper legal form, not that I doubt myself or doubt you – but as promising youths as I am have heretofore turned vagabonds – tell me then how much you could conveniently lend me, to be repaid in two years from the time of lending? you need not be in haste to answer my question just deal with it in the common course of correspondence.

I have been to Switzerland – paid Heidelberg[17] a visit and the sun shone so bright and clear that I felt an irresistable impulse to go further – trusting to my legs and mountain stick, repudiating guides, eating bread and milk and sleeping when possible in the country villages where nobody could detect my accent, I got through amazingly cheap – I have thrown my adventures into the form of a story, the incidents real, the plot fanciful, it will probably appear in the Chronicle,[18] just enquire, and try can you unravel a character or two – you will laugh at least twice over it, I lay a wager –

Guess what the packet was, a bit of sugar work from Ada.[19] why in heaven's name didn't you open it before you paid the extravagant postage it must have cost. Noll[20] and I echoed each others laughter as the mysterious bundle was unloosed and the little cherries of gum and lump sugar revealed themselves – I wrote to Ada but did not mention this – if she permits you kiss her for me. I shant be jealous because I know that Ada is true.

Poor Mrs Wright, my personal connexion with her seems to be once more revived, on account of both of you I am glad to see you there – all my lifetime with her flashed back upon me as you mentioned her name[21] – the hardworking little woman running up and down stairs at the tinkle of the bell, bearing folds of savoury beef or rectangular masses of Yorkshire pudding – the lively cachination[22] of John Hall and the lisp of Kuppers,[23] all are present with me; a word from you now and then may serve her son Alfred[24] –

You talk about a settlement. this is beggarly after your offer. I know what you mean – these paltry half dozen Punches[25] is it not? I'll quietly leave you and Mrs Fearnly[26] to settle the matter, for – I was going to swear By Heaven! – I wont pay a sous[27] for them!

Noll is in Cassell.[28] Frankland is in Geissen[29] – Frankland has engaged himself to one of the young ladies[30] we met at that village party – you noticed her, she sat nearly opposite and spoke English –

And now my dear Tom may the Lord prepare your pasture and the way of the reightious prove your everlasting delight.

Your affectionate friend | Tyndall

RI MS JT/1/HTYP/35–36
RI MS JT/1/T/862

1. *offer you have made me*: Hirst had offered to lend Tyndall part of his inheritance; see letter 0382 of this volume.
2. *'set your life upon a cast and stand the hazard of the dye?'*: W. Shakespeare, *Richard III*, V.v.9–10: 'Slave, I have set my life upon a cast, | And I will stand the hazard of the die'.
3. *'malice prepense'*: a wrong that is intentionally performed (*OED*).
4. *gripe*: the act of gripping or clutching (*OED*).
5. *'Whatever is is right'*: A. Pope, *An Essay on Man,* I.x.14.
6. *Alexander Pope*: Pope (1688–1744) was an English poet best known for his satirical work, including *The Rape of the Lock* (1712).
7. *good mother who thus watched over your welfare*: Hirst's mother, Hannah Oates Hirst, had recently passed away and left him an inheritance; see letter 0382 of this volume.

8. *my father*: John Tyndall Sr.
9. *'he being dead yet speaketh'*: Hebrews 11:4.
10. *'bone of my bone and flesh of my flesh?'*: Genesis 2:23.
11. *'be just and fear not'*: W. Shakespeare, *King Henry VIII*, III.ii.717.
12. *equipoise*: a state of perfect balance (*OED*).
13. *Marcus Antonius 'I have lost all except what I have given away!'*: not identified.
14. *antipodes*: the opposite of a given thing or person (*OED*).
15. *a man*: possibly Thomas M. Foy; see letters 0354, 0355 and 0357 of this volume.
16. *pericardium*: the sac or cavity enclosing the heart (*OED*).
17. *Heidelberg*: a city in the German duchy of Baden, best known for its university.
18. *the Chronicle*: probably the *Preston Chronicle*; see letter 0386 of this volume.
19. *Ada*: Ada Piercy.
20. *Noll*: probably Mr. Knoll.
21. *you mentioned her name*: see letter 0382 of this volume, in which Hirst mentions taking lodgings with Mrs. Wright.
22. *cachination*: cachinnation means loud laughter (*OED*).
23. *Kuppers*: Conrad Kupper.
24. *her son Alfred*: Alfred Wright, biographical details not identified.
25. *Punches*: *Punch, or the London Charivari* was a weekly satirical English periodical.
26. *Mrs Fearnly*: not identified.
27. *a sous*: a small denomination of money.
28. *Cassell*: Cassel, a town in the German electorate of Hesse-Kassel. The city now spells its name 'Kassel'.
29. *Geissen*: more commonly spelled Giessen (Gießen).
30. *engaged himself to one of the young ladies*: Sophie Fick, whom Frankland married in 1851 at St Martin-in-the-Fields (*ODNB*); see letter 0383 of this volume.

From James Craven 16 October 1849 0385

Halifax 16th. Octr. 1849

My dear Tyndall,

I should have answered your letter sooner had I not been very busy with my Effusion on Geology[1] which I mentioned to you in my last.[2] I found it a much more difficult task than I had anticipated – & as you will no doubt presume I placed but little confidence in myself & thus my trouble, in hunting in all the Works I could find on the subject, to find suitable paragraphs, was endless – however suffice it that through the kindness of Tom[3] it was completed in a tolerable shape – The History was the commencement – then the review of the various great divisions in the strata giving an outline of the forms of life which characterised them & then ended up with a review of the present causes that are in existence & of the effects taking place around us with a view to explain the previous phenomena – I then gave them the nebular theory[4] – & the facts which suggested it & finally adverted to the theory of the Vestiges of Creation[5] – On closing however this paper & when quietly seated by myself on the evening when it had been delivered I certainly felt happy – but in a few days I felt a sort of disgust creep over me – learning – reading – science & everything else seemed fanciful & vain – like a will o' the wisp[6] which may be followed but never to realize pleasure

I looked back on my past life – on the 2 or 3 years of my apprenticeship[7] – how I had spent my time principally on one object & I was perfectly disatisfied & miserable with the result of my labors & resolved since no comfort or happiness attended these *[researches]* to have no care for them in the future – I was low indeed. Everything which I had expected would follow necessarily from my studies was disappointed & I found nothing but vexation – After such reflections as these I resolved that it was better to devote ones attention exclusively to monetary concerns & inwardly dammed my sad lot that ever I was in a profession – give me thought I – the manufacturer, the speculator – anything to keep in activity the mind – there, there is something to show for all this labor – the reward is something which procures you at least comfort in one respect while the other is but one series of disappointments

Such were the reflections – & mentioning them to Tom he kindly comforted me by saying he, you, & all, have or must go through this stage & at the same time encouraged me to go on – Last night he lent me Emerson's Essays[8] as appropriate to my present disease which in time I will read well. – Yet I cannot say that yet I am quite resolved to go on in this apparently barren track – the excitement attending the other pursuit is great – & always furnishes something for the mind (if you can bring it down to that) while at the same time from the success which I have lately had in some transactions I have taken in hand is an additional incentive to such a pursuit – I am well aware of the sordid & worthless nature of the object – money – but perhaps after all there is more pleasure & enjoyment in gaining it than in the attainment of what has been called scientific knowledge – however to proceed to your letter –

You ask me if I 'ever think of death' & what my 'feelings would probably be were I certain of its approach'[9] Yes – often – I have often when some known friend has been suddenly called away asked myself – where is the being I once knew – his body is the same throughout but where is the spirit? & I often fancy to myself that perhaps there may be some truth in the doctrine of transubstantian[10] though of a much higher kind than that which transmits the soul from a higher to a lower creature – It is a curious subject however & in a future letter I will give you a dish on this subject for Tom has just told me that this is the wrapper sheet[11] so I must be brief – Talking of dishes – At the time you wrote yours dishes[12] seem to have made a great impression on you – Tom must have given you to understand that the powers I possess on this subject are prodigious or something of the kind – perhaps you dreamt it or rather as clairvoyance is the 'go' here at present was the means by which you arrived at the conclusion – however it is incorrect but more on all these subjects in my next – for somehow my letter has been sadly curtailed by the lengthened prelude of my own state however believe me dear Tyndall to remain

Yours very affectionately | James Craven

RI MS JT/1/HTYP/37–38
RI MS JT/1/H/139–140

1. *Effusion on Geology*: it appears that Craven wrote an essay on geology, though it has not been identified.
2. *my last*: letter missing.
3. *Tom*: Thomas Archer Hirst.
4. *the nebular theory*: the hypothesis, first developed in the eighteenth century, that solar systems are created when gaseous clouds collapse and flatten due to gravity. As the clouds flatten, they cool and shed material that eventually condenses to form stars and planets.
5. *Vestiges of Creation*: reference to the anonymous book *Vestiges of the Natural History of Creation* (London: John Spriggs Morss Churchill, 1844), which suggested the existence of stellar evolution and the transmutation of species. The author was later identified as Edinburgh publisher and journalist Robert Chambers (1802–71). On *Vestiges of the Natural History of Creation* and its authorship, see J. Secord, *Victorian Sensation: The Extraordinary Publication, Reception, and Secret Authorship of* Vestiges of the Natural History of Creation (Chicago, IL: University of Chicago Press, 2001).
6. *will o' the wisp*: a 'will-o'-the-wisp' is a mysterious object that appears suddenly and confuses its viewers (*OED*).
7. *my apprenticeship*: i.e. Craven's work for Richard Carter.
8. *Emerson's Essays*: probably R. W. Emerson, *Essays* (London: James Fraser, 1841). See letter 0380 of this volume. Later editions of this book were republished under the title *Essays: First Series.*
9. *'ever think ... its approach'*: the letter in which these quotes appear has not been identified.
10. *transubstantian*: probably a misspelling of 'transubstantiation', the Catholic doctrine which states that the wine and bread of the Eucharist are converted into the blood and body of Christ during Mass (*OED*).
11. *wrapper sheet*: the sheet of paper that encloses and protects a letter and is used to address the correspondent (*OED*).
12. *Talking of dishes – At the time you wrote yours dishes*: reference not identified.

From Thomas Hirst 16 October 1849 0386

Halifax Monday Evening | Oct[r] 16th 1849.

My dear Tyndall –

As Jemmy[1] is writing to you,[2] I will just send a word in the same package. And first of all I thank you sincerely for your two last Preston Chronicles. You have succeeded very well in the passage from Faust[3] & it is a splendid passage most assuredly. I have since seen a translation by Hayward[4] in Prose and I tell you honestly your translation of the passage alluded to is by far the better of the two. You have managed to convey the idea with greater clearness & vividness. Of the poetry, I can say that of all the readers of the Preston Chronicle, it had the most significance for me. You have many of my own thoughts there Tyndall, thoughts that I never expressed even to myself or knew I possessed until you told me, whether this is owing to our common soul or whether you may have remembered me during their composition I don't know, but I find it suits me *[also]* – These thoughts however with which I can sympathise are detached & after reading it over carefully I find myself unable to grasp the idea in its entirety. I dare say you will smile at my request but I should like you to give me a kind of commentary on it & try if by another mode of expression you can cause a brighter gleam to shine through that thick skull Tom Hirst's – that 'dummer Mensch'[5] Perhaps this analytical process of your own poem will not be exactly pleasant but between us of course there is no ceremony. I have read (hurriedly) Fichte's Characteristics of the Present Age,[6] a book I believe I never thanked you for but what matter, you know that I felt so if I did not express it. I prize it much & the fly leaf[7] of it the most of all – I found his philosophy though very attractive & interesting from its logical correctness & conciseness, somewhat strange to me & in order that I might better understand the Book in question I purchased his 'Vocation of Man'[8] which may be called more elementary. I am just in the middle of it now & trying if possible to understand thoroughly his metaphysical philosophy. Those questions on External objects, their properties &c being only matters of our own conciousness & sensation, all this I find startling & still attracting in the extreme. Its seemingly paradoxical assertions I am forced to admit & I begin to feel that my relation to the world & every other person is about to undergo some change Already I fancy my me is rising in the scale of importance & objective Nature becoming more plastic to its power. If it will but lead me to that eminence; if it will but convince me of the Godlike properties of my own Soul. Oh how shall I be rewarded! That dimly imaginative state will at last become a reality. Beauty & Truth I shall discover are in fact in me not out of me. The dictates of conscience will be followed with cheerfulness – Duty will at last be synonymous with desire – and I may realize that grand object, to walk the Earth a Man!

'Ah! Tom', I hear you say, 'your going to be dissapointed you are too enthusiastic'. I am afraid so John – in fact it is just half an hour since I wrote the above lines, I have cooled down & can tell you if it was not written it never would be. but if you will be kind enough to get into a fit of excitement yourself you will oblidge me, for then you wont laugh. on the contrary if you descend from the excited platform among the cold hearers as I have just done why you'll recommend a gag for the future to be applied when the same symptoms of this complaint make their appearance. I am so cozy now in my little lodgings. nobody troubles me & few seek me, for amongst my former jovial companions I am little missed, & every night after my work I shut myself up without regret for

> When I am safe in my sylvan (?) home
> I laugh at the pride of Greece & Rome[9]

Then Sundays, they are delicious days, so quiet, so peaceful aye and sometimes I think Holy for with my Bible before me which I love to read now, I read 'Ye must be born again'[10] & hope & fancy that my regeneration may be even now effecting, for 'The Wind bloweth where it resteth & thou hearest the Sound thereof, but canst not tell whence it cometh nor whither it goeth. So also is every one that is born of the Spirit'[11] –

God Bless you Tyndall, I am in one of those rare moments just now, when you feel at peace with yourself & with everybody, a brief moment of perhaps as pure happiness as we can hope often to realise – You have felt the same many a time. 'Nicht wahr'?[12] I could go on writing for hours; but then the postage, & besides I have said enough about myself. there are others in the world to whom we ought to lend a helping hand, among these Jemmy.[13] Since our talk together I have thought more about him and have had some anxiety. After mature consideration I feel convinced now is a crisis in his existence. You wrote to him about his diet[14] & he read me part of the letter, we may have done him some injustice on that point. perhaps its influence over him is not so great as we thought, at any rate he seems to think your remarks however good in themselves have none or very little application to him. A short time ago he was rather low spirited it appears he had been reviewing his own life, he thought of the many resolves he had made that had fallen to the ground, the number of books he had read & the little benefit he had received, and the poor little fellow was prostrate before such accumulated failures – Now this is a symptom which may lead to two very opposite results & the conclusion of it is what makes me anxious There is danger in his path. I feel great responsibility devolves on me, & I hope to let my actions be the results of this conviction – If he have but perseverance enough, if we can only excite him to a renewed effort the symptom is encouraging. But if on the contrary this cannot be done, if it leads to despair & finally carelessness you will fear with me for the result. For you know how superficial has been his training, as well as his tendency to little sensualities. We have had some earnest talk on the matter I have tried to encourage him by shewing him how we have all had to battle with the same fears & despondencies. I have tried to apply a stimulus to furnish him with an object that should act as an incentive and to rendering this object more permanent by placing it in himself. I have avoided pointing out any course of study as yet, it would be premature for him to make new resolves at present. This indifference he has for such studies this sense of deprivation from other pleasures must first be banished. He must love it for its own sake & find interest instead of effort in every step. I have at last persuaded him to read one chapter I thought appropriate in Emerson We have read part of it together 'Man Thinking'[15] & I have translated it into more popular phraseology. I think it will be of some service. I have thus spoken at length in order that your letters may bear upon this situation. nothing of course must be said of *[these]* remarks but if you have any suggestions for my guidance tell me. I have received no answer to that little d–d[16] question of ours – G.S. Phillips spent last Sunday with me. I find him a kind warm hearted friend –

Adieu, my dear Tyndall. yours affectionately – Tom Hirst

Remember me to Messrs Frankland and Knoll. Tell the former I have not yet received his dissertation – What have you been engaged with since your Examination Have you decided on the subject for your dissertation? Whats is it? We are just setting up our Laboratory in Mrs Wrights kitchen but have done nothing yet – Have you had a trip yet, you must do so before you begin another session –

Mr Tyndall | Care of Professor Bunsen | Marburg | Hesse Cassel | Germany

RI MS JT/1/HTYP/44–47
RI MS JT/1/H/139

1. *Jemmy*: James Craven.
2. *Jemmy is writing to you*: probably letter 0385 of this volume.
3. *the passage from Faust*: Tyndall translated a portion of Johann Wolfgang von Goethe's tragedy *Faust* for the *Preston Chronicle* and signed it with the pseudonym 'Wat Ripton'. W. Ripton, 'Goethe and Faust', *Preston Chronicle,* 29 September 1949, p. 3.
4. *Hayward*: Abraham Hayward (1801–84) published a translation of Part I of Goethe's *Faust* in 1833 (*ODNB*).
5. *'dummer Mensch'*: stupid human (German).
6. *Fichte's Characteristics of the Present Age*: J. G. Fichte, *Characteristics of the Present Age,* trans. W. Smith (London: John Chapman, 1847). The book was a translation of a lecture series first published in 1806. It discussed the five epochs of history through which Fichte believed humanity had passed.
7. *fly leaf*: a blank page at the beginning or end of a book (*OED*). It is likely that Tyndall wrote an inscription for Hirst on the flyleaf.
8. *'Vocation of Man'*: J. G. Fichte, *The Vocation of Man,* trans. W. Smith (London: John Chapman, 1848). The book was first published in German in 1800; it was intended to further explain Fichte's views on atheism.
9. *When I am safe ... Greece and Rome*: R. W. Emerson, 'Good-Bye', 4.1–2: 'O, when I am safe my sylvan home, | I tread on the pride of Greece and Rome'.
10. *'Ye must be born again'*: John 3:7.
11. *'The Wind ... of the Spirit'*: John 3:8.
12. *'Nicht wahr'?*: haven't you? (German)
13. *among these Jemmy*: see letter 0385 of this volume.
14. *You wrote to him about his diet*: letter not identified.
15. *'Man Thinking'*: This is a state of mind advocated by R. W. Emerson in 'American Scholar', an essay first given as a speech in 1837.
16. *d–d*: damned.

To Thomas Hirst and James Craven 21 October 1849 0387

Marburg Octr 26/49[1]

– 'For Jemmy'[2] –

'A lucky job that your letters came on Sunday when I have time to reply to them and an unlucky job they did not come on Saturday and thus spare me *[8]* silbergroschen[3] which I am ill able to afford. I sent a packet of documents off last night & in it scraps to both of you; inwardly protesting to have done with all correspondence for at least 6 weeks, when lo! you missives appear & like the shifting sands of Shorecombe Bay[4] my resolution is tossed aside to allow a reply to rush out of me.

'There is one mind common to all individual men' I actually laughed at the ridiculous truth of the adage as I read Jimmys letter.[5] Without ever dreaming of Jimmy & his experiences, without having the least idea that he was turning up his toes and kicking at the pricks of fate, he will find a transcript of his own mind at the end of that affair which I have just sent off to the Chronicle.[6] Arrak[7] Jimmy! as the people in Ireland say, did you imagine that God and Nature intended you to sit quietly on your rump after the achievement of a lecture on Geology? Nature cries aloud, 'I'd see you whipped first my little Sir' – Are you going to set up a mark on the door post of the little closet in Northgate and say in your heart 'I'll grow until I reach such a height then I shall be a fine fellow, but the devil an inch taller will I stretch myself! I shall then look down upon Richardson & Tidmarsh and that's enough for me.' Jemmy however reaches the proposed mark & finds to his infinite chagrin, that nature did not intend him to stop there – 'grow on you little cripple' she cries; Jemmy clenches his teeth, looks with the ferocity of ten Tigers into the face of Nature and cries 'I wont! I'm damn'd if I do! I'll put a stone on my head first, I'll warp my spine, turn my shin bones into circular ones, – do anything sooner than grow one inch higher!' –

– 'For Tom' –

I will have a random rap at both of you through this sheet. Every man is a microcosm; a world in miniature, and the world according to the Heathenish notion was a tree, what a profit it is both for Tom Carlyle & me that it is so, as by the age of this phantasmal vegetable we can climb like Jack on his beanstalk to a region far higher than the stilts of Johnson or Walker[8] would enable us to reach. Dear Igddrasil![9] fresh & beautiful you appear, yet is the sound of mourning in your branches as the wind passes over them. What says that little bough yonder? It peeps discontentedly upward and murmurs in its little timber heart. 'What a miserable devil of a branch I am, to be stuck here; why did not nature contrive that I should sprout out a little higher? would that I could mingle my leaves with those yonder, how happy should I be then' And thus whispers of discontent ruffle the sobriety of the vegetation from top to bottom. What says the highest branch? Is he contented? Infinitely so when he looks downward & compares himself with his brother twigs, but he turns his eye to heaven & begins to complain like the rest. 'O Stars', he cries, why did God set you out of my reach? Why do I not fold you like doves with silver wings in the umbrage of my leaves, – here I am fast rooted to this diabolical Earth, a prisoner to the laws of vegetable culture – Why Oh Heaven could I not spring up in a night like the gourd of Jonah[10] & find the everlasting stars like dewdrops on my leaves next morning? Why? Oh Why? – 'Patience, patience, dear disconted little mourner', whispers a soft voice from between the cherubim,[11] 'every branch of you is where it ought to be; the top branches may perhaps blossom in another age, yet without you they cannot exist: I will not say that you may not reach your tendrils higher, that depends upon your own chemical skill in the decomposition of CO_2, know then thy work, make use of the atmosphere around you this will lead you to results far more beautiful and blessed than looking upward in vain longing to a position which is in point of fact no wise more honourable than your own, be true to thine own position that is what is required of thee, in that must thy happiness consist.

'For Jimmy'

Let me take another turn over to Jimmy & ask him a random question or two as they occur to me – Are you sure Jemmy that manufacture and Speculation[12] are so happy in this excitement? All very fine if it be a game of gain, but it may be a game of loss; they have something to shew for their labour say you – yes I know a speculator who has something to shew, viz, a bare rump to the winds I doubt much whether solid satisfaction will be found here – if your genius leans that way, try it by all means, but beware of letting a few failures, which every man that breathes the breath of life has at one time or another experienced, have too much influence upon you – Speculation! the word brings to my mind the unspeakable agony of poor John Richardson during his first transaction with Jonathan Crowther[13] – The Scholar looks upon the healthy ploughman whistling all day long at his horses tails, as the happiest man; and the ploughman when he sees the scholar, damns his own thick head & hard fate which have shut him out from the enjoyment of the other. To many, the life of a German Student is very romantic, while in point of fact life is as prosy[14] in Germany as elsewhere. Lord Lyttleton[15] when at a theatre always fancied happiness in the next box to his own, but he never could make it his actual companion. Take care Jimmy. I say again that you do not find this speculation a mockery, a delusion, & a snare. I tell you plainly my mind, that if you had worked as hard at your profession, as I was obliged to do, you would by no means be disgusted with it – I'm afraid you have been playing the Engineer, and there is no peace on Gods earth for the man who does not handle his work in rough earnest. I know your proclivity in this respect, & I know my own, I have often smashed my shins over the same stumbling block, if you have failed & failed and failed, it is no more than others have done before you – 'Never mind the defeat, up again old heart there is victory yet in store for you'[16] I should like to know your state of mind during the preparation of that lecture,[17] if contented then was the labour its own reward, you have increased your stock of knowledge and had besides an evenings calm happiness – had you begun another lecture next day on the mountains of the moon it would have effectually shut out vexation – you will probably gyrate & root yourself finally upon the same old rock–

– 'Again I turn a new leaf for Tom'–

Here once for all I would set myself in 'Stabiles gleichgewicht'[18] and fix myself in that relation which I know I should occupy, even though I should descend sundry paces from the tip-toe eminence where friendship would place me. I will fix myself on something solid at all hazards, though I barter for it the glory of an occasional ride in a balloon – know then Tom that in this thing called Human Life I have a weight of things to bear – I'm not an egotist, I'm not cold, I'm not selfish, but my action when free is most effectual I hate responsibility, I would give my life for the life of my friend, but dont make me responsible for his life – I come to you then like the ochre slave which you see sometimes painted on tea trays, with

<*Handwritten page missing; next portion of letter is LT Transcript only*>

a chain on my arms and another round my ankles and beg of you to have the goodness to shake them both off – liberate me and I am not the less your servant – My first act afterwards will be to fold you in my free arms and kiss you, swearing inwardly, though you may not hear the adjuration, that I will fight by your side more manfully than if ten thousand responsibilities pricked me forward. I am willing to lend and borrow on equal conditions, right gladly I will take part in all your battles, but henceforth and forever I disclaim the leadership, I will fight as a volunteer by your side, giving insight when I have it, and accepting it when I have it not. In making this arrangement I obey a dictate of my constitution, which has more than once manifested itself, and in doing so shirk no dictate of duty, but on the contrary I give my arm a freer swing. Do you understand me? You are playing deep Tom, there are heavy stakes on the board, if played bravely and with entire selftrust the result must be a dead gain. I have handled and am handling the same dice myself and would wish for entire freedom in the game. I am fearless but would be free.

Fichte is no milk for babes Tom – he is meat for strong men, beware of conformity. I would rather for my part see you pitching him to the devil, than that he should come between you and nature – apply yourself to his solution, feed on that which you can appreciate, but beware of being warped too far by Fichte or anybody else. You will sometimes, as by accident, drop upon the precise point from which he viewed the subject – I have done so sometimes in taking

<*Handwritten letter resumes*>

a walk – sometimes in the act of putting on my nighshirt, but this eternity which embraces us can be regarded from many points and your interpretation of it may be far better for your own private purposes than Fichte's – I thought of writing to you before on this subject – Fichtes great gospel will I believe prove eternally true, which may be so condensed 'Want is not the source of vice, it is the motive to activity and virtue; indolence, sluggishness is the source of all vice – How to enjoy as much as possible – how to do as little as possible This is the question of a perverted nature and the various attempts to answer this question are its crimes. There is no salvation for man until this natural sluggishness be successfully combated – until he find all his pleasures & enjoyments in activity & activity alone'[19] – –

– Scrap for Jemmy –

A deal of unhappiness arises from the simple dislike to try a strange track. I never commenced to level a line without a feeling akin to reluctance; but once upon it I could get on very well. I feel the same now with my dissertation, but I shall soon be in the middle of it and make friends with my new thoughts – Conceive a screw with a triangular thread conceive one of the bevelled surfaces of this thread produced indefinitely. I am to investigate the properties of this surface and to determine the conditions of equilibrium upon it. There are some who have a dislike to go into company but find themselves agreeably surprised when the 1st formalities of the thing are rasped away. Jemmy appears to have the same reluctance in attacking a new subject where as if he did so he would in a few hours find himself among agreeable and entertaining friends. Oh Jemmy the power of Gold to produce happiness has been experimented upon since the world began and always with the same result. Would you change places with John Abbott? try it however and add your testimony the rest–

– Scrap for Tom –

I have just got a 'Truth Seeker'[20] which contains an article from January Searle[21] on Friendship – it is the best production of his that I have seen. I fancy however one recommendation of his at least would prove exceedingly perilous – an intimate friendship with a young lady without warmer feeling is a desideratum[22] – I wonder would January himself like to choose as a wife a young lady who cultivated such a platonic intercourse with another – there is length, breadth, & depth in the article however & it is worth reading – I would be inclined to abolish what he calls 'the scholar proper'. If I had a child I would teach him a trade, perhaps the use of boxing gloves to boot –[23]

RI MS JT/1/HTYP/40–42
RI MS JT/1/T/523
Handwritten copy in Thomas Hirst's hand

1. *Octr. 26/49*: Louisa Tyndall annotation to typescript letter: 'It was really written on Sunday Oct 21/49 | received by Hirst Oct 27th See the journal.'
2. *'For Jemmy'*: James ('Jemmy' or 'Jimmy') Craven.
3. *silbergroschen*: Prussian currency.
4. *Shorecombe Bay*: probably Morecombe Bay, notorious for its shifting sands and fast moving tides.
5. *Jimmys letter*: letter 0385 of this volume.
6. *the Chronicle*: probably the *Preston Chronicle*.
7. *Arrak*: not identified.

8. *Johnson or Walker*: not identified.
9. *Igddrasil*: Igddrasil (Ygddrasil), according to Norse mythology, is an enormous tree that is connected to the nine worlds.
10. *gourd of Jonah*: Jonah 4:6: 'And the LORD God prepared a gourd, and made *it* to come up over Jonah, that it might be a shadow over his head, to deliver him from his grief. So Jonah was exceedingly glad of the gourd'.
11. *Cherubim*: an angelic being (*OED*).
12. *manufacture and Speculation*: professions Craven was considering; see letter 0385 of this volume.
13. *Jonathan Crowther*: this may refer to the Mr Crowther referred to in Tyndall's journal. Tyndall's 24–5 January 1845 journal entry noted: 'My levels were also corroborated by Mr Crowther' (RI MS JT/2/13a/73).
14. *prosy*: tiresome, wearisome (*OED*).
15. *Lord Lyttleton*: possibly a reference to George Lyttleton (1709–73), first Baron Lyttleton, a politician and patron of the arts.
16. *'Never mind the defeat ... yet in store for you'*: R. W. Emerson, 'Essay II: Experience', *Essays: Second Series* (Boston: Phillips, Sampson & Co., 1844), p. 87: 'Never mind the ridicule, never mind the defeat: up again, old heart! – it seems to say, – there is victory yet for all justice'.
17. *that lecture*: see letter 0385 of this volume, in which Craven describes preparing a lecture on geology.
18. *'Stabiles gleichgewicht'*: Stable equilibrium (German).
19. *'Want is not ... activity* alone*'*: J. G. Fichte, 'Lecture V.: Examination of Rousseau's Doctrines Concerning the Influence of Art and Science on the Well-Being of Man', in *The Popular Works of Johann Gottlieb Ficthe,* trans. W. Smith, 4th edn, **000 vols** (London: Trübner & co., 1889), vol. 1, pp. 197–203.
20. *'Truth Seeker'*: a Transcendentalist publication edited by Frederic Richard Lees.
21. *January Searle*: a pseudonym under which George Searle Phillips frequently published.
22. *desideratum*: something needed or to be desired (*OED*).
23. *gloves to boot –*: Thomas Hirst annotation: 'Copy of Letter from John Tyndall | Marburg. to Messrs Hirst & Craven | Oct[r] 26 – 1849'.

From Thomas Hirst 25 November 1849 0388

Halifax Sunday evening | Nov[r] 25th 1849

My dear 'Tom' (as you have thought */*proper to*/* rechristen the man John Tyndall)[1]

Your travels have duly appeared in the Chronicle[2] & have been truly enjoyed by the namesake of the 'traveller'.[3] Would that his empirical self more truly resembled this delineation of the best parts of us both – But why you foolish man straitened & poverty stricken as you are why did you send that to the Preston Chronicle? I could have got you something handsome for that manuscript. Don't you do so any more – And this reminds me of that immense loan[4] that you make such fuss about as if I were about to invest in your hands my eternal as well as temporal welfare or even any part of the latter. I've not so much to spare as all that; besides the quantity I swallow, there will be a little purse which I wish you quietly to appropriate for your convenience inasmuch as I shall consider it the most satisfactory way of employing it & 'by George' if you don't take it without any more fuss or word of thanks or any thing else you shall not have it at all, I'll throw it into the streets instead or do some other rash act. I shall send you £20 as it is all I dare spare or trust you with. more all at once would make you *[frisky]*. I have taken duly into consideration all my wants & made ample allowances for them so don't be afraid, I can spare £5 as soon as you will write & tell me to whom I am to pay it here, the rest in January – Now tell me whether so little will do you any good, if the time is convenient, if not I can alter it, it makes little difference to me & finally to whom I am to forward it. I have not the slightest idea how to convey £ s.d[5] to the Continent or I should have spared myself the trouble of filling this page with such unnecessary remarks by quietly enclosing the amount. However you see the nature of the case & are not quite destitute of common sense, so as we say in Yorkshire 'Shut up'

Now Letter for letter No 2 – by far the more acceptable although it does give me a rub down. Advice is allways welcome Tyndall especially when it corresponds to the dictates of your own heart, & an error now & then even if it did but call forth the utterances of such didactic maxims I should think almost justifiable, but in this case I gave you fair warning I told you I was enthusiastic, & I don't regret it Your illustrated Igrdrasil[6] is beautiful *[is]* interesting but somehow it don't hit my nail on the head. I'm not discontented. In my moments of high hope (enthusiasm if you like) which I would have recur more often, I see but a point to which I may reach & that also by decomposing CO^2. You who are so very apt in your chemical & mathematical & Botanical allusions, illustrating intuitive feelings by Parkside Curves[7] & Goethes[8] profundity by infusible Platinum may also find one for this truth That the attainment of excellence of any sort worth the attaining, was ever accompanied & inspired by pure moments of exalted hope & longing – Have you not had such moments yourself? & would you not rather suffer what may be called a slight shock on once more touching Earth than forfeit the expansive vision & the refined atmosphere that your ride in the balloon of Hope rewards you with?

I know full well in principle that I must be my own guide my own parson, confessor & adviser in this world Tyndall but the remarks of your letter surprised me inasmuch as they discovered to me a great discrepancy in my action. I have relied on your Authority I have & do still put great trust in your word & advice: but why? because you have appealed to myself that advice has hitherto proved most acceptable to my own heart & I have followed it & will follow it God willing in the teeth of all consequences. But there is just this one chance that I might appropriate something true & honest to yourself, yet not to me in my different sphere, & putting faith merely in the former existing sympathy, appropriate something unquestioned, & this might draw on a flood of circumstances & consequences in which though you might swim I should undoubtedly sink from the fact that I had trusted to your ability to save us both – from thinking that your heart was worth more to me than my own. I see plainly the responsibility you would throw off &

under which in <u>such</u> a case you would place yourself – Duty calls me decidedly to release you & I do it willingly though with some traces of sorrow. It is natural to long for a staff on which to lean, a back on which to throw the excess of weight you groan under & it is *[wise]* that it is so, the time does come when you feel inwardly that your own legs may possibly be able to bear your own weight but at first you hesitate, like a baby learning to walk – to make the attempt, to trust their unknown strength. Tis, thus I am at present I feel inwardly that <u>I</u> *[am]* the best authority to myself, I feel that everything truly worthy that I can call my own is my own attainment But there come dark moments Tyndall when I would give the world for some support & like Peter the less faith I have the more I fear to sink & the more I cry for help.[9] I tremble at times at the responsibility of myself, that I a youth of 19 should disclaim all outward guide & authority, Both Priest Bible, & carve out my own path, amid the shrieks & warnings of older heads than mine who warn me that I am on the high road to perdition If my own heart be the veritable map, would that I had more faith in it, or could decipher it more readily If you cannot point the road Tyndall you can at least strengthen that faith without disobeying your love of Duty If you cannot decipher this map for me, Oh tell me convince me without fear of repetition that it is at least <u>decipherable</u> & when done so to be relied on.

My Last[10] was a letter of joy *[in]* exultation this of doubt & fear I grumble not at either, they are necessary stages of my progress, this I believe though cannot see – I was bred up in the belief of my parents & Oh now that I view it as a thing not mine it looks a treasure thrown away. Compared with these rags I wear this coat so full of holes how warm & comfortable seems that boyish jacket. I half mourn my stature & blame myself for growing when I see I cannot wear it now. This religion of my Fathers once I threw it off, I fancy now too hastily, it calls to me with a voice more plaintive more attractive now. I must listen to its appeal – I see another period of trial yet in store for me nay it has already come the contest will be fiercer than before, the result more lasting –

I will write again, for I do not yet plainly see my course. One thing I would fain preserve unquestioned I will *[strive]* to do, but it also will be tested sorely I mean this capability of myself. Did not the Almighty Maker of all, when he planted this reason of mine – this distrust in myself, give me also a support. True he has given me a desire that prompts obedience & a heart to obey, but has he not also placed in me this <u>distrust</u>, is it then without an object? You will understand me I feel sure & will show me some light if you can do so with comfort to yourself; if not why leave me to myself or what is the same tell me it must be so –

2 scraps accompany this

<Handwritten enclosures missing; LT Transcript only henceforth>

I have once said[11] 'I admit in <u>principle</u> that all trust is due only to myself', and yet I do not only feel so, this 'pale cast of thought' has 'sicklied o'er'[12] this truth I thought so firm and stable, I have laid aside my book and sit for hours to gaze on nothing, my eye turned inwards but with no satisfaction. Just at this time I have met a man of comprehensive mind, one that has tested these things ably, has doubted, thrown aside & at last accepted once again his orthodoxy. Is there not something suggestive here? I will trust no more to him than I would to any other human authority but the belief that has called him back must surely possess some force and true significance and must not be dealt with lightly. I refer to Hutchinson, a man you know something of – it was only last night we were engaged in conversation, for 4 hours debated earnestly. He is a man before whom you must throw off all deceit and mere appearance, he did not treat me with assumed superiority nor quench a thought by pure conventionality. He met me on my own ground with all the sincerity and honesty I could wish for. Coming on me whilst in such a state it is useless to deny he made a vast impression, not conviction though. As 'Christel' thought of 'Tom'[13] I thought of him, that he seems to feel the solid ground under him, while in my staggering walk I seem to tread on insecurity. As I said before, the matter has not been sufficiently tested by me yet, it will be done. I hope for strength and honesty, fearless of all consequences I enter on the struggle, the result and mode of combat you shall some time know. One thing we both agreed upon this purification and putting under abeyance our grosser nature, without this we can see nothing truly and will always be diffracted. 'Tis easy said all this, but devilish hard to do. The more we strive against temptations the more they pull, the more we shun the world the more overtures it makes. This fair exterior we shew the world covers sins as black as hell, Tyndall – the blacker the more we look at them. I have felt moments of real misery to think there are passages of my life I shun to bring to light, I have felt it many a time when with yourself and amidst that confidence we feel in each other's presence, those heart revealings, I have stopped, and half feared you would see the cloud gathering over my countenance at the memory of deeds I've done, which even you must never know. I'm not a murderer or a thief, although I use these accents of solemnity. The world would judge me lightly, would soon acquit me with a verdict for myself. Not so this higher tribunal within me. Before it I am cast down with shame, Tyndall, and see but one method of escape and comfort – the future, and by Heaven! if strength and life is given me it shall make some retribution.

Remember me kindly to Frankland. I received his Dissertation,[14] for which I thank him, & wish you to convey such to him for me. How is yours coming on?[15] You wanted to have done with writing for 6 weeks[16] – at the end of that time I shall expect to hear from you. Jemmy[17] desires to be remembered. He will write soon. Not being ready at present, this letter comes alone. Ada[18] often comes in to see me. I have never delivered your kiss yet:[19] I am too bashful & she is getting too big to be kissed with impunity by strange gentlemen. Mrs Wright is suffering from toothache and Mr Wright's excesses just now. I believe the poor woman is miserable sometimes. Alfred[20] is with a chemist in Leeds. The lad has some good in him, inclined to be rather gay though; he wants to come home again. I think he had better not, from all accounts.

But good-bye for the present. | Yours affectionately, | T. Hirst.
Mr Tyndall – | Care of Professor Bunsen | Marburg | Hesse Cassel | Germany

RI MS JT/1/HTYP/44–47
RI MS JT/1/H/141

---

1. *as you have thought proper to rechristen the man John Tyndall*: Louisa Tyndall annotation: 'this alludes to a story by Tyndall in Chronicle, where 'Tom' is the hero'.
2. *the Chronicle*: the *Preston Chronicle.*
3. *Your Travels have ... namesake of the 'traveller.'*: as Louisa Tyndall notes at the top of this letter, Tyndall had published an article (under the pseudonym 'Wat Ripton') about his experiences in Europe in which he named the traveller 'Tom'. W. Ripton, 'The Sisters of the Rhine', *Preston Chronicle,* 24 November 1849, p. 3.
4. *immense loan*: Hirst had recently offered to loan Tyndall part of the inheritance he received upon his mother's death; see letter 0382 of this volume.
5. *£ s.d.*: pounds, shillings, and pence.
6. *Your illustrated Igrdrasil*: Yggdrasil is a tree from Norse mythology; see letter 0387 of this volume.
7. *Parkside Curves*: not identified.
8. *Goethe's*: see letter 0386 of this volume, for discussion of Tyndall's translation of *Faust.*
9. *like Peter ... I cry for help*: a reference to the Apostle Peter's fear as he began to walk on water to meet Jesus, who was already walking on water. See Matthew 14:22–33.
10. *My Last*: possibly letter 0386 of this volume.
11. *I have once said*: Louisa Tyndall annotation: 'Scrap 1 continuing H. Let. Nov. 25/49'.
12. *'I admit in principle ...'sicklied o'er'*: source of quotes not identified.
13. *'Christel' thought of 'Tom'*: a reference to the *Preston Chronicle* article in n. 3.
14. *his Dissertation*: Frankland's doctoral dissertation at the University of Marburg discussed, among other chemical reactions, the isolation of ethyl (*ODNB*). Frankland was awarded the PhD in July 1849.
15. *How is yours coming on?*: Tyndall's dissertation at the University of Marburg focused on the mathematical properties of screw surfaces.
16. *done with writing for 6 weeks*: see letter 0387 of this volume.
17. *Jemmy*: James Craven.
18. *Ada*: Ada Piercy.
19. *never delivered your kiss yet*: see letter 0384 of this volume.
20. *Alfred*: Alfred Wright, son of Mr. and Mrs. Wright, biographical details not identified; Tyndall asked Hirst to speak with Alfred in letter 0384 of this volume.

From James Craven 29 November 1849 0389

Halifax, 29th Nov. 1849

My dear Tyndall,

My last letter[1] was really so meagre and so scant that I am sure you would have deemed me shabby were I to have renewed a similar one in return for your two very kind letters[2] and especially the last. It was in accordance with this idea that I resolved to defer this pleasure – for such really it is – until a time and fitting opportunity arrived, and now I sit down with the intent of giving you a full benefit which will most likely tire you.

The sentiments contained in my last letter were not very pleasing to you as is very evident. You ask did I think that the achievement of a paper on Geology was the end and object of all my existence? You seem herein to have misunderstood my letter – for I said on looking back on my past life – on the 2 or 3 years of my apprenticeship – I felt dissatisfied with the result. You say too 'There is one mind common to all individual men'[3] and you actually laughed at the ridiculous truth of the adage as you read my letter. I certainly cannot for the life of me make out what you are aiming at by the quotation at all – unless you mean to give your disapprobation to an occasional review of the past which is in part borne out by stating that had I immediately engrossed my attention on the mountains of the moon, on the completion of the Essay, the ennui would have disappeared. I certainly agree with this – to be engaged is to be happy, at least I find it so – but to question the policy of inspecting the course one has been pursuing – an idea which has the stamp of being natural – is certainly rather a difficult matter. Pray have you never found yourself sitting before a waning fire, with a faint glimmering light, or better still no light at all, silently meditating on the present – looking forward to the future and stealing away again into the past? If you have not I have and many an hour at time and time have I wiled away in this quiet manner. Probably however you call this '<u>dreaming</u>'. Nevertheless it is very pleasant though perhaps on this score it is avoided by Stoics, poets and philosophers. To be brief however I think I have misinterpreted your meaning – as peaceful quiet contemplation cannot but be dear to us all, and especially so to those who are of a meditative turn. I have lately been subject to that most enviable state – the life of Do-nothing & Co. I will explain: – My brother's[4] majority[5] happening on the 30th of Oct. I determined to pay him a visit at his lodgings in town for the proper celebration thereof. A kind old lady[6] – a friend of ours who is wealthy – invited us to go to Brighton[7] and in compliance with her wishes I remained with her for 10 days. Now during the whole of this period I enjoyed everything which one could wish for. I had but to mention my desire and it was satisfied (if in the reach of money). Consequently you would naturally suppose this together with the company of Madame and a mademoiselle should make me comfortable and happy enough. But no, though the change at first was certainly agreeable and pleasant I soon found it pall on me, until eventually I was positively glad to return to actual work again. Having made this experiment of the happiness which riches can afford I would certainly not give an iota for it and I can assure you that no one sits down to Emerson and other works with more pleasure than I do at present, so that on the whole I may say I have reaped a great benefit – tho' be it distinctly understood that my passion for such reading does not reach to such an extent that I am at all neglectful in my own personal concerns (for I think my bump of selfishness must be very large[8]) though perhaps not so keen as when I wrote you before. I find in fact, if I may so speak, that there is a mysterious law at work declaring that to be happy I must be supplied with books of various kinds to suit the fancy of the moment – and that without these time flags and passes drearily on its way. I have endeavoured to combat this but generally without effect for it appears to have become a habit having been so long and invariably practised. Stay! but what am I talking of? Certainly it is some outlandish egotistical blarney or other. So we will 'shift' the scenes – though what I have written tends to establish your doctrine of a 'great reader is like a great smoker'.[9]

I have got Emerson's Essays.[10] I often read the one on Self Reliance,[11] it is one of which I felt in need being as you are aware too much given to place reliance in others than in myself. Emerson of all the writers I ever read impresses you with ideas of your own dignity and importance. All his Essays lead you to entertain nobler views of the objects and aim of your existence, while the most noble and the most glorious ideas are poured forth in every sentence. I often think when reading him how often the self-same thoughts which he has transcribed here occurred to me – but yet the many and various points from which he takes his survey – the association of ideas which he possesses – make him the Man. I wish I had known you had not his Poems, I would have sent them you instead of Dr Millingen[12] – however perhaps you will be aware that he wrote one on the death of his son[13] – a most touching and heartrending one it is – fully showing the agony of his mind at the time it occurred. Thus we see that in Philosopher, Poet, or whatever else you like, this same Death produces a feeling of sorrow – nay even (as Combe[14] has it) the clergyman who knows that his friend is removed to 'a heavenly and better place' laments his absence from this 'vale of sorrow'.[15] After all there is something so solemn, so searching and appealing to one's inmost self by the death of a dear friend that forces upon us – Well can I remember a faint indistinct whitey face which often comes across me even in my gayest moments – it is the one of my own poor mother[16] as she lay immediately before being conveyed to the earth – to that to which she was about to return – and often in my own undertakings do I imagine her fair face in the distance beckoning me on in the path of virtue and uprightness. Oh yes – surely there is something in all this – though a mystery conceals it. You asked me in a late letter[17] if I ever thought on these subjects, and what would be my ideas were I certain of its approach? I must say I am too much inclined to regard these matters as questions for discussion rather than for actual practice. One sees so much dishonesty in this all important matter Religion that one looks on its propagators with distrust. Another flagrant instance has just occurred in this little town, viz. Mr Obrey[18] the minister at Harrison Road Chapel who has mulcted his creditors to some £4000,[19] and when one sees so much of this hypocrisy going about – prevalent alike among Churchmen, Roman Catholics and Dissenters, no wonder we begin to question and discuss rather than to practise. On the question of Belief I have in former letters given you a tolerable dose – viz. that matters which one is required to be perfectly acquainted with are merely points of history and therefore by no train or system of reasoning can their truth be arrived at – or if so of the foolishness and absurdity of such a belief as an all important necessary to salvation, etc. etc. together with the proposition that matter of fact people like myself cannot be satisfied with anything less on a subject of the highest importance than to arrive at a result plainly and logically. No doubt you laugh at this. I believe myself it is impracticable in this case, else why our different sects? and opinions? Notwithstanding all this I feel daily a certain consciousness of the depth and intensity of this subject creep over me – the Bible seems more noble and glorious than it was wont – when used as the Latin and French task and the school book. The ideas which it conveys in every word – the proverbs and advice which it throws out – all suggest as you have stated the Divine Authorship. Seeing then that my tendency is, at present, in favour of arriving at a more just and definite idea on these matters, I leave off paying any very particular notice to them, trusting that by observation and experience I may eventually be able to clear up many of those passages which at present are obscure. At present my belief is – in an inward and instinctive idea common to all, which if followed up to, notwithstanding all the allurements cast in our way, we shall be all right. I am not one of those who in the wayward frolicsome child see the creature of the Evil One, but on the contrary I perceive him gifted with an inward idea more pure, more holy (because free from worldly cunning and worldly desires) than he can be at any other period of his existence.

I think now by this time you will be almost tired of the subject – perhaps of me too – but I must and will fill up the sheet, though the matters may be rather desultory. As a matter of course then the weather must be introductory – so you must know that our hills are crowned with snow – our ponds are slightly frozen over and everything around represents the misty, dark and chill November. Yet amidst all this I am glad to say we find plenty of work – in fact we are extremely busy what with Assize,[20] Arbitration and other cases – as usual we are out on the most wintry days and not on the pleasant ones; however this system having been renewed for such a length of time I am almost indifferent on the matter. By the bye, in a late letter[21] you intimate that you fancy[22] I have merely been <u>playing</u> the Surveyor all this time – what could be the <u>reason</u> of your supposition I know not – certain it is whether play or work I would not hesitate to take in hand any survey of ordinary difficulty notwithstanding my general hesitation and non-self-reliance. I cannot see that being diffident to the merits of the profession because so devoid of interest and of a general requirement that you should arrive at such a conclusion – besides, you must remember that the idea of change was occasioned by the thought of altering a certain course of action – which change was in some measure by the additional excitement to counterbalance the thing given up. Such was merely an epitome of my thoughts; at the moment – suffice it that they do not occupy any place with me now, though certainly occasionally a twitch as to the ways and doings of the present lot of hosts of Surveyors has occurred and which as yet I have failed to answer. For my own part however I look forward without much uneasiness to the future, confident that by application and attention I can at least at any time command a tolerable practice here. Meanwhile I look forward to the time when my articles will be out and as to what course I shall then pursue will depend much on circumstances.

I have read up to Article III which is the latest published of your production of the 'Sisters of the Rhine',[23] and certainly I have been very much pleased with it. Methinks that truths and ideas conveyed in this style are much more palatable than when drawn up and moulded by a heap of dry reasoning, and trust that some day or other I may have the pleasure of reading a similar but a more extended work of yours. I am afraid on the whole this has not been such a very agreeable letter as usual. Were I to reindite it I would alter it in many places. However you will probably make all allowances and round off some of the sharp edges of it. And now I conclude with my very, very best wishes for yourself, trusting you may continue in the enjoyment of good health – In all of which John, Tom and Father[24] etc. all unite, – and hoping a speedy return to Yorkshire,

Believe me, | A sincere and grateful friend, | James Craven.

P.S. I have no news to communicate whatever. Halifax is extremely dull – at least I find it so after London and Brighton. However time passes swiftly over which shows I am not very very miserable.

We are at present in want of a room for a laboratory in consequence of which our studies in this respect are rather curtailed. This defect however we trust to overcome shortly when we shall resume with renewed vigour.

RI MS JT/1/TYP/11/3534–3538
LT Transcript Only

---

1. *My last letter*: letter 0385 of this volume.
2. *your two very kind letters*: one is probably letter 0387 of this volume; other letter not identified.
3. *'There is one mind common to all individual men'*: this quote appears in letter 0387 of this volume.
4. *My brother's*: John Craven; no biographical details identified.
5. *My brother's majority*: John Craven must have been turning twenty-one, the age at which men reached legal majority in Britain.
6. *a kind old lady*: not identified.
7. *Brighton*: a popular seaside destination in southern England.
8. *my bump of selfishness must be very large*: Craven is referring to the phrenological belief that the size and shape of each person's skull correlates with their character. The statement suggests that Craven considered himself selfish.

9. *'great reader is like a great smoker'*: the letter from Tyndall containing this quotation is missing.
10. *Emerson's Essays*: see letter 0386 of this volume; Hirst had encouraged Craven to read Emerson.
11. *Self Reliance*: R. W. Emerson, 'Self-Reliance', in *Essays* (Boston: James Munroe and Company, 1841). The book was later reprinted as R. W. Emerson, *Essays: First Series* (Boston: James Munroe and Company, 1847).
12. *Dr Millingen*: probably a book by John Gideon Millingen (1782–1862), a retired surgeon and author best known for works such as *Adventures of an Irish Gentleman* (1830) and *Recollections of Republican France from 1790 to 1801* (1848).
13. *wrote one on the death of his son*: Emerson wrote about his son's death in the poem 'Threnody' and in the essay 'Experience'. J. Porte, S. Morris (eds), *The Cambridge Companion to Ralph Waldo Emerson* (Cambridge: Cambridge University Press, 1999), p. 145.
14. *Combe*: possibly George Combe [Comb] (1788–1858), Scottish proponent of phrenology (*ODNB*).
15. *'a heavenly and better ... vale of sorrow'*: not identified.
16. *my own poor mother*: not identified.
17. *in a late letter*: letter missing. Craven also addressed the subject of death in letter 0385 of this volume.
18. *Mr Obrey*: not identified.
19. *mulcted his creditors to some £4000*: a mulct is a fine imposed for an offence or an unfair tax (*OED*).
20. *Assize*: a legislative court sitting (*OED*).
21. *in a late letter*: letter 0387 of this volume.
22. *you fancy*: in this context, 'fancy' means 'imagine' (*OED*).
23. *your production of the 'Sisters of the Rhine'*: W. Ripton, 'The Sisters of the Rhine', *Preston Chronicle,* 24 November 1849, p. 3. 'Wat Ripton' was one of Tyndall's pseudonyms. See also letter 0388, n. 3 of this volume.
24. *John, Tom and Father*: 'John' is possibly John Richardson. 'Tom' refers to Thomas Archer Hirst. Craven's father was John Craven of Halifax.

To Thomas Hirst 2 December 1849 0390

Dec. 2nd, 1849.

My Dearest Tom

The sadness of an autumn evening is brooding over my soul as my pen moves – but I wont be poetic, in fact I cant – during the week my brain has been wrapped up in the intanglements of my dissertation,[1] and of other similar matters and you know that under such circumstances a man's gift of speech is often crippled – I will write and write on however, stating my thoughts as they occur and leaving it to your loving kindness to pardon my short comings – your last was a letter of exultation,[2] I saw that, and I foresaw the probable reaction, I saw it and dreaded it, and in a moment of weakness wrote as I did[3] – I have often said, or half said since, 'I ought not to have written <u>so</u>' but a moments thought convinces me that I ought; for Tom must know at all hazards that I am a struggler – not a hopeless one thank God – like himself – I say that I marked your exultation and foresaw the probable reaction, how was this possible? simply because I have passed through the same ordeal myself and I know how sad it is to have the peachbloom of hope rudely brushed away – but let not Tom think that the bloom is gone for ever, that there is no balm in gilead and no physician there;[4] – so Tom has released me from my responsibilities – does he imagine that I am therefore the less his? at the present moment I feel forcibly the terrible energy of paul's[5] words when he said 'I could wish myself accursed from Christ for the sake of my brethren'[6] – would that I could share your burden Tom – by Heaven I think I should prefer sinking under it myself to seeing you sinking – am I therefore more free than I was before? Think it not Tom – I did not mean that my letter should be an isolation between you and me – It is not good for man to be alone, the herd of elephants swims the river with their trunks knotted and the life of every man is intertwined with that of the human race – when I say that you must trust yourself, I intended my words to bear an interpretation that you have not yet given them –

Remember Tom I write incoherent, my lungs at the present moment are full of the fumes of Salpetrige Säure[7] and my eyes are smarting from the same, I have been all day experimenting with a galvanic battery on the influence of magnetism on crystalline bodies, and you know as I said before how disjointed my expressions on the present subject must be – as I read your letter many experiences of my own and of others flashed upon me – how vividly Faust[8] rose to my mind as you spoke of your fathers[9] creed. Tired of the world he is about to quit it when the *[easter]* him[10] breaks in upon him

> 'Christ is arisen | Joy to the mortals'[11]

The memories of his youth bend sweetly over his soul and he exclaims

> O tönet fort ihr Süssen Himmelsglieder
> Die Thräne quillt die Erde hat mich wieder.[12]

I remembered that Shakspeare wrote Hamlet and that to write Hamlet Shakespeare must have been Hamlet and I remembered an incident in my own life indicative of a state of feeling which you well know – I went to the polytechnic institution in London[13] – at the end as usual were a series of dissolving views – one an old church before the dark portal of which a monk was prostrate on the paving stones. – Would to God I exclaimed that I could go and do likewise – amid my hearts sterility I felt that even superstition would be a relief – in later times I have applied this little experience to the solution of such men as Pusey[14] – the religious barrenness of our age is intolerable to such men, and they turn with loving eyes to the traditions of other times hoping by them to revive the dying embers – I often think however that the salvation which Pusey finds in these traditions lies though all unconscious to himself in the effort that he makes to restore them, and that were his wishes in the matter fulfilled he would by no means find the same enjoyment in their realization as in the exertions which he made to realize them – This truth is common place – almost every body has experienced it – and where do these facts point? I interpret them as hinting that employment – action; as mans proper tendency, that in it alone is he to find health and strength bodily and spiritually – I turn to an old journal and find the following under the date, Tuesday 30th January 1844 – I had received a letter from a pious young friend of mine[15] whom I loved and love and who loved and loves me – 'George's letter as I predicted contains a lecture – poor fellow I sometimes wish to be like him for I feel the want of an anchor for my opinions – he begs of me not to trust altogether to my own reasoning and philosophy – surely the mind of man requires a firmer support than either philosophy or Reason can furnish and a higher principle of action than a love of morality – this latter is very good as regards the wellbeing of society, but the man who is moral merely for the sake of preserving an unblemished reputation in the eyes of the world takes low ground, fixes his principles upon a quicksand and is a stranger to the comfort drawn from the assurance that his actions are regarded and approved of by a higher authority than that of man – How often is Paul's case ours notwithstanding! – 'To will is present with me but to perform that which is good I find not[16] – For the good that I would I do not and the evil that I would not that I do'[17] A thousand times have I reasoned regarding this principle of distrust and the necessity of an exterior support exactly as you do – though I question whether I ever felt the matter so deeply as you for two years I would have given anything to have been a christian – and read books calculated to bring about this consummation – The reading of Dr Burnets account of the death of Rochester the infidel[18] set me on my feet as a believing christian for some time – this want within myself was near driving me to join the methodists, expecting that their prayings and groanings and religious excitements would arrest the dryrot of my soul – all this have I gone through Tom, this has been the forge in which my present creed has been hammered into shape.

And do not think Tom that it is men of devotional feeling alone who are thus tormented – Byron perhaps suffered as much in this way as any man ever did – and I read that Tom Moore, the muse's curled darling, who spent his youth in love, and his manhood in the breath of fame is impatient of life; Charles Lamb[19] talks as follows –

By myself walking
To myself talking
When as I ruminate
On my untoward fate
Scarcely seem I
Alone sufficiently
Black thoughts continually
Clouding my privacy[20]

x x x
Dreams of Antipodes
Night riding incubi
Troubling the fantasy
All dire illusions
Causing confusions
Figments heretical
Scruples fantastical
Doubts diabolical
Abaddon vexeth me
Maker perplexeth me
Lucifer teareth me![21]

And for all such men as Charles Lamb and other men besides, you and me inclusive Tom, is there I believe a blessed truth in the following words of Carlyle –

'Consider how even in the meanest sorts of labour the whole soul of man is composed into a kind of real harmony the instant he sets himself to work! Doubt, Desire, Sorrow, Remorse, Indignation, Despair itself, all these like hell-dogs lie beleaguering the soul of the poor day worker as of every man: but he bends himself valiantly against his task and all is stilled – all these shrink murmuring far off into their caves'[22]

I have already given you an extract from my old journal I will give you another it is dated the 30th of Nov. 1849

'Visited Faber[23] as the clock struck 6 – the moon pushes back the infant daylight and still holds sway in heaven – it is beautiful beyond description the frost as hard as flint underfoot and the air as clear as crystal all round – to protect me during the mornings I have a large amorphous catskin coat – a regular Esquimaux[24] article – it is pleasant to rise early – long experience of both enables me to decide between it and the pleasure of lying in bed – the former is infinitely preferable – if I had a son he should start early to his legs – yesterday morning hewed the last difficulty of my dissertation to pieces – I have pored rather long over this matter, but could have got on more speedily had I availed myself of assistance, but I sat down to the affair as my proper work, determined to do it myself through much entangled cogitation, by patient abiding when my way was dark as night, I became acquainted with my subject – by degrees difficulties began to crumble down, and now they are all conquered and my work is done – Let me here set my seal to the word Patience nothing is done without it, looking patiently into the darkest subject a thousand hidden beauties reveal themselves – I must tell this to Tom.'[25]

No man knows the value of work as an educator, and as an administrator of happiness better than Emerson, still is there danger in rising from the perusal of his essays that an underestimate of the importance of action has been imbibed – I dwell upon this subject Tom because I believe nothing is more calculated to combat your present despondency than this very thing employment – What drives the fashionable world to theatres and thousands to vice and infamy – that very feeling which now preys upon your heart – and it would drive you to the same had you not sufficient iron virtue in your constitution to resist it – you see the matter is to direct this inherent activity, and let not your daily avocations Tom be rated too low – done in a right spirit they are as beautiful and exhilarating as any others – There is no stability on earth Tom – life is a flowing river, dam it up and it will create festering sedge marshes and unwholesome fens – let it flow out in honorable effort and it is beautiful as the figure I have used to express it – in a word Tom it will never do to sit for hours looking disconsolately inwards for by this manner of looking there is nothing found –

I am glad you have met Mr Hutchison and from your character of him I could squeeze his hand this instant with brotherly love – there seems to me something holy in that yearning which he has inspired in you for the creed of your fathers; but I think you will hardly find it satisfying, I think Mr Hutchison has not doubted after your fashion or rather believed after your fashion – I would not throw a dust particle in your way Tom, but if ever I spoke sincerely to mortal man it must be to you and I therefore state my belief that you will not find satisfaction in this change – try it my son, try it if you can, but do not despair if it also slips through your fingers – What are the experiences of those who have held this old religion? of the best and most celebrated among them? They have had their hours of exstasy like you and their hours of doubt and pain like you also – can you expect to find a city of refuge here? – as I write the hymns and prayer of my childhood rise before me hymns the spirit of which my later life has often reeochoed, for instance

> {'Oh for a closer walk with God
> A calm & heavenly frame'}[26]
>
> {A poor wayfaring man I go
> In restless wandering to and fro
> Fightings without & fears within
> Tempt my desponding soul to sin}[27]
>
> 'Could we but stand where Moses stood
> And view the landscape o'er
> Not Jordans streams nor deaths cold flood
> Could fright us from the shore.'[28]

'Could we but stand' make your own inference Tom, the man was not satisfied with his <u>stand pünkt</u>[29] –

I know there are many others more to the point but I cannot remember them –

<u>Sunday morning</u> – I have just had a dash of frosty water over my lank physiology – I have reread my last paragraph commencing with 'No man knows &c' I have mentioned the fashionable world being driven to theatres, but here you

may misinterpret me and that gravely, you may think that my advice merely furnishes you with a subterfuge to enable you to fly from yourself – undoubtedly in the example that I have cited, in the extravagances of gaming fox hunting &c men find subterfuges to cheat their taskmaster into repose, I dont harp on this string Tom, I lay bare my motive when I say that I believe it to be a law of your nature against which it is useless to fight and conformity to which will draw a thousand satisfactions in its train – I do not urge you to employment for the sake of any <u>results</u> you ever will or ever can accomplish tho' these have also a neccessary place – I urge you to it as the most practical act of worship, the best heart purifier that man possesses – and here I imagine we may solve the apparent inconsistency of Emerson when he sneers at action – he is often the apologist of those who retire in disgust from the field of worldly enterprise, and no wonder that in defending them he should often treat the important things of the world with ridicule – men of the world regard their achievements as important, they mistake the true office of action and work for the result alone and if they can get the result without working many do not work at all; and this view corresponds I fancy with another expression of his which I half remember – in speaking of sugar he says the negros have got all the profit. I receive the commodity merely – they the education. – I think there is a great deal of sap in this sentence Tom. You talk of sins as black as Hell Tom[30] – I would not treat them with willful levity but I cannot help saying that I care nothing about them; I am not going to judge of you by any act or any ten or twenty acts you ever performed – If men were measured by their acts what paltry maggots the best of us would appear – I measure you not by your performances but by your tendencies – 'This fair exterior we shew the world covers sins as black as hell'[31] – Shall I take this categorical proposition as it stands? Cannot I indulge in an inference of the following shape – that this proposition instead of being received as the statement of a fact merely proves the purity and rigour of that tribunal by which those sins are judged – Dick Charlesworth[32] would commit all your sins ten times over without any feeling of remorse – There is no necessity whatever of bringing all the passages of your life to light – nay it might be very impertinent to do so, and still less necessity that you should feel miserable about them.

I have often felt angry with myself when not able to enjoy the beauty of a starlight night, deeming it a kind of duty that I should be entranced – I have gone out among the fields with this feeling that I ought to be delighted and have returned without profit. I have gone out with a thought in my heart, with a problem in my head, it might be mathematical it might be philosophical, it might be an endeavour to solve some obscurity in Emerson or Fichte, and I have strolled along busily engaged with my subject and as I raised my head from the ground and looked among the woods I might fancy angels peeping from the branches or the stars of heaven stooping down to talk to me – but to go out with my eyes open saying in my heart 'I must enjoy this', has ever been a futile job with me – This action and this reward appear to be something like complementary colors – dont you think that Emerson has taken the same wild goose journeys when he is able to say of the moonshine – 'go out to look at it and it is gone'? [33]

There are foolish people in the world to whom if I wrote this long letter they would thank me for the trouble I had taken on their account – what are their thanks to me compared with the sweet and salutary exercise of my own thoughts in writing it – my dear Tom be of good heart your fate is not the black and cheerless thing which your last letter[34] would lead one to suppose –

I could continue writing but I think I have written enough – buckle on your armour then my son and test the blessedness of work – do not expect too much, angel visits are always gratuitous they do not come at our invitations; they are not to be run after, nay they like us best when we appear most independent of them –

Thus far Tom I have repeated no mans words – I testify that which I have known and speak of that which my own eyes have seen.

Tell Hutchinson that I would share my breakfast with him – With regard to the money[35] Tom I shall not call upon you till February or March and perhaps not even then, but put a little by as I may be truly hard up at the time mentioned and 15 or 20 pounds might prove a deliverance from many inconveniences

Jemmy[36] perhaps thinks me too hard and has got tired of my maxims – why did not the little fellow drop me a line? – hard as I am however I have still a soft little dewdrop of affection for Jemmy.

Remember Tom Adam and Eve were placed in Eden, not for the purpose of gathering nosegays, and inhaling the perfume of gilliflowers[37] but 'to dress it and to keep it'–[38] The Israelites in the desert ate angels food,[39] but they had to <u>gather</u> it – and it fell in atoms not larger than a grain of coriander seed – they once got an abundant mess of quails <u>without labour</u> but a tremendous purging was the consequence – I often think of the last verse of the last chapter of John. 'And there are also many other things that Jesus did the which if they should be written every one I suppose that even the world itself could not contain the books that should be written'[40] – Here is a picture of a life of action

With regard to keeping our grosser nature in abeyance I think the life of man consists in something more than this this is at best a negation and I believe that human life has a positive element in it – you sometimes strengthen an adversary by rating his prowess too highly – and a danger is often exaggerated by dwelling too much upon it – It is a question whether on the whole it would not have been better had you been left destitute as you anticipated – my poverty appears before me in a far more tangible shape at present than it did a year ago, but I am not a whit unhappier now than then; I know struggles await me, but I know also that struggles rightly interpreted, rightly wrought out, are rich blessings and therefore I fear them not – I have accustomed myself to look at dark possibilities so that if they come they shall not take me unawares – you offer me money – yet you would do me a far greater service by selling a manuscript for me – a shilling thus earned would be worth half a crown borrowed – I sometimes feel thankful to the Gods for this poverty inasmuch as by it I am allotted my share of healthy effort – it marries me to my work.

This 'self capability' Tom is a thing which wrongly interpreted may be wrought into the sorriest caricature imaginable you are no self subsistent point in space, shut out from all exterior relations – you are a man, or a big boy rather, with an intellect and a heart both of which imply like verbs active objects acted upon – you have intellect I say and the cold beautiful nature around you as its field of exercise – you have a heart, and moral feelings – the human race your own neighbourhood, your friend – your flower garden, the midnight stars and the sailing moon are all its legitimate objects –

I have blotted this out because I was wandering into theory, and theory is not suitable at present – something plainly tangibly practical is what you want & I perhaps have stepped aside from this in one or two cases when I ought to have kept it constantly in view

I will not close this letter yet as my mood is too intellectual and the white snow on the housetops and the stiff walls of the castle on the hilltop tend to render this mood predominant – I have a Southampton letter[41] to reply to in which the theory of Newton as to planetary motion is virulently attacked I will turn to my thoughts to this for the present, nothing doubting that the wave of love and sympathy will in due time return to my soul – not long ago I should have regarded this as a painful defect – I now begin to discern that it is part of a law and bow to it accordingly – nor must you therefore infer that my love is impulsive, a thing of fits and starts, not to be trusted – it is within me now as certainly as the moment I finished the reading of your letter, and any act I should have ever done at love's prompting I would do <u>now</u> – feelings sometimes retire from expression for the purpose as it were of collecting strength for a fresh avowal – you see Tom in the illustrations furnished by his present moods what manner of man your friend John Tyndall is, and the chances are that these moods only shadow forth the experiences of John's friend Tom. There is tranquility in the recognition of a fact and the ability to allot to each its proper place – when therefore my intellect predominates and my heart seeks repose I will by no means call myself cold and fret myself accordingly. I recognize a harmony in the working of both principles and give place in due season to each – when writers endeavour to carry the expressions of one mood into another their <u>untruth</u> reveals itself at once – and in this way I could write a very noisy but a very false letter if I pleased.

Do you imagine Tom that a poet finds that bliss in the <u>visual</u> contemplation of nature which his expressions would lead you to imagine? not at all Tom. What he sees becomes his property, and he builds starcastles and weaves rainbow draperies from the materials furnished by nature; in this action of the poet consists his blessedness, and as a kind of repayment to nature for furnishing him with such rich materials he is wont to give her all the praise – but he is the true artizan of his own delights – nature gives the blocks but he builds and harmonises –

A thought sometimes flashes on me Tom in the course of my work that I should like to communicate to you and I had begun to make hasty memoranda of such & intended to send them to you all broken and incoherent as they were whenever I should write to you – your letter however called for a special reply and I therefore refrain from sending any of my fragmentary cogitations at present. how glad I shall be to see you in Halifax, and to see Jimmy and to sit with you both for 3 long hours before that bright phenomonon so long invisible to me an English fire and mingle our little experiences together into a fricassee of thought ten times richer than Giblet soup – I should be tempted to propose a deputity to Miss – I declare I forget her name – Saxton,[42] is it not, of the Upper George[43] for a bottle of her best port I should not care whether at my expense or yours, and to propose further that we should get tipsy every one – now dont open your great eyes in wonder – I positively am of opinion that all this could be nicely managed, and that the recording angel instead of knitting his brow and setting down a black mark in the judgment book would indulge in a good natured chuckle at our audacity – I further believe Tom that a time is in store for each of us when our yoke shall be easy and our burden light, when like free men we can handle implements which would now be perilous as edged tools to children

– The pharisees[44] called Him a gluttonnous man and a winebibber,[45] the time advances when we shall practically know how high was his liberty compared with that of the Baptist who dared not venture beyond locusts and wild honey[46] –

A wealthy man can afford to wear a bad coat, and Jesus Christ on a similar principal could afford to eat and drink –

There are times Tom when a man is disposed to see something humorous in the solemnest things, when even grief and sorrow put on such comic faces that one does not know whether to laugh or to let it alone – I verily believe I have dropped upon me one of these moments this evening – I often do in fact, and if I spoke my feelings outright, you would deem me a blasphemous cur,[47] in which you would be totally wrong as blasphemy or irreverence is a thousand miles from my thoughts – I *[verily]* believe a man might pray to God and laugh at the same time, but if you have never got into one of these moods you will see no way of pardoning me –

So Ada[48] calls to see you – well I should not blush to kiss my little girl ten times over were she here beside me – I just fancy her here at my elbow on the sopha,[49] I declare I think I should treat her as a sugar plum and eat her up in the extravagance of my love – I look around these four bare silent walls, upon that black surly looking stove, I look at myself here half buried in papers

<Handwritten pages missing; LT Transcript only henceforth>

and all alone, and just imagine the change that Ada's presence would make in all these things!

I have mentioned 'Faber'[50] in a former extract. I will continue the extract, as it will furnish you with a bit of biography, 'a young man of high parts but totally devoid of energy. Without external pushing he would turn into a morsel of the poorest animalism[51] and spend his time eating, smoking, and sleeping. I have known him now nearly a year, attended the same lectures and worked in the same laboratory. He has not perhaps during that time derived 5 shillings worth of benefit. I thought him an admirable mathematician when I came to Marburg, he talked so glibly of the Differential and Integral calculus; but now I see his utter shallowness in this as well as on most vicious trait – a wish to appear a man of genius without making the trouble to earn the reputation – a nature given to self puffery,[52] and by this he has signalized himself in the University. I have got hold of him, but how long the gripe will continue I know not; bad habits are not uprooted in a day. He now starts at 5 o'clock every morning, shews me his lamp in the window to testify that he is at his post, and works well in the laboratory all day. People laugh at the idea of his continuing thus, but we will try in spite of this laughter. Skill is doubtless necessary in the treatment of such a case; but by skill alone such a case was never cured. Hence the faithless laughter of those who know of no higher influence. Shall I attempt to cure a patient by advice? ridiculous! I must endeavour to weave his life into my own. My chief influence must be a silent one, he must feel me without hearing me and to accomplish this what is necessary?'[53]

The answer to this question Tom has some bearing on this matter of self-capability, and I have therefore made an extract. I said that your life is intertwined with that of the human race, and so is mine Tom, and in an especial manner with yours; you often speak of my influence over you, but forget altogether of your influence on me. Now Tom, properly speaking, both are in equilibrium. And Faber reacts upon me too, though in a less degree. In fact I have stated in my journal that the terms Debtor and creditor are abolished between us.[54] This interaction of individuals is one of the sublimest laws of human nature and we must be careful lest a wrong headed self-capability shut us out from its participation. I thank the gods, not for rendering me independent of Tom, but for meeting one of the necessities of my nature by giving me such a friend. If ever man was independent of these influences, it was he whose birthday now approaches, yet what says he? Look to that inimitable prayer in the 17th chapter of John – in the 19th verse he says, 'And for their sakes I sanctify myself'.[55] There was action and reaction even here, communicativeness and receptivity – no rejoicing in isolation. It would seem as if the disciples were as a necessary to the nature of Christ as he to their natures. They were undoubtedly a means to his improvement.

And now I must close Tom – I shall wait patiently and hopefully for your next letter. Always write your heart to me my son and I will try to do the same. Let us endeavour to be friends Tom in as high a sense as any pair of fellows as ever lived. I find this feeling growing within me – will time perfect it? How pleasant it will be to realize, in our own actual lives, that principle which every body deems either fabulous or perished out of the world. I am not too anxious for the consummation. If I feel it growing within me, I'm content. It will be no hothouse plant, no Jonah's gourd,[56] when perfected. Isolation! aye – If the gods have a blessing in store for me, may they grant it by making our lives one. But where is this imagination of mine leading me? Shall I say into a land of impossibilities? What say you?

This letter has drawn me into a third morning, and now, amid the silence of the world, I bid you goodbye.

John Tyndall

RI MS JT 1/HTYP/48–55
RI MS JT/1/T/524
Portions of letter are LT Transcript Only

---

1. *my dissertation*: Tyndall's dissertation at the University of Marburg was a mathematical investigation of screw surfaces (*ODNB*).
2. *your last was a letter of exultation*: possibly letter 0386 of this volume, as the tone of 0388 does not seem to fit this description.
3. *wrote as I did:* possibly letter 0387 of this volume.
4. *there is no balm in gilead and no physician there*: Jeremiah 8:22: 'Is there no balm in Gilead; is there no physician there?'
5. *paul's*: the Apostle Paul, to whom fourteen of the twenty-seven books of the New Testament are attributed.
6. *'I could wish myself accursed from Christ for the sake of my brethren'*: Romans 9:3.
7. *Salpetrige Säure*: nitrous acid (German).
8. *Faust*: J. von Goethe, *Faust* (1787), one of the most celebrated works in German literature; see also letter 0386 for a discussion of a passage from *Faust* that Tyndall translated for the *Preston Chronicle*.
9. *your fathers*: Hirst's father was Thomas Hirst (1797–1842), a wool-stapler (*ODNB*).
10. *easter him*: Easter hymn.
11. '*Christ is arisen | Joy to the mortals*': J. von Goethe, *Faust: The First Part of the Tragedy,* iv.384–385.
12. *O tönet fort ihr Süssen Himmelsglieder | Die Träne quillt, die Erde hat mich wieder*: J. von Goethe, *Faust: Der Tragödie erster Teil,* iv.430–431. Translation from the German: 'Sound forth your sweet celestial music! | The tears well forth, the earth has me again!'
13. *the polytechnic institution in London*: the Royal Polytechnic Institution, founded in 1838.
14. *Pusey*: Edward Bouverie Pusey (1810–82), a Church of England clergyman and theologian and founder of the Tractarian movement (*ODNB*).
15. *a pious young friend of mine*: George Latimer. On 30 January 1844 Tyndall mentioned in his journal that he 'Sketched a long French letter wrote part of one to Latimer and retired to bed' (RI MS JT/2/13a/f. 14).
16. *'To will is present with me but to perform that which is good I find not*: Romans 7:18.
17. *For the good that I would I do not and the evil that I would not that I do.'*: Romans, 7:19.

18. *Dr Burnets account of the death of Rochester the infidel*: Gilbert Burnet (1643–1726), a theologian and writer. He authored *Some Account of the Life and Death of John Wilmot, Earl of Rochester*, which was published in 1812 (*ODNB*).
19. *Charles Lamb*: Charles Lamb (1775–1834), an English writer (*ODNB*).
20. *By myself walking ... Clouding my privacy*: C. Lamb, 'Hypochondriacus', 1.1–8: 'By myself walking, / To myself talking, / When as I ruminate / On my untoward fate, / Scarcely seem I / Alone sufficiently, / Black thoughts continually / Crowding my privacy'.
21. *Dreams of Antipodes ... Lucifer teareth me!*: C. Lamb, 'Hypochondriacus', 2.5–15: 'Dreams of Antipodes, / Night-riding Incubi / Troubling the fantasy, / All dire illusions / Causing confusions; / Figments heretical, / Scruples fantastical, / Doubts diabolical, / Abaddon vexeth me, / Mahu perplexeth me, / Lucifer teareth me –'
22. *'Consider how ... into their caves'*: T. Carlyle, *Past and Present* (London: Chapman and Hall, 1843), chapter 11.
23. *Faber*: William L. Faber, a young American doctoral student at Marburg. Tyndall was attempting to mentor him. See also C. A. Russell, *Edward Frankland: Chemistry, Controversy and Conspiracy in Victorian England* (Cambridge: Cambridge University Press, 1996), pp. 70–1.
24. *Esquimaux*: the aboriginal people who inhabit northern Canada, Alaska, Greenland, and eastern Siberia (*OED*).
25. *'Visited Faber ... this to Tom.'*: this appears in Tyndall's 30 November 1849 entry in his journal, where he refers to Faber as 'Yankee' (RI MS JT/2/13b/469).
26. *'Oh for a ... heavenly frame'*: W. Cowper, 'O for a Closer Walk with God', ll. 1–2.
27. *A poor wayfaring ... soul to sin*: 'Hymn 341', 2.1–2, in E. Smyth, *A Choice Collection of Hymns, Psalms, and Anthems* (Dublin: B. Dugdale, 1776), p. 318.
28. *'Could we ... from the shore.'*: I. Watts, 'Hymn LXVI', 6.1–4. In Rev. G. Burder (ed), *The Psalms and Hymns of Isaac Watts* (London: C. Whittingham, 1806), pp. 360–1.
29. *Stand punkt*: standpoint (German).
30. *You talk of sins as black as Hell Tom*: see letter 0388 of this volume.
31. *'This fair exterior ... black as hell'*: this is a quote from letter 0388 of this volume.
32. *Dick Charlesworth*: not identified.
33. *'go out to look at it and it is gone'*: R. W. Emerson, 'Beauty', *Nature* (Boston: James Munroe and Company, 1836), p. 19: 'Go forth to find it, and it is gone'.
34. *your last letter*: see letter 0388 of this volume.
35. *regard to the money*: see letters 0382, 0384 and 0388 of this volume for Hirst and Tyndall's discussion of the possibility of Tyndall borrowing money from Hirst.
36. *Jimmy*: James ('Jimmy' or 'Jemmy') Craven.
37. *perfume of gilliflowers*: Gilliflower (gillyflower) is a term applied to plants that are scented like a clove (*OED*).
38. *'to dress it and to keep it-'*: Genesis 2:15.
39. *The Israelites in the desert ate angels food*: Psalm 78:25: 'Man did eat angels' food'.
40. *'And there are ... should be written'*: John 21:25.
41. *a Southampton letter*: not identified.
42. *Miss ... Saxton*: not identified.
43. *Upper George*: a hotel and hostelry in Halifax.
44. *pharisees*: members of a religious party within Judaism (*OED*).
45. *winebibber*: a winebibber (wine-bibber) is a person who drinks frequently (*OED*).
46. *the Baptist who dared not venture beyond locusts and wild honey*: this refers to John the Baptist. Matthew 3:4: 'And the same John had his raiment of camel's hair, and a leathern

girdle about his loins; and his meat was locusts and wild honey'.

47. *cur*: a poorly bred, cowardly man (*OED*).
48. *Ada*: Ada Piercy.
49. *the sopha*: a couch or sofa (*OED*).
50. *Faber*: see n. 24.
51. *animalism*: the notion that humans are animals, or the exercise of animal-like behaviors (*OED*).
52. *self puffery*: undeserved self flattery or praise (*OED*).
53. *'a young man of high parts … necessary?'*: a quotation from Tyndall's 30 November 1849 journal entry (RI MS JT/2/13b/469–470).
54. *Debtor and creditor are abolished between us*: Tyndall made the following remarks in his 30 November 1849 journal entry: 'If I act on Faber, Faber reacts on me to the same extent, the term of debtor and creditor are forever abolished between us' (RI MS JT/2/13b/470).
55. *'And for their sakes I sanctify myself'*: John 17:19.
56. *no Jonah's gourd*: see letter 0387, n. 10 of this volume.

From William Faber 17 December 1849 0391

Marburg 17th Dec. 1849.

My dear John

I can stand it no longer. I must go to Cassel[1] tomorrow, my Elise[2] is expecting me. Don't scold, I shall soon be back and work as hard again for it.

Your thankful and affectionate | Yankee.[3]

RI MS JT/2/13b/471
LT Transcript Only

1. *Cassel*: a city in the German electorate of Hesse-Kassel. The city now spells its name 'Kassel'.
2. *Elise*: not identified.
3. *Yankee*: a nickname for William L. Faber; see letter 0390, n. 23 of this volume.

BIOGRAPHICAL REGISTER

This register contains the names and biographical details of people who are mentioned three or more times in the letters included in this volume. Biographical information about people mentioned once or twice is contained in the notes appended to the appropriate letter(s).

Conventions

In transcribing the letters the editors have generally retained the spellings of family names given in LT's transcripts. However, in a number of cases a variety of spellings were used by correspondents; for example, the family name of Richard Boyle Barnard was often spelled 'Bernard'. Where there is a preferred spelling – for example, as used in official documents or biographical dictionaries – the alternative is placed in '( )' brackets. Thus, 'Barnard (Bernard), Richard Boyle' is employed below. Where there is no preferred spelling, the alternatives are given in alphabetical order, separated by a '/': thus, 'Steuart/Stewart, William Richard'. Where known, the dates of birth and death are given in '( )' brackets to the right of the part of the name for which the nickname is given, i.e. 'Frankland, Edward ('Trismegistus', 'Tris') (1825–99)'.

Nicknames and pseudonyms were widely used by Tyndall and his friends. For example, Tyndall was known as 'Alpha' and 'Purybingle', Phillip Evans was called Jim and William Ginty styled himself 'The Attorney General'. Where such nicknames occur in the letters, the person they refer to is given in an accompanying note. These nicknames are given in this register; they are placed in '( )' brackets.

Ranks in the British army mentioned in letters, with abbreviations (in ascending order):

Private	Pt.
Lance Corporal	LCpl.
Corporal	Cpl. or Corp.
Sergeant	Sgt. or St.
Second/First Lieutenant	Lt.
Captain	Cap.
Major	Maj. or Mr.
Colonel	Col.
Major General	MG.
General	Gen.

Abbot, John, a friend from the Ordnance Survey who was planning to emigrate from Ireland to America with his family in the spring of 1845. He was either from, or lived in, Clonakilty, County Cork.

Alexander, John (1764–1843), a wealthy and successful mill-owner who had developed a thriving business milling wheat, oats, and other grains at Milford, 3 miles north of Leighlin Bridge close to the River Barrow. He also purchased an estate at Milford and in 1799 constructed an impressive manor house. In 1840 the Milford estate occupied 1600 acres and accommodated 90 tenants. He was widely appreciated as a good employer and an improving landlord, he was highly respected, both by fellow Protestants and by Catholics, and was considered by John Tyndall, Snr, to be a suitable patron for his son (S. Kinsella, 'Landlords, Politicians, Entrepreneurs: The Alexanders of Milford, Co. Carlow in the Nineteenth Century', *Carloviana*, 58 (December 2009), pp. 70–83).

Allen, Robert (Bob) C., a civil assistant on the Ordnance Survey of Ireland. He worked in the A District under the command of Henry Tucker. Allen joined the Survey on 1 June 1840. He worked in the District Office at Waterford, and seems to have worked closely with Joseph Bickett. His main duty was plotting, and during his later months on the Survey he was examining plans. He left the Irish Survey, along with most of the workers in A District, at some point between August and October 1841; the records for these months are missing so there is no exact date of departure. Tyndall's journals and later letters indicate that Allen and his wife (name unknown, first initial M.) settled in Preston in the mid-1840s.

Bailey, John (Captain B., C.B.), a Royal Engineer stationed at Yorkshire. Little is known about Bailey, but he was likely the Captain who led Archibald McLachlan's division of the Ordnance Survey. McLachlan referred to his superior officer only as 'Captain B.' or 'C.B.'. McLachlan respected Captain B., but he was evidently transferred to another division in late 1843.

Banister, Frederick Dale (1823–97), a friend of Tyndall's from Preston. He was educated at the Grammar School of Preston and, beginning in 1838, had his first job at the Preston firm of Myres and Newton, a civil engineering and architectural company. In 1847 Banister began to work for the London, Brighton and South Coast Railway as a Civil Engineer.

Banister, Mrs, likely Mary Norris Banister (*c.* 1796–1864), mother of Frederick and Emily and wife of Samuel Banister, the High Constable of Preston. In the letters, she is mentioned as a patroness of Tyndall's.

Banister, Miss, likely Miss Emily Banister (*c.* 1820–74), Frederick Banister's elder sister. She was the daughter of Samuel Banister, the High Constable of Preston. Miss Banister was a social acquaintance of Tyndall's in Preston.

Barnard (Bernard), Richard Boyle (1787–1850), the second son of Francis, Earl of Bandon. After studying at Cambridge and being called to the Bar, he returned to the family seat and served as MP for Bandon Bridge, County Cork, from 1812 to 1815. He subsequently vacated his parliamentary seat and was ordained as a priest. In 1820 he became rector of Glankeen, County Cashel, and two years later he was appointed Dean of Leighlin and was based at the Cathedral Church of St Laserian in Old Leighlin. Tyndall and his father considered Barnard a useful and reliable patron.

Blamire (Blanmire), William Esq. (1790–1862), farmer, Whig MP for Cumberland and Tithe Commissioner. In August 1836 Blamire, Thomas Wentworth Buller and Richard Jones were appointed as Tithe Commissioners after the Tithe Commutation Act of 1836, which permitted citizens to make their Tithe payments in kind instead of requiring monetary payments. The Tithe Commissioners were given the authority to settle boundary issues for farmers affected by the change in Tithe payments.

Buller (Bullen), Thomas Wentworth (1790–1862), Tithe Commissioner. The Tithe Commutation Act of 1836 authorized the replacement of payment in kind for monetary payments of the Tithe. In August 1836 Buller, William Blamire and Richard Jones were appointed as Tithe Commissioners after the Tithe Commutation Act of 1836, which permitted citizens to make their Tithe payments in kind instead of requiring monetary payments. The Tithe Commissioners were given the authority to settle boundary issues for farmers affected by the change in Tithe payments.

Bunsen, Robert Wilhelm (1811–99), Professor of Chemistry at the University of Marburg. Bunsen taught both Tyndall and Edward Frankland during their time at Marburg. Bunsen was regarded as one of the greatest chemists of his day. During the late 1840s Bunsen worked with Frankland on organometallic compounds, but would soon leave organic chemistry behind entirely. He later became Professor at Heidelberg, where he worked with Gustav Kirchhoff on the characteristics of light, making a major contribution to the field of spectroscopy.

Byron, Lord George Gordon Noel (1788–1824), Sixth Baron Byron, English Romantic poet. Byron was one of the most famous and influential poets of the nineteenth century, even though the suggestive nature of his poems often ran counter to Victorian high moralism. Tyndall was deeply influenced by Byron's poetry and by the figure of the independent, defiant Byronic hero.

Carlyle, Thomas (1795–1881), Scottish philosopher, essayist and historian. Carlyle had delivered a series of public lectures in London in 1837, 1838, 1839, and 1840. These lectures made Carlyle a well-known figure in England by the 1840s. He was concerned with the state of contemporary society, and took great interest in Irish poverty and the state of Ireland. CaTyndall was particularly influenced by Carlyle's works on history, including *On Heroes, Hero-Worship, and the Heroic in History* (1841) and *Past and Present* (1843).

Carter, Richard (1818–95), an employer of Tyndall's. After a yearlong tour of the United States, Carter settled in Halifax, Yorkshire in 1840. In the early 1840s Carter was occupied in surveying Halifax with the aim of enabling the construction of a railroad that would connect the major manufacturing towns of Yorkshire. He went on to become the surveyor of the proposed West Yorkshire Railway. He hired Tyndall as a leveller in December 1844. Tyndall briefly left Carter's firm in early 1845, but otherwise worked more or less steadily for Carter from December 1844 to July 1847. In 1846, Parliament rejected both Carter's West Yorkshire Railway and a rival plan, leading to a period of financial uncertainty for Carter's firm. However, Carter would later become the surveyor for the West Riding Union Railway. Later in life, Carter purchased significant property in the South Yorkshire town of Barnsley, where he was eventually elected mayor. In the 1880s he retired to Harrogate, where he passed away in 1895 (J. Forrest, ed., *Minutes of the Proceedings of the Institution of Civil Engineers*, vol. 123 (London: Institution of Civil Engineers, 1896), pp. 445–8).

Chadwick, John C., a civil assistant on the Ordnance Survey who worked in the 4th division in the C District under the command of Lieutenant Fenwick. He worked on the Survey between 1838 and 1842, at which point he resigned from the Survey and moved to the town of Kinsale in the County Cork.

Colby, Thomas Frederick('Carbon') (1784–1852), whose father was a major in the Royal Marines, trained at the Royal Military Academy, Woolwich, before joining the Royal Engineers. Despite an accident, which led to the amputation of his left hand, he had extensive experience as a surveyor, including extending the meridian line from Dunnose into Scotland. Following the death of William Mudge in 1820, Colby lobbied and obtained the post of Superintendent of the Survey, which he sought to improve. After Parliament's decision in 1824 to initiate the Survey of Ireland, Colby took charge of the project and decided against employing civil surveyors, instead giving the task to the Royal Corps of Sappers and Miners. He began by carefully measuring an eight-mile baseline at Lough Foyle and by making improvements to the surveying instrumentation. Colby was a stickler for accuracy; progress on the Survey was slow and expensive and the first county map – Londonderry – was only published in May 1833. From 1833 he joined Henry de la Beche in making a geological map of the west of England and by 1838 the Irish Survey was mainly overseen by Thomas Larcom. Under pressure to complete the Survey of Britain and to trim its budget, Colby became increasingly involved in arguments over the scale of maps produced by the Survey. He left the Survey in April 1847, having been promoted to Major-General, and died six years later. Archibald McLachlan and Tyndall appear to have assigned Colby the nickname 'Carbon' in their correspondence; they had a low opinion of his work and viewed him as one of the Survey leaders responsible for the surveyors' mistreatment.

Conwill, John (1802–80), was educated in his father's school at Rathornan and taught for eleven years at the National School at Bagenalstown School, where he employed the Lancasterian system of teaching. Following the opening of a National School at Ballinabranna in 1833 he was employed as its schoolmaster, a position he held until September 1847, when he transferred to the National School at Leighlin Bridge. In 1840 the school at Ballinabranna attracted 236 pupils (both male and female), mostly Catholics, and Conwill's annual salary was £20. As his letters to Tyndall demonstrate, he possessed an impressive range of attainments, including poetry, mathematics and English grammar, a subject on which he sternly opposed the local school inspectors. Although Conwill was a Catholic, he and Tyndall developed a close personal relationship; Conwill was Tyndall's academic mentor and Conwill viewed Tyndall as his chosen pupil.

Craven, James ('Jemmy', 'Jimmy') (1829–86), a young protégé of Tyndall's and an early friend of Thomas Hirst's. Tyndall met Craven's father John Craven while working in Halifax and took an interest in Craven's education. Under Tyndall's guidance James Craven studied philosophy, geology, and chemistry; however, Tyndall did not feel that James Craven was as dedicated to the work as Hirst.

Cuddy, Martin ('Blackthorn'), a civil assistant who worked with Tyndall in the 5th Division of C District of the Survey. He joined on 30 April 1838 and was mainly involved in registering and examining work on the ground. In May 1841 he was transferred to the 2nd Division under Lieutenant Wilkinson in Bandon, where Phillip Evans later joined him. On 8 April 1842 Cuddy was transferred to the British Survey.

Cunningham, Colonel, the private secretary to George Murray, Master General of the Ordnance; first name and biographical details not identified. According to Tyndall, Cunningham informed the employees of the Ordnance Survey that they had the right to petition the Master General of the Ordnance Survey.

Edmondson, Anne (née Singleton), wife of George Edmondson. Anne was the daughter of her husband's former schoolmaster; the couple married in 1820. She played an integral, supportive role in her husband's work as Headmaster of Queenwood.

Edmondson, George (1798–1863), headmaster of Queenwood College in Hampshire. Edmondson attended Ackworth School in Yorkshire and was a Quaker. From 1817–24 Edmondson lived in Russia; Czar Alexander I invited

Edmondson's mentor Daniel Wheeler to advise on the formation of new agricultural institutions there, and Edmondson accompanied Wheeler as his assistant and as a tutor to Wheeler's children. After his return to England, Edmondson opened a school at Blackburn in 1825. In 1841 he opened another school at Tulketh Hall in Preston. In 1847 he became the Headmaster of Queenwood College, a position he held until his death.

Eivers, Lawrence (Larry), a civil assistant working in the C District of the Ordnance Survey. He worked on the Irish Survey between 1829 and 1842, at which point he was transferred to the English Survey. Due to his long service on the Survey, he was one of the highest paid civil assistants. By June 1842, Eivers was one of just twelve men, including Tyndall, left in the 3rd division of the C district (also remaining were George Latimer, William Hunter and Private Foy, while the division was led by James Sinnett). At this point the division was engaged in work in the town of Kinsale.

Emerson, Ralph Waldo (1803–82), an American poet and essayist and a leader of the Transcendentalist movement. Tyndall greatly admired Emerson's writings and urged his protégés Thomas Hirst and James Craven to read Emerson's work. Emerson typically wrote his stories as lectures before publishing them. He published two large series of essays in the 1840s, the first in 1841 and the second in 1844.

Evans, Phillip Deighton (Phil, 'Jim') (*c.* 1822–1906), the son of Richard Evans of Carlow, joined the 5th Division of C District of the Irish Ordnance Survey on 28 July 1838 in Leighlin Bridge. He appears to have already known Tyndall when he joined the Survey and may also have been a pupil of John Conwill. His primary activity within the Survey was drawing and his salary rose from 1*s.* to 2*s.* 9*d.* in the summer of 1840. He and Tyndall were close friends both at work and in their leisure time and they lodged together. In the winter of 1841–2 Evans was moved to the 2nd Division and on 20 May 1842 he was transferred to the English Survey, together with William Ginty and others. He joined Tyndall in making the complaint of mismanagement against the Survey's commanders; both men were dismissed from the Survey in November 1843. At the time of his marriage in 1861 Evans was described as a 'Temporary Foreman of Works Warr Department' and he gave an address in Limerick (Irish Church Records). He died in Liverpool in 1906.

Fichte, Johann Gottlieb (1762–1814), a German philosopher who is considered one of the founding figures of post-Kantian German Idealism. Fichte termed his Idealism *Wissenschaftslehre*, which translates to 'Doctrine of Scientific Knowledge'. Fichte based his version of Idealism on subjectivity. In 1849 Hirst and Tyndall engaged in a discussion of Fichte; Tyndall warned Hirst that 'Fichte is no milk for babes Tom – he is meat for strong men, beware of conformity' (letter 0387).

Foy, Thomas M. (b. 1814/15), enlisted in the Corps of Royal Sappers and Miners in February 1833 and commenced employment in the Survey in February 1834. He worked in the 5th Division, C District, under Lieutenant Wynne and Corporal Davy. He was later charged with being absent without leave, being married without permission and having embezzled £19 11*s*. 3*d*. of public money. For his misdemeanours he was reduced to the rank of Private and on 27 January 1841 he was sentenced to a fortnight in prison in Cork Barracks. In May 1842, he was promoted to Lance Corporal. Much of Tyndall's correspondence with Foy in this volume deals with an outstanding debt Foy had to Tyndall.

Frankland, Edward ('Trismegistus', 'Tris') (1825–99), a chemist and teacher at Queenwood College. In 1848, he accompanied Tyndall to the University of Marburg in Germany, where they both studied for doctoral degrees. In 1849 he became engaged to a young German woman named Sophie Fick. They were married in 1851 and would have several children, including Margaret (b. 1853), Frederick (b. 1854), Sophie (b. 1855) and Percy Faraday (1858).

Fresenius, Carl Remigius (1818–97), a German chemist who trained in Justus von Liebig's laboratory. Two of his chemical textbooks were translated into English as *Elementary Instruction in Qualitative Analysis,* trans. J. L. Bullock (London: John Churchill, 1843) and *Instruction in Chemical Analysis (Quantitative),* trans. J. L. Bullock (London: John Churchill, 1846). Thomas Hirst and James Craven used Fresenius's books as a reference in their self-instruction in chemistry; Tyndall oversaw and aided Hirst and Craven in their studies of Fresenius.

Ginty, Margaret ('Peb') (née Roberts) (*c.* 1826–?), married Tyndall's close friend William Gilbert Ginty on 18 May 1846, in Killucan, Westmeath, Ireland. In 1847 Conrad Kupper met Margaret Ginty, remarking in a letter to Tyndall: 'Mrs Ginty has been here some time ago. By George, she is one of the most agreeable women I ever saw in my life. Had she not been married I would have fallen in love with her, there is almost no doubt in it' (letter 0335). The Gintys later relocated to Manchester, where they appear on the 1851 English census along with three children: Kathleen, William Roberts, and Fanny.

Ginty, William Gilbert (Bill, 'Alpha', 'The Attorney General') (*c.* 1820–66), born at Carrick-on-Shannon, County Leitrim, Ginty joined the 5th Division, C District of the Irish Survey on 1 February 1836 as a civil assistant, serving under Lieutenant Wynne. His duties included plotting, registering, laying down lines, tracing for ground examinations, and writing and drawing. He left the Survey in the spring of 1839 (probably to work on the English Tithe Survey) but re-joined in October 1840. A close friend of Tyndall's, they shared lodgings until he was transferred to the English Survey on 20 May 1842. He was dismissed from the Survey in November 1843. On 18 May 1846 he married Margaret Roberts in Killucan, Westmeath,

Ireland; the couple later relocated to Manchester, where they appear on the 1851 English census along with three children: Kathleen, William Roberts, and Fanny. Ginty became a prominent Manchester businessman, served as honorary secretary of the Manchester Athenaeum and was involved various civil engineering projects, including the construction of the Liverpool-Bolton railway. In 1854 he moved to Rio de Janeiro, where he worked as an agent of the Baron de Maua, and contributed to such projects as construction of the imperial gasworks and a railway that ran from Rio to Tijuca Serra. For his contributions to civic projects in Rio he was awarded a knighthood in the Order of the Rose of the Brazilian Empire.

Graham, Sir James Robert George (1792–1861), a member of the Conservative party and Home Secretary from 1841–6. During this time he was involved in certifying the maps of the Ordnance Survey, though Tyndall and his correspondents did not consider him trustworthy. He also held the post of First Lord of the Admiralty from 1830–4 and again from 1852–5.

Griffith, Richard John (1784–1878), Irish geologist and Commissioner of Valuation for the Ordnance Survey. He was in charge of the townland valuation survey of the Ordnance Survey of Ireland from 1830 to *c.* 1842, a survey that became known as 'Griffith's Valuation'. Griffith was also known for publishing the first geological map of Ireland, based in part on the data he had gathered on the Ordnance Survey. In 1858 he was made a baronet for his work at the Ordnance Survey.

Hall, John, an engineering surveyor who lived in Mrs Wright's boarding house concurrently with Tyndall. He appears to have engaged in religious conversations with Tyndall and Hirst. In 1846, Hall lost over £100 through his speculations. He worked with Mr Sherrard on surveys, hiring John Tidmarsh in 1847. Sherrard and Hall conducted surveys of the London and York Railway.

Hamley, William George (1815–93), appointed as a Royal Engineer in 1833 and worked on the Irish and English Ordnance Surveys between 1840–6. The civil assistants on the Survey often addressed their complaints about pay to Lieutenant Hamley. On 29 September 1843 Hamley gave a speech to the civil assistants on the Ordnance Survey in which he addressed their complaints about pay, encouraging them to leave the Survey if they were unsatisfied.

Hebdon, John, a friend of Tyndall's who lived in Preston and worked at the local bank. Possibly John Henry Hebdon (*c.* 1822–?) who appeared on the 1851 census of Preston along with his wife Alice Hebdon (1824–83); his occupation was listed as 'Banker's clerk'.

Hirst, Hannah Oates (1804–49), mother of Thomas Archer Hirst. She was born in Yorkshire and was the daughter of John Oates, a ship-owner and wool merchant. She married the Yorkshire businessman Thomas Hirst in 1824 and was widowed in 1842. Her husband's death left her in somewhat difficult financial circumstances; however, she managed to save enough money to leave her children an inheritance upon her death.

Hirst, Thomas Archer (1830–92), a mathematician and one of Tyndall's closest friends. Hirst was the son of a successful Yorkshire businessman, also named Thomas Hirst, and Hannah Oates Hirst. He was twelve when his father died and seems to have been deprived of most of his inheritance under questionable circumstances. Tyndall and Hirst met in 1845 when Hirst moved to Halifax to become an apprentice to Richard Carter on the West Yorkshire Railway. Tyndall took an interest in Hirst's development and served as a mentor to him, even contacting Thomas Carlyle on his behalf. Hirst later studied mathematics and chemistry, first at the Halifax Mechanics Institute, and later at Marburg, Germany, where he followed in Tyndall's footsteps.

Holland, John, a civil assistant on the Ordnance Survey. He joined the Irish Ordnance Survey on 1 September 1836. He was plotting and drawing for a daily rate of 2 shillings in January 1839. His pay was increased to 3 shillings in April 1842. Holland was transferred to the British Survey on 5 August 1842. Captain Tucker dismissed him, along with Tyndall and seventeen other civil assistants, in November 1843. In 1844 Holland returned to the Survey on the reduced wages of 2/9 per diem.

Holmes, George, a civil assistant who joined the Irish Survey on 26 June 1839 and worked in the 5th Division, C District, initially in Corporal James Mulligan's surveying party under the command of Lieutenant George Wynne. During the spring of 1841 he worked closely with Tyndall and was transferred to the English Survey in May 1842. In August 1843 Holmes had a temporary job working for Mr Johnson, Civil Engineer, in Manchester.

Hunter, Richard (Dick), a civil assistant who joined the Survey on 7 May 1838. He worked in the 4th Division, District C of the Irish Ordnance Survey, under the command of Lieutenant Robert Fenwick. His duties included plotting and drawing, on which he worked closely with his brother William. He was transferred to the English Survey on 20 May 1842. He worked with Tyndall and Ginty in an effort to expose the poor pay of the civil assistants.

Hunter, William (Willy), a civil assistant on the Ordnance Survey who worked in the 4th Division in the C District, under the command of Lieutenant Robert Fenwick. He worked on the Survey between 1839 and 1842, at which point he transferred to the British Survey. In 1843 Hunter was involved in the civil assistants' agitation for more pay. He was the brother of Richard Hunter.

Hutchinson, Mr, a friend with whom Hirst discussed religious matters. Hutchinson had questioned his own beliefs, but had returned to his orthodoxy by 1849 (see letter 0388). Hirst considered him trustworthy.

Jones, Richard Rev. (1790–1855), a political economist with a focus on taxation, and Tithe Commissioner. In August 1836 Jones, William Blamire and Thomas Wentworth Buller were appointed as Tithe Commissioners after the Tithe Commutation Act of 1836, which permitted citizens to make their Tithe payments in kind instead of requiring monetary payments. The Tithe Commissioners were given the authority to settle boundary issues for farmers affected by the change in Tithe payments.

Knoll (Noll), Mr, a kind German whom Tyndall knew during his time at Marburg. Hirst was also fond of Mr Knoll, writing that 'I shall long remember the kindness shewn me by Mr Knoll, his kind, honest & manly conduct will often recur to me' (letter 0382).

Kupper (Kuppers), Conrad, a German friend of Tyndall's who was fluent in French and German but struggled with the English language. Tyndall and Kupper appear to have met in 1847 when they were both living in the boarding house of Mr and Mrs Wright in Halifax. In 1847 Kupper and Tyndall discussed religion with one another. In 1848 Kupper was living in Rheydt, Dusseldorf and in 1849 Tyndall noted that Kupper was in Leipzig.

Latimer, George, a civil assistant on the Ordnance Survey. He worked in the 4th Division in the C District, under the command of Lieutenant Robert Fenwick. Latimer worked on the Irish Survey between 1837 and 1842, at which point he was moved to the English Survey. Latimer had a brother named William who also worked on the Survey in the same district. In 1844 he was considering emigrating to America with Tyndall, but changed his mind at the urging of his family.

Latimer, William, a civil assistant on the Ordnance Survey. He worked in the 4th Division in the C District, under the command of Lieutenant Robert Fenwick. Latimer worked on the Irish Survey between 1840 and 1842, at which point he was moved to the English Survey. Latimer had a brother named George who also worked on the Survey in the same district.

Lees, Frederic Richard (1815–97), a Spiritualist and temperance advocate who worked as an editor of the Transcendentalist publication the *Truth Seeker* from 1844 to 1850. Lees and Tyndall corresponded regarding the *Truth Seeker*. In 1838 Lees married Mercy Joanna Jowett (d. 1870). Together, they had a son, Frederic Arnold Lees (1847–1921), and a daughter, Mary Eleanor Whatmoore. In 1878 he remarried, to Sarah Barnsley, but he had no children with her.

Lindsay, James S., a civil assistant on the Ordnance Survey of Ireland. Lindsay was among the civil assistants dismissed from the Survey on the orders of Henry Tucker in November 1843.

Marquis, William, a civil assistant from Dublin who worked in the A District headquarters under the command of Captain Henry Tucker, having joined the

Irish Ordnance Survey on 1 December 1837. His primary duty was drawing but he also undertook examining work at various times during 1839. In this he appears to have worked closely with Archibald McLachlan. He left Ireland in August 1841 when the work in the A District was all but complete, and was transferred to the English Survey. He was stationed at Preston.

Martin, Robert (Bob), a civil assistant in the A District of the Irish Ordnance Survey, commanded by Captain Tucker. Having joined the Survey in February 1827 he became one of its most senior assistants and an accomplished draughtsman. In the spring of 1840 he was based in the Waterford region in the 4th Division, A District, under Lieutenant Thomas R. Lyster, and later that year the Division was transferred to Clonmel in County Tipperary. In August 1841 Martin left Ireland for the British Survey. In 1847 Martin was anxious about losing his employment and Tyndall provided him with reassurance on this matter.

McGowan (McGowen), John, a civil assistant on the Ordnance Survey. He worked in the 4th Division in the C District, under the command of Lieutenant Robert Fenwick. He joined the Survey on 18 November 1835. His primary duty on the Survey was drawing. By the beginning of 1842 he was working in the 3rd Division with Tyndall, Sinnett, Foy, O'Neill, Doig Junior, Eivers, William Hunter, George Latimer and Bloomfield. He was transferred to the British Survey on 5 August 1842. Captain Tucker dismissed him, along with Tyndall and seventeen other civil assistants, in November 1843. In 1844 William Ginty reported to Tyndall that McGowan had re-joined the Survey (letter 0288).

McLachlan, Archibald (Archy, 'B.'), a civil assistant who worked on the A District of the Ordnance Survey, commanded by Captain Henry Tucker. McLachlan was employed on the Ordnance Survey between 1827 and 1837, where his duties included comparing plans and drawing. He returned to the Survey in February 1841, before being moved to the British Survey in August of the same year. However, he quickly became embittered by the poor treatment of the civil assistants on the Survey. He was one of Tyndall's most important sources of Survey intelligence as Tyndall prepared the 'Spectator' letters.

Moore, Thomas (1779–1852), Irish poet, songwriter, and entertainer. Between 1808 and 1834, Moore wrote lyrics to accompany a large number of Irish folk tunes; Moore's lyrics were printed by the publishers William and James Powers in London and Dublin, along with arrangements of the folk tunes by the composer Sir John Stevenson (1761–1833). The collected songs became known as *Moore's Irish Melodies*. Moore was also known for the epic poem *Lalla Rookh* (1817). William Ginty frequently quoted Moore's songs in his letters to Tyndall (see, e.g., letter 0249 or 0266).

Mulhall, Margaret, a Preston resident.

Murray, George (1772–1846), born in Crieff, Perthshire, he entered the army in 1789, when he was commissioned ensign in the 71st Regiment of Foot, and had reached the rank of Captain by 1794. He excelled as a Quartermaster-General and was appointed to the Duke of Wellington's army in the Peninsular War against Napoleon, although a brief posting in Canada meant that he did not fight in the Battle of Waterloo. In 1819 Murray was made Governor of the Royal Military College, Sandhurst, and was elected MP for Perthshire in 1824, also becoming Lieutenant-General of the Ordnance in the same year. He was made Master-General of the Ordnance during Robert Peel's first ministry between December 1834 and April 1835, and again in the second Peel ministry of September 1841 to June 1846. Perhaps the most respected soldier in Britain after Wellington, he died only a month after leaving the post of Master-General.

Newton, James W. (Jim), a Private in the 13th Company of the Corps of Royal Sappers and Miners at Leighlin Bridge, in 5th Division in the C District of the Ordnance Survey.

O'Connell, Daniel (1775–1847), former MP (1828–41) and Lord Mayor of Dublin (1841–2), known as 'the Liberator'. O'Connell was the oldest of ten children in a large Irish Catholic family. He set his sights on a legal career in the early 1790s and made a name for himself in Ireland as a skilled and sought-after barrister; many of his followers would continue to call him 'Counsellor O'Connell' even after his entry into politics. O'Connell joined the cause of Catholic emancipation around 1805 and quickly became one of its most passionate voices. He was widely credited with persuading Peel to enact Catholic emancipation in 1829, removing many restrictions on Catholic voting rights and ability to participate in British public life. He was elected to County Clare's Parliamentary seat in 1828 and would subsequently be elected MP for County Waterford, County Kerry, Dublin City, and Kilkenny City, reflecting his tremendous popularity in Ireland. In 1841 O'Connell was elected to a one-year term as Lord Mayor of Dublin, the first Irish Catholic Lord Mayor since 1688. In the 1840s, O'Connell campaigned vigorously for the repeal of the 1801 Act of Union. To push for this, he held a series of well-attended political rallies, called 'Monster Meetings'. The Prime Minister, Sir Robert Peel, banned one proposed monster meeting at Clontarf, just outside Dublin, in October 1843. O'Connell was arrested, charged with conspiracy and sentenced to a year's imprisonment in London with a fine of £2,000, although he was released after three months by the House of Lords, which quashed the conviction and criticised the unfairness of the trial. The final years of O'Connell's life coincided with the onset in 1845 of the Great Famine in Ireland. O'Connell's last years saw serious fractures within the Catholic emancipation movement, including the rise of the Young Ireland movement, which included both Catholics and

Protestants. At this stage, O'Connell was seeking to form a political alliance with Lord John Russell's Whig Party. He died in Genoa in May 1847, en route to see Pope Pius IX, before this alliance was formalized.

Parker, John, a civil assistant on the Ordnance Survey. He worked with Tyndall in the 5th Division in the C District of the Ordnance Survey. Parker joined the Survey on 5 November 1838. His duties were plotting. He was transferred to the English Survey on 8 April 1842. He lived in Preston along with Tyndall in 1843, but was not among those dismissed from the Survey that November. He went to Liverpool under Lieutenant Hamley's command in 1844.

Peel, Sir Robert (1788–1850), Second Baronet and Prime Minister of the United Kingdom 1834–5 and 1841–6. Born in Bury, the son of a printed calico manufacturer, he became an MP in 1809. He soon entered the cabinet, in May 1812, as Irish Secretary, in which role he strongly supported the Protestant ascendancy. Peel returned to England in 1818, serving as Home Secretary from 1822 to 1827, and again from 1828 to 1830, overseeing a transformation of policies for law and order. He also enacted the emancipation of Catholics in 1829, a measure forced upon him by unrest in Ireland led by Daniel O'Connell, but which left Protestants feeling betrayed by their former champion. Following the upheavals of electoral reform, Peel became Prime Minister in December 1834, although his ministry lasted only until April of the following year, with his Irish Tithe Bill having just been introduced to Parliament. After six years in opposition to successive Whig governments, Peel became Prime Minister again in September 1841, moving quickly to tackle the economic recession and the huge deficits run up by the Whigs. In the budget of 1842, Peel announced the reintroduction of income tax at 7*d.* in the pound. Ireland, however, was excluded because of concerns over the collection of such a tax, and instead the Irish duty on spirits was raised. In the same budget, he reduced the sliding scale in the duty on imported corn, and in 1846, partly prompted by the Irish famine, repealed the Corn Laws altogether. In Ireland, Peel's appointment in 1841 of an ultra-protestant Lord Lieutenant gave new momentum to O'Connell's campaign to repeal the Act of Union, peaking with the so-called 'Monster Meetings' of 1843, which led to the passing of the Irish Arms Act restricting the manufacture and importation of arms and ammunition. Peel, however, refused to enact further coercive legislation in Ireland, fearing that it would only increase support for O'Connell, and instead introduced conciliatory measures including the creation of three new Irish universities. With both the repeal of the Corn Laws and his distaste for coercive measures in Ireland enraging reactionary Tories and Unionists within the Conservative Party, Peel resigned in June 1846, dying four years later after falling from his horse. Tyndall's 'Spectator' letters were published during Peel's second term as Prime Minister and were addressed to his attention.

Phillips, George Searle ('January') (1816–89), a journalist and writer who befriended Thomas Hirst in 1849. Phillips worked as editor of the *Leeds Times* beginning in 1844. He often published under the pen name 'January Searle'. He also edited the Transcendentalist publication the *Truth Seeker* along with F. R. Lees, and was a lecturer at the People's College in Huddersfield.

Piercy (Peircy), Ada, a young girl whom Tyndall knew in Halifax. Tyndall met her in January 1847 and described her in his journal as 'the sweet little girl Adah Piercy ... she promises to become mine if I wait eight years and at the end of that time am not an old man with grey hair' (RI MS JT/2/13a/174). The references to an engagement or attachment between the two (see, e.g., letter 0380) seem to have been made in jest.

Richardson, John, a mutual friend of Tyndall and Hirst who also worked for Richard Carter's firm.

Richardson, Mr, a lecturer in Halifax whose lectures on chemistry and pneumatics Thomas Hirst often attended. It is possible, but unlikely, that Mr Richardson was Tyndall and Hirst's friend John Richardson (see above).

Sayers, Edmond B. (Bob), a civil assistant on the Ordnance Survey of Ireland. He joined the Survey on 1 November 1839. He worked in the B District's 1st Division, under the command of Lieutenant Stothend. This Division was based at Limerick during Sayers's time on the Survey. His main duties were surveying and plotting, and his daily rate of pay had risen to 2 shillings by the time he left. The records state that he set sail for Britain on 13 August 1842. It seems that he left the Survey for a period before this; the records are a bit sketchy during 1841, but he is listed as having re-joined on 30 November 1841 after a period away. He was one of the civil assistants dismissed by Henry Tucker's 7 November 1843 order (letter 0262). After his dismissal Sayers remained in the Preston area and accepted a few small surveying jobs. He later moved to London, where he resided at 47 King Street.

Scott, William ('Sir Villainy'), a Corporal in the Corps of the Royal Sappers and Miners. He was an acting Lance Corporal on the Ordnance Survey in Ireland until he was promoted to Corporal on 1 July 1840. He was transferred to the British Survey in 1842 and appears to have been Ginty's direct supervising officer. Ginty had a low opinion of him and nicknamed him Sir Villainy.

Scott, Sir Walter (1771–1832), a Scottish poet and novelist. One of the most popular writers of his time, Scott is best known for works such as *Ivanhoe* (1809) and 'The Lady of the Lake' (1810). He was born in Edinburgh and worked as a solicitor and as an editor before making his name in poetry and as a novelist. Scott was also a respected literary critic who wrote for such publications as *The Edinburgh Review* and *Blackstone's Edinburgh Magazine*. The 1826 economic crash wiped out Scott's fortune, and his attempts to repay his creditors led to a series of strokes that would eventually end his life.

Sinnett, James (1807–44), enlisted in the 13th Company of the Corps of Royal Sappers and Miners on 21 April 1826 and was promoted to the rank of Corporal in March 1839. He worked in the 4th Division, C District, under the command of Lieutenant Robert Fenwick, and was respected as a skilled draughtsman – Major Waters described him as 'attentive and useful' and of 'very good' conduct. Before being transferred to England on 5 August 1842 he worked closely with Tyndall. He died in Liverpool in August 1844.

Stanley, Charles Edward (1819–49), a Royal Engineer on the Ordnance Survey and Tyndall's Division Officer in Preston. He was the son of Bishop Edward Stanley FRS (1779–1849), the Bishop of Norwich. He was originally posted to the Ordnance Survey of Wales and became a Lieutenant in 1838 before being transferred to the English Survey. Tyndall thought highly of Stanley, but told John Tyndall, Snr that unfortunately 'he is young, and has no power. If he had, things would change' (letter 0267). In 1846 Stanley married Eliza Dolly Clayton in Lancashire. In 1847 Stanley was transferred to Tasmania, where he was the Private Secretary to the Lieutenant Governor. Stanley died in August 1849 after contracting gastroenteritis.

Steuart/Stewart, William Richard (1798–1852), descended from an established family of Scottish ancestry that settled in Ireland *c.* 1719, having obtained estates in Counties Carlow and Meath. The family's wealth was further increased in the mid-eighteenth century by a marriage settlement that included 2,000 acres of land abutting the south of Leighlin Bridge, where Steuart's grandfather (also named William Steuart) built Steuart's Lodge. William Richard later inherited the estate and in 1820 married Elizabeth Dawson Duckett, from another very prosperous Protestant land-owning family. He held the rank of Captain, served as a local magistrate and was appointed High Sherriff of Carlow in 1821. He owned many of the properties in Leighlin Bridge, including the house where the Tyndalls lived and both John Tyndall, Snr and Caleb Tyndall were at various times in the employment of Steuart. Tyndall and his father looked to the Steuarts for patronage and approached them for support when Tyndall was considering various forms of employment. Tyndall also designed a porch for Steuart's Lodge.

Steuart/Stewart, (Mrs) Elizabeth Dawson (née Duckett) (1802–93), born into a very wealthy family with an estate of 12,000 acres and a gothic mansion (Duckett's Grove) situated several miles northeast of Carlow. She married William Richard Steuart on 18 August 1820 and they lived at Steuart's Lodge, Leighlin Bridge. She took a keen interest in Tyndall's career and was probably his most consistent and long-lived patron, her husband having died in 1852.

Taylor – most likely Alexander Taylor, a civil assistant on the Ordnance Survey who joined the Irish Survey in July 1838. Taylor was transferred to the British Survey on 8 April 1842 and worked in the same Liverpool division as William Ginty. Taylor appears to have been fired from Corporal William Scott's division alongside Ginty and several others in November 1843 (see letter 0263). He returned to Ireland where he later obtained a position at the Land and House Agents office in Dublin.

Tidmarsh, John Francis (Jack; Tid) (1824–1906), was born in Cork and educated at Dr Radcliffe's School before becoming an assistant draughtsman at an architectural office in Dublin. He joined the Irish Ordnance Survey on 11 May 1840, working in the 5th Division, C District under the command of Lieutenant George Wynne, where his principal duty was plotting, although he also undertook drawing and tracing for examination on the ground. He worked closely with Tyndall in Youghal, where he lodged with him and William Ginty, and again in Kinsale and Cork, where Tyndall enjoyed the hospitality of Tidmarsh's family. He was separated from Tyndall after being transferred to the English Survey on 20 May 1842, where he worked in the 5th Division and undertook fieldwork in Lancashire, Yorkshire and Westmoreland. In the summer of 1843 he contemplated emigrating to the Cape of Good Hope to take up a job with relatives, but instead remained in England where, by 1845, he was employed, alongside Tyndall, making surveys for the West Riding Union Railway. At this time, he and Tyndall were once more lodging together in Halifax. In March 1849 Tidmarsh emigrated to Australia, settling in Adelaide, where he initially worked as a surveyor and valuator for the city's first building society and later prospected during the Australian gold rush in 1852. He established a successful candle and soap manufacturing company, although overwork impaired his health. In November 1906 Tidmarsh ended his own life by firing a revolver into his head ('Death of Mr. John Tidmarsh', *Register* (Adelaide), 12 November 1906, p. 6).

Todd, John, Civil assistant who worked in the 2nd Division in the C District of the Ordnance Survey, under the command of Lieutenant Wilkinson. Todd joined the Survey on 1 August 1837, and performed duties that included plotting and drawing. Todd remained on the Survey after most of his division had left, and was working on the parishes of Cork. He was transferred to the British Survey, along with Tyndall, when the work in the C District was completed on 5 August 1842. He was not among those dismissed in November 1843, and he remained on the Survey in Preston.

Tucker, Captain Henry, Local Director of Preston Ordnance Survey. Possibly born in Kent, where his father Stephen Tucker was vicar of Borden from 1797 to his death in 1838. He enlisted in the Royal Engineers in *c.* 1815, and was made 2nd Lieutenant on 1 March 1816, and then 1st Lieutenant on 9 April 1825. He joined the Ordnance Survey in 1824, and was in Ireland by at least 1832, when he was stationed in County Antrim. He was in Ardee, County Louth in early 1836, where he was in charge of the E District, and was promoted to 2nd Captain on 25 December 1837. By January 1838 Tucker was stationed in Dublin, and was in command of the A District, which was operating in Counties Kildare, Wicklow, parts of Carlow, Queen's County and Dublin. In January 1840 he was stationed at Waterford, where he remained throughout the year and into 1841. He was then transferred to the English Survey, where he superintended much of the surveying of the six northern counties, commanding the 2nd and 3rd Divisions, which he ran from the Divisional Office at Wakefield. Tucker was subsequently stationed at the Ordnance Survey headquarters in Southampton, where he supervised the preparation of special plans of Windsor for Prince Albert. He became Brevet Major on 11 November 1851, and Lieutenant-Colonel on 29 March 1853, and in the same year returned to Ireland where he was put in charge of the Survey headquarters at Mountjoy House, Phoenix Park, Dublin. In 1854 the superintendency of the entire Ordnance Survey of Great Britain fell vacant, and Tucker was widely regarded as the most appropriate candidate. However, his profound deafness, alluded to by William Ginty in letter 0185 (Volume 1), debarred him from the post, which went instead to the less experienced but well-connected Major Henry James. Outraged at being passed over, Tucker unsuccessfully tried to have the Irish Survey made independent of the rest, and then 'left the Survey in disgust to die a few years later in the West Indies' (J. H. Andrews, *A Paper Landscape: The Ordnance Survey in Nineteenth-Century Ireland* (Oxford: Clarendon Press, 1975), p. 246). He presumably died between 1856, when he received his final promotion to Colonel, and 1858, after which he no longer appears in the annual *Army List*. Tucker was one of Tyndall's most disliked superiors on the Ordnance Survey; he is probably the 'Acid' that McLachlan and Tyndall criticize in their correspondence, although the identification is not certain.

Tyndall, Emma (1817–1904), Tyndall's sister whose education he repeatedly encouraged. She visited her brother in Preston in August and September 1843. She does not appear to have married and later in life suffered from a mental illness which was manifested as an extreme form of religiosity.

Tyndall, John, Snr (1792–1847), grew up in Leighlin Bridge and married Sarah McAssey in 1813. He probably served in the Royal Irish Constabulary at the police station in Nurney from 1828 to 1833. The family then moved to Castlebellingham, County Louth for the next three years before returning to Leighlin Bridge, where he worked as a boot and shoemaker. He may also have served as a land agent for William Steuart, one of the principal landlords in Leighlin Bridge and also the main patron of the Tyndall family. Although self-educated he was a keen reader and sought to expand his own education and encourage his children's. A staunch Orangeman, he was a member of the Brunswick Club which sought to advance the Protestant cause and oppose the social and political advance of Roman Catholics. Louisa Tyndall described him as 'a vigorous man, [with a] tall and powerful frame, [who] suffered much from sleeplessness'. Plagued by a 'delicate stomach', he was 'fastidious in various ways ... and was very particular about the habits of his children' (LT, 'Biography', vol. 1, pp. 4–5). His 1847 death was a blow to his son, who relied on his father for advice.

Tyndall, Sarah (née Macassey/Macasey/McAssey) (1793–1867), descended from a prosperous farming family with lands in Ballybrommell, County Carlow, that included in earlier generations Quakers by the name Malone. According to Louisa Tyndall, she 'must have been a brave, loving, devoted soul ... Tyndall used to say that he had rarely seen such gentleness combined with so much courage. She was of a type occasionally met with in the south of Ireland, tender poetical, naturally refined, and well educated for her class' (LT, 'Biography', vol. 1, p. 5). She was also remembered for having encouraged the education of her children and for reciting poetry to them.

Tyndall, William (b. 1804), an uncle of Tyndall's who emigrated to Philadelphia.

Walkem, Charles J. (*c.* 1805–?), a Corporal in the 14th Company of the Royal Sappers and Miners. He enlisted to the army in September 1822. He was a carpenter by trade. He worked in the A District of the Ordnance Survey of Ireland, and as a Private was involved primarily in surveying duties. He became Lance Corporal in August 1838, and he later became the Superintendent at the District headquarters at Waterford. In this post he was Captain Tucker's right-hand man. He left the Irish Survey, along with most of the workers in A District, at some point between August and October 1841; the records for these months are missing so there is no exact date of departure. He was transferred to the Survey in Preston, but following the disputes between the Preston civil assistants and the Survey leadership, Walkem was transferred to Quebec, Canada in 1844. His departure, according to Tyndall's friend John Todd, was 'universally regretted' (letter 0294). He appears on the 1871 Canadian census living in Quebec along with a wife, Mary Anne Walkem, and four children.

Wright, Mary Ann, a sister of William Wright who lived for a time at No. 12 Nichols Street, Dublin. She may be one of the Wright sisters who emigrated to America to be with her brother in 1844 (see letter 0304).

Wright, Sarah/Sally Maria (née Neale/Neill) (1815–88), was from a prosperous family in Roscrea, County Tipperary. She married William Wright on 7 February 1841 and emigrated to America in early March 1841.

Wright, William (1818/9–92), from Leighlin Bridge, a close childhood friend of Tyndall's, with whom he subsequently maintained a long-term correspondence. In early 1841 he was working in Roscrea, County Tipperary, and courting Sarah Maria Neale, whom he married on 7 February 1841 at St Nicholas Parish Within, Dublin (Irish Church Records). Possibly due to opposition from her family, the newlywed couple rapidly emigrated to America, sailing from Liverpool in early March 1841. They settled in Cincinnati, Ohio, where William worked in the dry goods and grocery trades. He frequently urged Tyndall to join him in America.

Wright, Mr, the overseer of a lodgings house on Ferguson Street in Halifax and a notorious local drunk. Tyndall's journal entry for 24 January 1847 described him as 'a soul in ruins – An evident appreciation for the highest and noblest conceptions – a love for the grand and beautiful overcome only by a love for grog ... An exalted genius united to drunken depravity' (RI MS JT/2/13a/174).

Wright, Mrs, matron of a lodgings house on Ferguson Street in Halifax, where Tyndall and many of his friends stayed. Tyndall consistently praised Mrs Wright's hospitality and good care of her lodgers.

Wynne, George (1804–90), born in Dublin, he enlisted in the Royal Engineers in *c.* 1825, becoming 2nd Lieutenant on 26 September 1826, and 1st Lieutenant on 12 March 1831. He joined the Irish Ordnance Survey in September 1835, working in County Galway and at Tyrellspass in County Westmeath, before being stationed at Leighlin Bridge from April 1838. By this time, he was in charge of the 5th Division, C District, under the command of Major Marcus Antonius Waters. In April 1840 he was transferred to Youghal, before handing over command of the 5th Division to Lieutenant Paul Whittingham on 18 September of the same year. He requested to be taken off the Survey altogether, apparently because of his wife's illness. Wynne later served with distinction in Greece, China and the West Indies, rising to General by the time he retired in 1877. In a letter to *The Times* in 1890, Tyndall recorded: 'In 1839, I being fresh from school, Lieutenant Wynne became my chief. A few years afterwards he was sufficiently my friend to offer me the use of his purse for the prosecution of my studies in Germany. Our friendship, afterwards, was of the most intimate and cordial character; and notwithstanding my heterodoxy, and his Christian piety, it was never for a moment shaken. Let me briefly say that a spirit more erect and pure I have never known ... The earth's noblest ones are sometimes found among its noiseless ones. A conspicuous illustration of this truth is furnished by the life and character of General George Wynne' ('The Late General Wynne', 12 July 1890, p. 11).

APPENDIX: JOHN TYNDALL'S TIMELINE OF THE 1843 ORDNANCE SURVEY CONTROVERSY[1]

I make my appearance[2] before Sir R. Peel or – as a delegate chosen to represent the sentiments of a number of men lately dismissed from the Ordnance Survey – and as their case is one of peculiar interest I beg his consideration of the circumstances which led to their dismissal.

Memoranda

Sept. 20th order read 1st time.	Heard the order[3] read and requested time for deliberation – this was granted, met in the evening – concluded that it would not be safe to sign either. Determined on sending a Memorial to the Master General[4] in which our reasons for declining to sign the form should be given.
Sept. 22nd order read 2nd time.	Order read by Sergt. Penny[5] – each man's name was called out and he answered 'yes' or 'no' according as he was willing or not willing to sign.
Sept. 23rd Mem. forward	Memorial laid before Lt. Stanley our Division officer, who was requested verbally to forward it through the proper channel.
Friday Oct 13th Met by Capt. Tucker	Were met by Captain Tucker who spoke to us relative to our applying to the Master General instead of to Colonel Colby – said that he knew the consequences of our proceeding would be that half the men would be discharged and the other half increased. Though the good men suffered from a contact with the bad, still were they unwilling to see one of the latter dismissed even if it afforded them an increase – the universal feeling was that the matter should be referred to the M.G.[6] – Spectator.[7]
Oct. 14th letter forwarded.	Forwarded a letter to Captain T.,[8] in which our reasons for appealing directly to the M.G. were given: it contained a request to forward the Memorial to its original destination – this was signed by 12 men.
Original list made out.	About this time a list of men to be discharged had been made out – it contained a number of inefficient hands and bad characters – it did not contain the name of a single individual who had signed the letter to Captain T.

Wednesday Nov. 8th	Order which contained our dismissal read, in the number discharged were many who had not appeared on Capt. T's original list – and many were omitted who did appear in that list.
Nov. 8th adduce specimens and testimonials.	Allusion is made to previous notice – this notice referred to the inefficient men – adduce characters and specimens to prove that we should not be ranked among this number.
Capt T's statement in L'pool	Quite in keeping with Captain Tucker's former statement at Liverpool, that he was not bound to keep the good men and send away the bad, that he could discharge which he pleased.[9]

RI MS JT 1/TYP/10/3903

Notes

1. *John Tyndall's Timeline of the 1843 Ordnance Survey Controversy*: This document appears to be notes that Tyndall took for himself in the event that he was granted a meeting with Prime Minister Robert Peel or with George Murray, the Master General of the Ordnance Survey. We have included them as an appendix because they provide a useful outline of the major events in 1843 that led to Tyndall's dismissal.
2. *I make my … their dismissal*: Louisa Tyndall included this document in the notes for her biography and wrote: 'Notes of what to say to the M.G. if granted an interview' (LT, 'Biography', 117). By 'M.G.' Louisa is referring to George Murray, the Master General of the Ordnance Survey.
3. *the order*: the civil assistants were ordered to sign a document indicating that they were satisfied with their current rate of pay and would agree to remain on the Ordnance Survey with the understanding that their pay would not be increased. See letter 0236 of this volume, in which a copy of this order was enclosed.
4. *a Memorial to the Master General*: the letter of protest sent by the workers of the Ordnance Survey of England to George Murray, Master General of the Ordnance on 23 September 1843; see letter 0236 of this volume.
5. *Sergt. Penny*: James Penny (*c.* 1796–?) was a Sergeant in the 14th Company of the Royal Sappers and Miners, and one of Captain Henry Tucker's most trusted subordinates. He enlisted in the army in December 1814. Penny was a miner by trade. He worked in the A District under a number of different officers including Lieutenant Andrew Beatty, Lieutenant Leach, Lieutenant Henry James, and Captain Henry Tucker. He was primarily involved in supervising duties within various divisions, and he received 2 shillings per diem, the highest salary to which a member of this Corps could aspire. He was stationed in County Mayo in 1837, before moving to County Carlow, and then to Fermoy in County Cork in 1840. Captain Tucker thought very highly of him; he regularly stated that Penny was worthy of promotion in his monthly comments. Penny left the Irish Survey for the English Survey at some point between August and October 1841; the records for these months are missing.
6. *M.G.*: Master General of the Ordnance Survey, George Murray.
7. *Spectator*: Tyndall's *Liverpool Mercury* pseudonym; possibly a reference to letter 0248 of this volume.
8. *a letter to Captain T.*: see letter 0245 of this volume.
9. *he could discharge which he pleased*: Louisa Tyndall annotation: 'see Ginty Oct 13/43'; see letter 0244 of this volume.

INDEX

The index refers to the letters using their serial numbers. The letter 'n' denotes a footnote reference and appears if an item is not otherwise mentioned in the letter to which the footnote pertains. A number in bold signifies a letter to or from a correspondent.